MW00807685

Cartographic Japan

Cartographic Japan

A HISTORY IN MAPS

EDITED BY Kären Wigen,
Sugimoto Fumiko, and
Cary Karacas

THE UNIVERSITY OF CHICAGO PRESS
Chicago & London

Kären Wigen is the Frances and Charles Field Professor of History at Stanford University. Sugimoto Fumiko is professor of history at the University of Tokyo's Historiographical Institute. Cary Karacas is associate professor of geography at the College of Staten Island, CUNY.

The University of Chicago Press, Chicago 60637
The University of Chicago Press, Ltd., London

Printed in China

25 24 23 22 21 20 19 18 17 16   1 2 3 4 5

ISBN-13: 978-0-226-07305-7 (cloth)
ISBN-13: 978-0-226-07319-4 (e-book)
DOI: 10.7208/chicago/9780226073194.001.0001

Library of Congress Cataloging-in-Publication Data

Cartographic Japan : a history in maps / edited by Kären Wigen, Sugimoto Fumiko, and Cary Karacas.
pages cm
Includes bibliographical references and index.
ISBN 978-0-226-07305-7 (cloth : alkaline paper) — ISBN 978-0-226-07319-4 (ebook) 1. Cartography—Japan—History. I. Wigen, Kären, 1958– editor. II. Sugimoto, Fumiko, 1958– editor. III. Karacas, Cary, editor.
GA1241.C37 2016
911'.52—dc23

2015006383

♾ This paper meets the requirements of ANSI/NISO Z39.48-1992 (Permanence of Paper).

Contents

A Note on Japanese Names and Terms

Macrons are used in the text to indicate long vowels in Japanese, except in the case of frequently used names and terms (daimyo, shogun, Tokyo, Kyoto, and the like).

Japanese personal names are indicated in the Japanese fashion: surname first and given name following. The names of Japanese and Japanese-American authors writing in English are given in the reverse order, typical of English. For the sake of clarity, the surnames of all scholars who contributed to this volume are rendered in capital and small capital letters.

Introduction

Kären WIGEN

FIGURE 0.1 Map section in a bookstore. Photograph by Sakakibara Kazutoshi.

The Japanese people today are voracious consumers of cartography. In bookstores and libraries across the country, miles of shelf space are devoted to travel guides, walking maps, topical atlases, and works of historical geography (fig. 0.1). Schoolchildren are taught to map their classrooms and schoolgrounds, and retirees pore over old castle plans and village cadastres. Pioneering surveyors have been the subject of television shows and popular exhibits. Avid collectors covet exquisite painted scrolls depicting sea and land routes, while students and scholars help sustain a market for less expensive woodblock reproductions of city maps and bird's-eye views. On a more practical level, maps are mounted everywhere from subway walls to shopping malls. Recreational hikers snap up topographical sheets by the thousands, while citizen groups churn out digital charts showing the precise location of algae blooms, radiation levels, and other public health hazards.

If maps are ubiquitous in Japan today, however, it was not always so. Mapmaking in East Asia is an ancient practice, to be sure. But until the seventeenth century, maps were few and far between, a private privilege of the ruling elite. While no provincial maps from the era survive, commissioning geographic information on the provinces was one of the founding acts of the imperial state in the 600s CE. Images of the imperial realm as a whole came along much later. The earliest map of all Japan, in the "fish-scale" style named after its legendary designer, a monk named Gyōki, dates to the 1300s (fig. 0.2). Landholders over succeeding centuries occasionally commissioned property maps, and a few priests labored to locate Japan in a Buddhist cosmos. But for nearly a millennium, mapmaking remained rare. And even those fitful episodes ground to a halt during the bitter civil wars of the fifteenth and sixteenth centuries.

Only when the Tokugawa reunified the archipelago in 1600, laying the foundation for a 250-year cease-fire under the watchful eye of the samurai, did Japanese cartography truly take off. As we explore in part I, the ruling class took the lead. Tokugawa Ieyasu summoned massive new maps of the provinces, revealing the sinews of agrarian power in unprecedented and magnificent detail. The Buddhist establishment was not far behind, commissioning images of sacred sites from the level of the temple compound to the cosmos as a whole. Soon, warriors and priests were joined by ranks of commoners—scholars, artists, and eventually even satirists—who began producing woodblock prints and views for sale to ordinary people. As part II reveals, by 1800 the map market was saturated. Cartography consumers in the later Tokugawa period could choose from hundreds of styles and subjects, at scales ranging from the neighborhood to the nation and beyond. Map conventions were sufficiently familiar that humorists could use them to poke fun, mocking the excesses of Edo-era culture through cartographic parodies and puns.

The opening of the Pacific to global trade and warfare in the mid-nineteenth century only magnified maps' importance. Part III shows how the rush to modernize Japan's military, tax its countryside, plan for a growing population, and mitigate disasters provided much new work for Japanese mapmakers. So did the push to expand the frontiers of the state. From the 1870s to the 1940s, Japan's aggressive drive to acquire an empire of its own made Tokyo a voracious producer and consumer of cartographic information. Although that empire ultimately collapsed, postwar life generated ample demands of its own, as we see in part IV. Bureaucrats pursuing development projects needed maps every bit as much as generals preparing for battle. So too did leisure seekers and local businesses. And with the dawn of the digital age, public access to mapmaking technology exploded as well. Appealing to everyone from boosters to scholars, from quiet archives to bustling street corners, mapping in the postwar period became the booming practice that it remains today.

The present volume is designed to introduce non-Japanese readers to the resulting treasure trove of colorful materials that makes this one of the world's most diverse and spectacular cartographic archives. Our intent is twofold: to use individual maps as a window on particular moments in Japan's history, and to showcase contemporary cartographic research. Scholarly interest in maps has never been higher. Drawing on new work

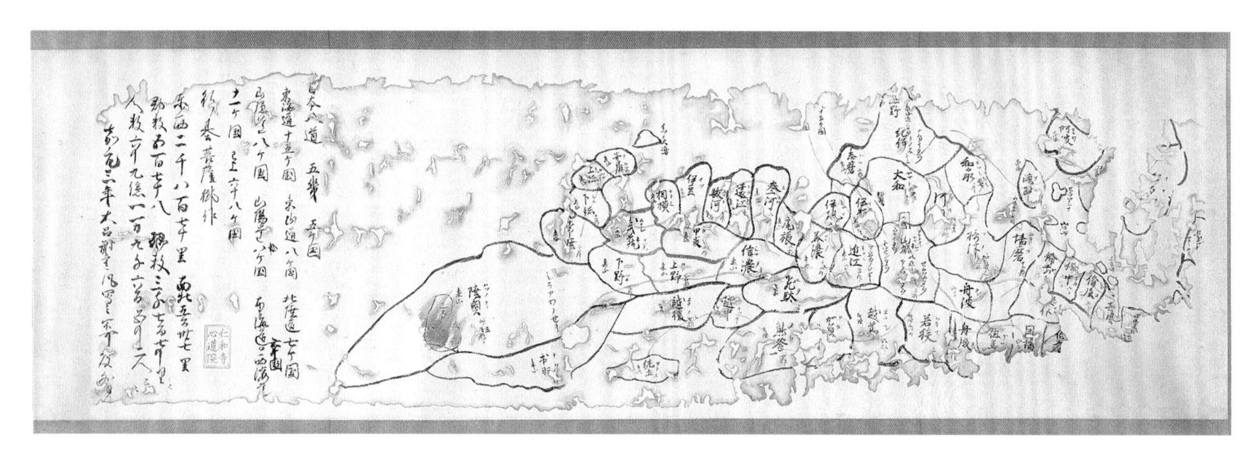

FIGURE 0.2 "Ninnaji Gyōki Map" [Ninnaji-zō Gyōki zu 仁和寺蔵行基図], Edo-period copy of a fourteenth-century original. Manuscript, 38.5 × 134.4 cm. Courtesy of Kobe City Museum.

in visual studies and material culture, historians in the early twenty-first century are alert to maps as both art form and commodity form to an unprecedented degree. One result is heightened expertise for dealing with this corpus of challenging material. Geographers, historians, and art historians on both sides of the Pacific have contributed to a growing number of specialized books and articles interrogating Japanese maps with considerable subtlety. Most of their findings, however, have not been accessible to a lay audience. *Cartographic Japan* aims to change that.

The book features fifty-eight short essays, each focused on one or two maps related to the contributor's specialty. Opening at the dawn of Japan's early modern cartographic explosion in the late sixteenth century, it ends with the great Tohoku earthquake and tsunami of the early twenty-first. Between those epochal moments, Japanese society—and the Japanese landscape—changed almost beyond recognition. Since maps documented those changes at every turn, the essays presented here offer a fresh approach to four centuries of tumultuous history. We hope this piques your interest in a fascinating archive.

Suggested Readings

Berry, Mary Elizabeth. *Japan in Print: Information and Nation in the Early Modern Period.* Berkeley: University of California Press, 2006.

Cortazzi, Hugh. *Isles of Gold: Antique Maps of Japan.* New York: Weatherhill, 1983.

Hubbard, Jason. *Japoniæ Insulæ, The Mapping of Japan: Historical Introduction and Cartobibliography of European Printed Maps of Japan to 1800.* Houten, The Netherlands: Hes & De Graaf, 2012.

Unno Kazutaka. "Cartography in Japan." In *History of Cartography*, vol. 2, bk. 2, edited by J. B. Harvey and David Woodward, 376–90. Chicago: University of Chicago Press, 1994.

Walter, Lutz, ed. *Japan, a Cartographic Vision: European Printed Maps from the Early 16th to the 19th Century.* Munich: Prestel, 1994.

PART I

Visualizing the Realm

SIXTEENTH TO EIGHTEENTH CENTURIES

Introduction to Part I

SUGIMOTO Fumiko 杉本史子

The Japanese archipelago is surrounded on all sides by the ocean, just as the planet itself is a world enveloped by the sea. It is here, in the great depths, that our journey through Japanese maps begins.

Human history became a global phenomenon in the sixteenth century, when economic activity began to integrate the world on a massive scale. Deemed the Age of Discovery on account of the Europeans who ventured outward from their peninsular landmass into the Atlantic and beyond, this period likewise found an unprecedented flow of people and things churning up the waters on the eastern side of Eurasia as well. East Asia was becoming an ethnically plural sphere of commerce that included people hailing from various places, including Europe. Pirates and merchants, military hegemons and resolute rulers all competed for profit in this vibrant sea trade.

Out of this sixteenth-century swirl emerged a regional balance of power among four distinct centers in East Asia: the Qing dynasty in China, the Tokugawa shogunate in Japan, the Yi dynasty in Korea, and the quasi-independent Ryukyu kingdom in Okinawa (caught between the Qing and the Tokugawa). The governments of Japan, Korea, and Ryukyu rigorously controlled the movements of people across their borders, and the East China Sea became a highly regulated body of water by the eighteenth century. There was even a de facto system in the region for helping castaways and repatriating them to their home countries.[1] It was in this globalizing context that cartography took off in Japan. As human interactions spilled over the confines of discrete political communities, rulers and merchants alike craved the synchronic overview that only maps can provide. This was indeed the era of the "mapped society," when princes and traders throughout the world rushed to make use of many different kinds of maps for an equally diverse range of purposes. The Japanese archipelago was no exception.

As a result this period of Japan's history has left us with a dizzying diversity of maps, each one made with particular goals in mind. There are maps projecting particular views of the entire world; maps of Japan itself; maps for administering regions, villages, and cities; flood control maps; maps to use while traveling by road, river, or sea; map-like architectural plans; maps emphasizing a certain party's claim to land and water rights; maps that made known boundary decisions; maps that publicized current events like natural disasters or military conflicts; maps reflecting the historical research of scholars; and sightseeing maps for pleasure.[2] Several conditions had to be met before this robust cartographic culture could emerge. The planning know-how required for making maps needed to be present, as did the vision to summon them into being. Also necessary were the cultivation of raw materials for paper, the circulation of materials for making color pigments, and the development of information circuits through which geographic knowledge could flow. How were these basic conditions satisfied under the Tokugawa? A brief history of Japan up to the founding of the shogunate will make the answers to that question easier to grasp.

In the eighth century, a state centered on the emperor was established in the Japanese archipelago. Over time, the class of people responsible for military affairs grew steadily in authority. By the twelfth century, although political legitimacy was still concentrated in the emperor and his court (*chōtei*), administrative power had been seized by a federation of samurai (*bakufu*). In the fifteenth century, powerful regional samurai lords went to war with each other throughout the archipelago; some launched themselves into smuggling and piracy in the East China Sea. In this context, the charismatic Toyotomi Hideyoshi mobilized the warlords of western Japan to invade the Korean Peninsula in a campaign to attack the Ming court in China. Hideyoshi's invasion failed, and his clan soon died out, setting the stage for the Tokugawa family to gather the political powers throughout the archipelago under a single umbrella.

Following five centuries of samurai tradition, the Tokugawa made a military corps the basis of political organization. Samurai lords with large territorial holdings were called daimyo, and the greatest of these were

FIGURE I.1 "A Picture of *Higaki* Boat Speed Race at the Mouth of Ōsaka's Aji River" [Higaki shinmen bansen Kakō Shuppan no zu 菱垣新綿番船川口出帆之図] by Gansuite Yoshitoyo 含粋亭芳豊, ca. 1850s. Woodblock print, 36.2 × 24.7 cm. Courtesy of Osaka Castle Collection.

the Tokugawa shoguns. The Tokugawa in turn made the other samurai lords their personal retainers. Theirs was the first government in Japan to regulate imperial and religious institutions by law. The shogun made his capital in Edo (today's Tokyo) whose population grew to over one million people by 1800, making it the world's largest city at the time. The Tokugawa government linked its headquarters in Edo both by land and by sea to the historical heartlands of central Japan, including Kyoto, the seat of the emperor, and Osaka, the historic commercial center. A symbol of this expanded network occurred every year in autumn with the *shinmen bansen*, a race for boats rigged with large sails carrying loads of freshly harvested cotton from Osaka to the mouth of Edo Bay (fig. I.1).[3]

The Edo-Osaka trunk line anchored a network of sea routes that reached around the entire archipelago to support the circulation of goods. At the same time, land routes radiating from Edo developed to service the ruling class (see the chapter by Constantine Vaporis below). These turnpikes facilitated the shogun's command for all daimyo to maintain their wives and children as permanent hostages in Edo, while they themselves were required to travel to and from their home domains. Official documents and letters could be dispatched quickly between Osaka and Edo along the same roads. The various highways each had fifty or more checkpoints by 1745, and since they emanated from the areas surrounding the castle and neighborhoods of Edo, they constituted lines of defense around the shogun's capital.

As lively as this interregional network was for those who used it, it was nevertheless controlled by the authorities and circumscribed to areas within and immediately around the archipelago. Points of contact with foreign lands were officially restricted to four gateways, the most important of which was Nagasaki, where most goods and books imported from China entered. Nagasaki was also a critical regional point of entry for Western goods, since it was the only harbor in Japan, Ryukyu, or Korea that was open to trade with Europe. The remaining three official gateway ports were Matsumae in the north (for trade with Ezo), Tsushima in the west (for trade with Korea), and Ryukyu in the south (for trade with China).

Life inside Japan's closely guarded shores was strongly shaped by the principle of socio-spatial segregation. An entrenched status system ensured disparity between people on the basis of territorial bonds, blood ties, and occupation. At the end of the medieval period, local communities (*sōson*) throughout the archipelago were largely self-regulating, maintaining their own laws, property, police, and courts. The political powers uniting the archipelago divided commoner settlements into two types: urban settlements (*chō*) and farm villages (*mura*) (chapter by Komeie Taisaku). In the early modern period, these communities could even submit lawsuits to the shogunal court upon securing permission from their regional lord.

The farmers in charge of leading the affairs of their villages were highly literate men, many of whom acquired surveying and mapmaking skills between the 1650s and the 1750s. Regional lords often enlisted their assistance to make maps of their domains. Such was the case with the celebrated civilian engineer Inō Tadataka, who, after retiring from a successful career in his family's business, led a team that spent sixteen years surveying the entire coastline of Japan.

Early modern samurai rulers, ensconced in cities far away from the sites of agricultural production, coveted something that would reveal in visual terms the territories that they ruled. For those at the top of the political pyramid that meant commissioning maps of the entire nation. On five separate occasions, daimyo from throughout Japan were ordered to submit detailed provincial maps and cadastral surveys to the shogun. These magnificent maps drew on a long tradition of political cartography in East Asia. For example, a world map made in Korea at the beginning of the fifteenth century had represented military positions using a visual system of round and square geometric labels (fig. I.2). Provincial maps made in Japan two centuries later made free use of similar labels as a way to present territory ruled by both the shogun and the daimyo (chapters by Sugimoto Fumiko).

The Korean world map is of interest to us in another way as well. Occupying what we might call the cartographic center of fifteenth-century East Asia, Koreans were in a position to draw on geographic information from China, Korea, Japan, and Ryukyu to compile an unprecedented picture of the world. Even though it predated the Age of Discovery, their map encompassed Africa as well as Eurasia, and it depicted the territories and capitals of successive emperors over a vast stretch of space and time.[4]

The world maps we are familiar with today are produced according to a way of thinking completely unlike that which inspired this Korean world map. The maps we normally think of as representing the whole world derive from sea charts that followed the global advance of European power. *Portolanos* originally made for voyages across the Mediterranean Sea were stretched over time to trace Europe's advance into new continents and seas between the 1450s and 1700s, incorporating the geographic knowledge of local peoples as they proceeded (chapters by Peter Shapinsky and Oka Mihoko).

The massive scale of mapmaking in the Edo era required a sturdy yet affordable physical medium, one that would be appropriate to the construction of large quantities of maps. In early modern Japan, the solution was readily at hand in the form of indigenous paper made from the pith of various tree species. Not surprisingly,

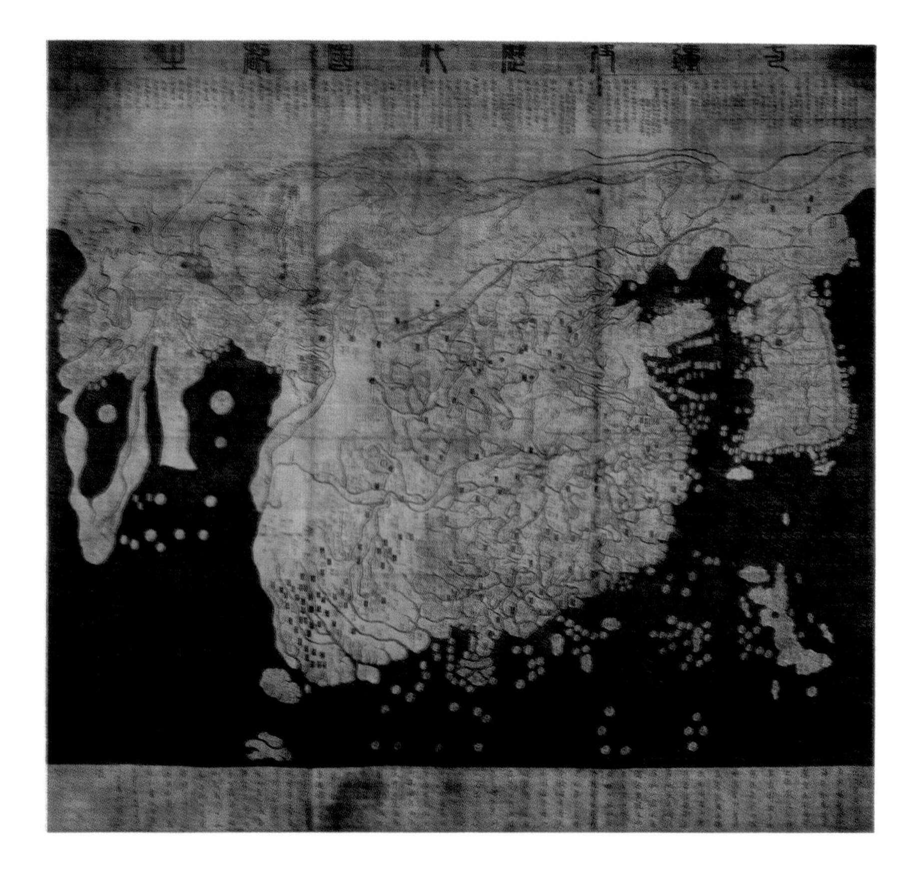

FIGURE I.2 "Map of Integrated Lands and Regions of Historical Countries and Capitals" [Kon'itsu kyōri rekidai kokuto no zu 混一疆理歴代国都之図], 1402. Manuscript on silk, 147.9 × 163.4 cm. Courtesy of Ryūkoku University.

mulberry and *mitsumata* trees, the raw materials for Japanese paper, came to be cultivated widely during these centuries. With its characteristic strength and lightness, Japanese paper could be used to make maps of many different shapes, and not a few mapmakers came up with clever designs to show three-dimensional spaces. Architectural renderings, for instance, could comprise several sheets of paper pasted together to indicate both the floor plan and side views of built structures. Engineers' maps, likewise, might reveal the invisible depths of mines by joining different slips of paper together so as to indicate underground tunnels (fig. I.3). The conical map of Mount Fuji introduced by Miyazaki Fumiko in her essay is another good example of a design that took advantage of the special characteristics of Japanese paper to show a three-dimensional object.

The medium of paper offered a range of viewing options as well. As was typical in Europe, luxurious cartographic artifacts were often designed for display on vertical surfaces. Beautiful maps composed with gold could be mounted on lavish screens for viewing by people assembled at ceremonial sites (fig. I.4). The world

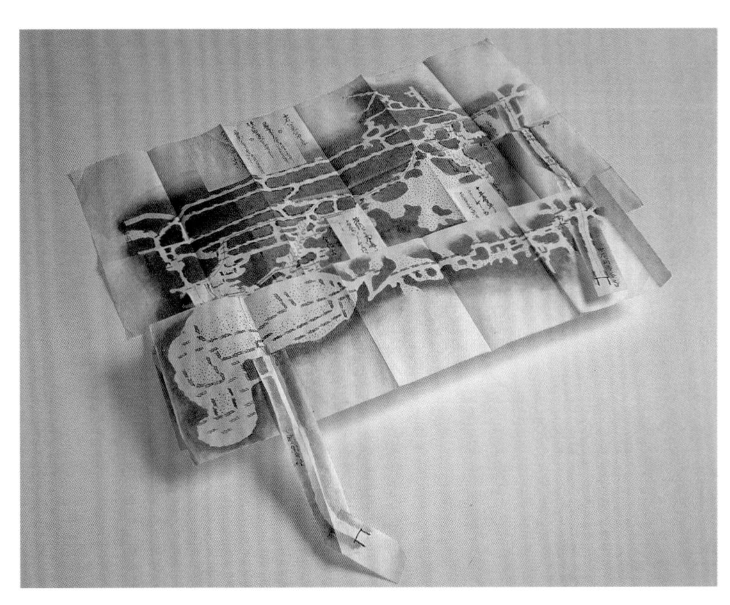

FIGURE I.3 "Underground Map of Nakao Shaft from Kamanokuchi Entrance" [Nakaomabu Kamanokuchi no zu 中尾間歩釜ノ口図], ca. 1832. 55.9 × 41 cm. Courtesy of the Historiographical Institute, the University of Tokyo. This shows part of a tunnel in the Sado gold mine, which was under shogunal administration.

FIGURE I.4 "Map of Bitchū Province" [Bitchū kuniezu 備中国絵図], date unknown. Folding screen: Chinese ink, color, and gold on paper, 187 × 626.4 cm. Courtesy of Okayama Prefectural Museum.

map discussed by Joseph Loh is such a screen-mounted map. Buddhist maps might be mounted on hanging scrolls and displayed on temple walls to explain religious ideas to adherents. But most maps in the Edo era circulated as portable paper sheets. This in turn gave cartographers considerable flexibility as to size. Broadly speaking, maps made in Japan during this period came in three formats. The smallest could be opened and held in both hands or folded up and slipped into the sleeve of a kimono, making them ideal for travel. A middling category, which might reach one meter on a side, was designed to be viewed on a tatami-mat floor. The third category was vastly larger, often exceeding three meters in length. A map this large was itself a singular space that gave the viewer who stood before it a commanding sense of power. Provincial maps and those made to promulgate the decisions of the shogunal court are prime examples of such giant cartographic documents. We can surmise that shoguns viewed these maps while standing on top of them in the large four-hundred-tatami-mat ceremonial rooms in the inner citadel of Edo Castle.

Another distinctive feature of early modern Japanese maps was their liberal use of colored symbols to depict spatial and social phenomena. This distinguishes them from both medieval Japanese maps (which used black or perhaps two or three colors) and modern maps (which rely principally on lines for the delineation of terrain). For such richly colored maps to materialize, it was necessary for trade to extend beyond the archipelago, for whereas paper was domestically produced, most of the minerals required for color pigments were not available in Japan at the time. For example, early modern cartographers routinely used red paints made from vermilion, a compound of mercury and sulfur, to indicate roads. While mercury had previously been mined and even exported from Japan, starting in the sixteenth century its importation became the norm, with the best vermilion coming from Fujian Province in China.[5] By granting one trade association monopoly rights to vermilion imports, the shogunal government attempted to regulate this key material, albeit not without competition from smugglers.[6]

When considering the cartographic culture of this period, we also must not overlook the phenomenon of commercial publishing, examined at greater length in part II. The Tokugawa shogunate, except in its last days, did not attempt to standardize the contents of published maps, so inconsistent images of Japan and the world circulated in print. Some cartographers continued to work in the simple "fish-scale" style that had been in existence for centuries (see fig. 0.2 above). The map analyzed by Marcia Yonemoto participates in this tradition. In these images, the sixty provinces appeared almost as if they had been glued one on top of the next, yielding "Japan" as an assemblage of provinces with no distinct boundaries of its own. This type of map was originally created by the ancient state for the use of officials, and as late as the medieval period, it would have been familiar only to a handful of aristocrats and literate elites. It was only in the early modern era that woodblock prints introduced such a picture of Japan to a wide readership. At the same time, much more accurate silhouettes of the nation—incorporating the rigorous coastal detail that emerged out of the maritime *portolano* tradition—also came into circulation (as discussed by Matsui Yōko). But whatever their style, all the published maps of "Japan" had one thing in common: they were crowded with up-to-date geographic information. The names of daimyo, castles, famous sites, checkpoints, sea and land routes, and toponyms of all types filled their lively surfaces.

While hand-drawn maps were kept by lords and various communities as proof of their carefully guarded rights and claims to rule, printed maps broke free from their producers, reaching far-flung places throughout the archipelago and beyond. Some of those maps slipped through official regulatory controls on information, crossed oceans, and were read by people on the far side of the world. In this way, Japanese maps gradually became drawn into the vigorous global cartographic cultural exchange of the early modern era.

(Translated by Robert Goree)

Notes

1. Kishimoto Mio, *Kinsei-ka ron to shinchō (bessatsu kan 16)* (Tokyo: Fujiwara Shoten, 2009); Kojima Tsuyoshi and Haneda Masashi, eds. *Higashi ajia kaiiki ni kogidasu 1: Umi kara mita rekishi* (Tokyo: Tokyo Daigaku Shuppankai, 2013).

2. Sugimoto Fumiko et al., *Ezugaku nyūmon* (Tokyo: Tokyo Daigaku Shuppankai, 2012) is the first comprehensive introduction to maps made in Japan from the seventeenth century through the first half of the nineteenth century. It is the result of a collaborative effort among specialists in history, geography, the history of science, architectural history, cultural history, the science of cultural

assets, Japanese-style painting, and the reproduction and restoration of cultural objects.

3. In 1859, a boat with a crew of twenty sailors and cargo weighing 270 tons made it to the finish line 650 km away in just 66 hours. This color woodblock print (fig. I.1) shows the start of the race. At the upper right, the boats are moored at the mouth of Osaka's Aji River, ready to depart. At center, magnificent warehouses are depicted just upstream from the mouth of the river; within this area, a brilliant bunting has been stretched for the race. Here is where the ship captains awaited the start signal. Once the captain of a crew received the required permit, he hurried out in a small boat to board the ship he would sail to Uraga.

4. For a long time, the only extant copy of the map was thought to exist in Japan, but another copy has recently been found in China. Sugiyama Masaaki, "Tōzai no sekaizu ga kataru jinrui saisho no daichihei," in *Daichi no shōzō*, edited by Fujii Jōji (Kyoto: Kyoto Daigaku Gakujutsu Shuppankai, 2007); Miya Noriko, *Mongoru teikoku ga unda sekaizu* (Tokyo: Nihon Keizai Shinbun Shuppansha, 2007).

5. A historian who entered the Yamato mercury mine in Nara Prefecture describes being surrounded above, below, and on all sides by an unimaginably beautiful red color. Matsuda Hisao, *Kodai no shu* (Tokyo: Chikuma Shobō, 2005).

6. Satsuma domain, which officially oversaw trade with Ryukyu in western Japan, was caught in the nineteenth century engaging in the illegal importation of vermilion off the coast of Niigata.

Suggested Reading

Kuroda Hideo, Mary Elizabeth Berry, and Sugimoto Fumiko, eds. *Chizu to ezu no seijibunkashi* (Tokyo: Tokyo Daigaku Shuppankai, 2011).

1 Japan and a New-Found World

Joseph Loh

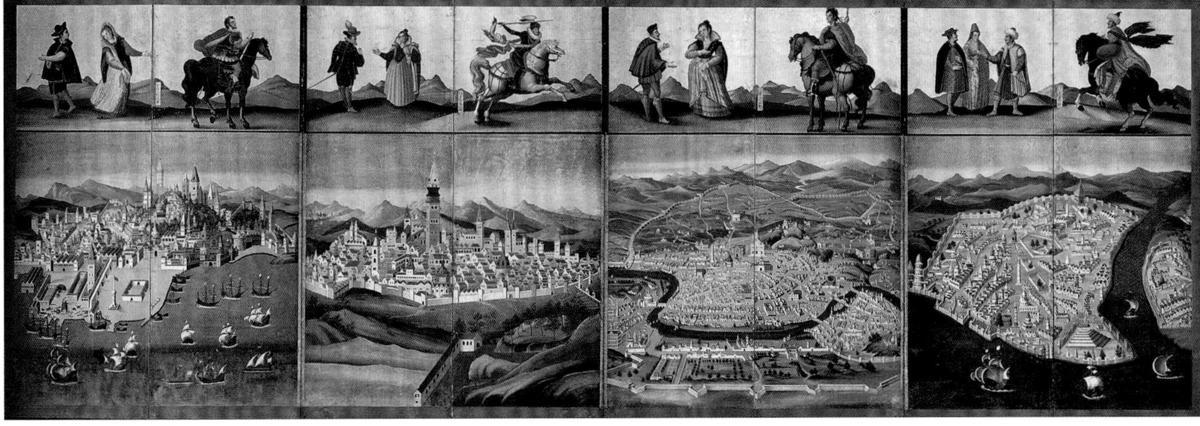

FIGURE 1.1 "Map of the World/Four Large Cities" [Sekai zu, yonto zu 世界図 · 四都図], early seventeenth century. Pair of eight-panel screens: ink, color, and gold on paper, 158.7 × 477.7 cm. Courtesy of Kobe City Museum.

Sumptuous folding screens provided elite Edo-period viewers with a map of a newly discovered world and a wondrous glimpse of places and peoples in faraway lands. On the pair shown here, one screen reproduces a European map of the world. A multitude of colors demarcate continents, countries, and regions. Palace-shaped cartouches mark cities and places of perceived inhabitation; European galleons traverse quilt-patterned seas. Spherical inserts show the polar regions, lunar and solar eclipses, illustrations of wind and compass roses, and latitudinal lines marking the equator and the tropical zones. Spouting whales share the high seas with mythic sea creatures and others that are half human, half monster. In the upper region of the North American continent, the artist has rendered mountain ranges in hues of green, brown, and gray to convey atmospheric distance and depth. The Japanese archipelago appears on the right edge of the map; below it, a circular insert expresses Japan's placement in relation to China, Korea, and North America. The paired screen shows views of the four cities of Lisbon, Seville, Rome, and Constantinople (Istanbul). Along the top are images of aristocrats in fancy dress and noblemen on horseback. Both screens make luxurious use of rich pigments and gold leaf.

This pair of monumental screens from the collection of the Kobe City Museum represents a hybrid art form: one that took pictorial elements from European *portolanos* (nautical charts), printed maps, and book illustrations and used them to embellish a traditional Japanese medium, the paired folding screen. Before he was finished, the designer of this particular set had drawn on at least six separate printed or painted sources. His view of Rome is based on the 1610 *Vita beati patris Ignatii Loyolae*, for instance, while the other three cities are apparently adapted from a famous late sixteenth-century Latin compilation called *Civitates Orbis Terrarum* or Cities of the World. Another European source provided decorative motifs and embellishments, while a domestic map must have been consulted for Japanese geographic features.

This lively mingling of Japanese and Western forms resulted from the unprecedented trade, religious engagement, and cultural exchange between Japan and Europe during the century from 1542 to 1641. The European traders and Christian missionaries who visited Japan during this time introduced a wide range of European visual imagery and pictorial techniques into the repertoire of Japanese artists. Historians have designated the resulting works as Nanban, or "Southern Barbarian," art: that is, Japanese art bearing a connection to European sources through visual design, subject matter, or context of production. Roughly twenty of the surviving multipanel folding screens from this period feature Western maps of the world. In making them, Japanese artists adapted material from European atlases and printed maps by such pioneers as Ortelius, Mercator, and Blaeu. As a genre, the Nanban map screens are among the earliest examples of Japanese visual culture shaped by European cartographic science, geographic knowledge, and overseas trade and exploration.

Following conventional Japanese practice, artists typically painted two screens to create a paired set. Their themes vary greatly, ranging from a map of Japan to European city and town views to depictions of foreign battles or Portuguese trading ships in Japanese ports. Despite being reminiscent of European models, these painted works are not mechanical copies of European maps. Rather, they are pictorial displays of stunning invention, in which Japanese artists confronted new subjects, motifs, and ideas and imaginatively transformed them to suit Japanese sensibilities. Being so radically new, and with patrons probably willing to pay handsomely for such works, the genre had no rules and few limits.

Three main groups of artists appear to have been responsible for the surviving screens from this era. One group was evidently connected to the Jesuit seminary that was active in Japan from 1590 to 1614, a second belonged to the more traditional Kano school (an influential painting lineage with origins in the fifteenth century), and a third consisted of anonymous town artists. In the case of the screens considered here, the artist must have had access to the Jesuit community, given the large number of sources from which the composition was built. He may also have received some measure of direct instruction from a European artist. This is suggested by the delicate brushwork in richly applied pigment intended to simulate European oil painting, the application of Western techniques such as modeling and shadowing, the distinctive quilted wave pattern, and the rendering of ships and topography.

The only patrons able to afford such luxurious objects were members of the ruling elite: warriors, abbots, court aristocrats, members of the imperial house, or rising merchants in the growing urban centers. In the case of Nanban art, some screens are known to have been commissioned by Christian missionaries. They might serve as lavish gifts for influential warlords or Japanese merchants, or as commodities to gain favor or mark the closure of significant business transactions. But wealthy Europeans appreciated these impressive objects as much as their Japanese counterparts did, and not a few were made as export objects destined for the palaces and salons of Europe. (This pair, in fact, only returned to Japan after a Japanese art dealer discovered them in Paris in the 1930s.) Monumental and impractical for travel, they were probably not meant to serve as a proper map or to be used anywhere outside the confines of an official reception room. Their main purpose would have been to impress viewers as dramatic centerpieces or backdrops for conversation in carefully controlled display environments and social situations.

Images like these capture an early Japanese comprehension of Western knowledge and an acceptance of a new cartographic vision of the world. But acceptance did not rule out creative adaptation. Artists routinely increased the size and prominence of Japan in proportion to the Asian continent and rendered the Japanese archipelago in greater detail. On some screens (as here), Japan is given further emphasis through special insets; others situate the Pacific Ocean or East Asia at the dramatic focal point of the composition, effectively putting Japan near the center of the globe. While offering stunning views of faraway places, such images assured seventeenth-century Japanese viewers of their own privileged place in the world. As objects of high culture, map screens such as these served simultaneously to locate Japan in a newly expanded world, legitimize the authority of their proud owner, and articulate Japanese identity in relation to peoples of other lands.

Suggested Readings

Bailey, Gauvin. *Art on the Jesuit Missions in Asia and Latin America, 1542–1773*. Toronto: University of Toronto Press, 1999.

Jackson, Anna, and Amin Jaffer. *Encounters: The Meeting of Asia and Europe 1500–1800*. London: V and A Publishers, 2004.

Okamoto Yoshitomo. *The Namban Art of Japan*. Translated by Ronald K. Jones. New York: Weatherhill/Heibonsha, 1972.

Santorii Bijutsukan, Kobe Shiritsu Hakubutsukan, and Nihon Keizai Shinbunsha, eds. *Nanban bijutsu no hikari to kage: Taisei ōkō kibazu byōbu no nazo* [Light and shadows in Namban art: The mystery of the Western kings on horseback]. Tokyo: Nihon Keizai Shinbunsha, 2011. (In Japanese with English-language chapter summaries.)

Shirahara Yukiko, ed. *Japan Envisions the West: 16th–19th Century Japanese Art from the Kobe City Museum*. Seattle: Seattle Art Museum, 2007.

2 The World from the Waterline

Peter D. Shapinsky

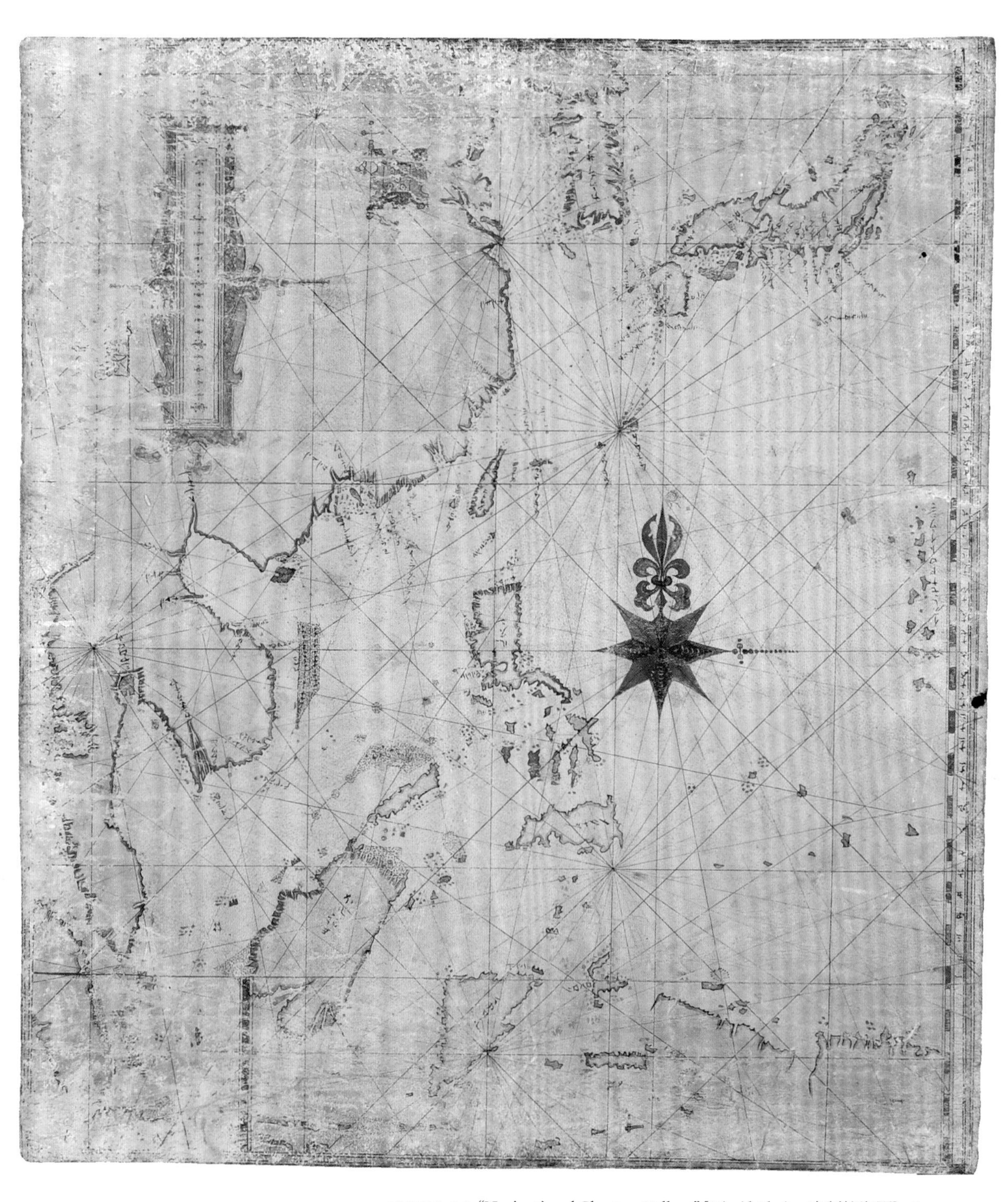

FIGURE 2.1 "Navigational Chart on Vellum" [Hisei kōkaizu 皮制航海図], 1630s. 43.5 × 38.7 cm. Courtesy of the Jingū Chōkokan Museum, Ise, Japan.

Only a handful of sixteenth- and seventeenth-century Japanese nautical charts of Asia survive, but their rarity should not blind us to their immense historical significance. These hybrid maps represent the dawning of a global, sea-centered view of the world. They emerged out of an environment of unprecedented cross-cultural experimentation, one where ship captains and cartographers alike integrated knowledge from Chinese, European, Japanese, and Southeast Asian sources to produce maps that reflect mariners' geographic perspectives and concerns. Most of the extant nautical charts made in Japan during this period originated in the so-called vermilion-seal trading system: a regime of chartered trade, created by Toyotomi Hideyoshi and continued by the early Tokugawa shoguns, that used official red seals to license and regulate commercial traffic with China and Southeast Asia between the 1590s and 1630s. A few copies of what may well be earlier Japanese nautical charts exist in modified form in Korean and Chinese compilations, and we have several copies of lost vermilion-seal maps that were made later in the Tokugawa period. But after the Tokugawa outlawed overseas travel in the late 1630s, charts only survived if they could take on new functions for their owners. The map shown here is a Japanese nautical chart once owned by the Kadoya family of sea merchants who were active in the vermilion-seal trade with Southeast Asia (fig. 2.1). It exemplifies both the processes of cross-cultural cartographic conversations that produced sea-centered visions of the world and the contingencies that enabled a few such maps to survive.

The Kadoya chart, like other vermilion-seal maps, was based on contemporary Portuguese portolan charts used for navigation. *Portolanos* first emerged in the Mediterranean world in the thirteenth century, about the time the compass began to see widespread use in Europe, and remained a dominant form of European maritime mapping well into the seventeenth century. Portolan charts are best understood as repositories of seafaring knowledge with no particular national allegiance. The Kadoya therefore invites a global reading of the charts as cartographic tools that could meet the needs of pilots worldwide. These were maps attuned to contemporary navigational practices, grounded in the experience of local pilots and constructed from the waterline.

To make the Kadoya chart, the anonymous drafter first acquired a bleached piece of vellum and lightly sketched a large "hidden circle" on it using a lead pencil. Around that circle he spaced sixteen equidistant starbursts radiating color-coded compass direction lines (known as rhumb lines): black for primary compass directions, green for secondary, and red for tertiary. Next, the mapmaker overlaid red representations of coastlines atop the rhumb lines. Reversing the land-centeredness of many maps, here the inland areas were either left empty or given over to ornamentation. Seas, coasts, and littoral regions were the primary focus. Particular littoral regions were exaggerated and drawn well out of scale with the rest of the map, either to highlight locations of interest to the patron or to show areas that required complex navigation. After that, the drafter wrote place-names perpendicular to the coasts in black ink. This removed any need for fixed orientation, allowing a navigator to simply trace itineraries from point to point following the relevant compass headings. Navigational information was encoded in various ways, by adding dots for shoals, a Spanish scale of miles, and lines of latitude, including the equator and tropics. Finally, the reverse side of the map reveals an unfinished attempt to translate the Ptolemaic cosmological system of celestial spheres to Japan. Someone has inscribed eleven white concentric rings. Inside each ring should have been inscribed a particular astronomical feature, beginning with the orbits of the moon, several planets, and the sun, followed by the firmament (the abode of the stars), and then three heavens. But the diagram on the reverse of the Kadoya contains only a solid white circle between the first and second rings (to indicate the moon), and a solid red circle between the fourth and fifth rings (the sun). Contemporary European navigational materials were often paired in a similar way with explications of Ptolemaic cosmology.

It is likely that a single, no-longer-extant Portuguese

portolan chart provided the base map for all known Japanese charts. One nineteenth-century copy even bears the Portuguese inscription "made by Sebastio" (Sebastio a fez). On the Kadoya map, Portuguese influence can be seen in the design of the compass rose, the scale of miles, the latitude depictions, and the flag indicating possession of Macao (from 1557). This last has been given an impressive pole extending from the island to the top of the map, perhaps signifying the importance of this colonial entrepôt for Portugal and regional trade.

Nevertheless, portolan charts were flexible tools that could easily be adapted to local circumstances and integrated into different navigational cultures. In contrast to modern scaled and gridded maps, they lack a single unifying projection and do not attempt to account for the curvature of the earth. Instead, drafters assembled local navigational lore gathered piecemeal by pilots (some mapmakers worked as pilots themselves) with the understanding that their charts would be used in conjunction with written itineraries, astrolabes, prayers, and other navigation tools.

Such characteristics made portolan charts compatible with contemporary East Asian navigational and cartographic practices. When European travelers reached Asian waters in the early sixteenth century, they joined sophisticated preexisting trade networks. These sea-lanes teemed with Chinese, Japanese, Southeast Asian, and Ryukyuan seafarers whose navigational cultures in some ways resembled their own. Like their Portuguese or Spanish counterparts, East Asian crews traced routes on maps in accordance with written itineraries, followed compass headings, measured distance with time, and found prayers efficacious at sea. Cross-cultural mixing continued aboard vessels commissioned in the vermilion-seal trade, for which the Tokugawa issued licenses to Chinese and Europeans as well as Japanese shippers. The resulting cosmopolitan environments produced creolized languages, hybrid shipbuilding practices, and cross-cultural fusions in the fields of navigation and cartography.

The Kadoya and other Japanese portolan charts thus reveal a fascinating mixture of Japanese, Korean, Chinese, Southeast Asian, and Portuguese influences. Placenames, written primarily in Japanese phonetic script, include Portuguese toponyms alongside Chinese, Japanese, and Southeast Asian names that were in common use throughout the Asian trading world. Comparing the depiction of Chinese, Korean, and Japanese coastlines here with those on contemporary European charts reveals the integration of considerable local knowledge. Subtle differences among extant Japanese charts also attest to continued efforts by cartographers involved in the vermilion-seal trading system to update and refine their geographic content over time.

Although it is not clear how or when the Kadoya family acquired this map, it was produced in the same vermilion-seal spheres of interaction in which they actively participated as traders. Rising to prominence as shipping merchants based in the Ise port of Ōminato in the late sixteenth century, they were thereafter enabled by Tokugawa patronage to add overseas trade to their domestic connections. Their commercial network linked their home in Ise with the Japanese ports of Sakai and Nagasaki and eventually the southern Vietnamese entrepôt of Hoian. One scion of the Kadoya house, Eikichi, traveled to Hoian in the 1630s and married into the Nguyen family, a powerful clan that made a practice of adopting and intermarrying with Japanese merchants in order to promote their commercial connections with Japan. Eikichi eventually became head of the "Japan Town" in Hoian. There, as in Nagasaki, as well as on the sea-lanes connecting the two ports, he would have encountered Chinese, Dutch, English, Portuguese, and Spanish pilots and merchants. Only a few years after his arrival in Hoian, the Tokugawa shogunate abruptly put an end to the vermilion-seal trade, and Kadoya Eikichi was only able to maintain connections with Japan by entrusting his goods and letters to a network of Chinese shipmasters and merchants who carried them to specific Japanese agents in Nagasaki.

The Kadoya chart was physically modified to fit this Japan–China–Southeast Asia trade network. Whereas most Japanese portolan charts follow the Portuguese original and extend from Japan in the east to the Arabian Peninsula and Africa in the west, the designer of this particular version has cut off everything west of the Straits of Malacca, focusing on the areas frequented by the vermilion-seal trade. The region bounded by the southern Chinese province of Guangzhou and the Mekong Delta in Vietnam is particularly magnified. At least one pilot found such detail helpful. Closer examination of the Kadoya chart reveals two lines of pinholes, suggesting the use of a divider compass to chart a course between Nagasaki and Hoian. Eventually the chart was remitted to the Kadoya in Japan, where the family labeled it for posterity as "a map used for crossing to foreign countries, made of bleached cowhide."[1]

After 1635, when the Tokugawa ceased issuing overseas trade licenses and forbade Japanese to travel overseas, maps like the Kadoya chart lost their function as navigational aids. Some of their owners adapted portolan cartography to fit the new land-centered era by turning it into both a tool and a certificate for the work of surveying the land of Japan; others seem to have interpreted portolan charts simply as repositories of information related to foreign lands. For Kadoya Eikichi's descendants, this particular chart may have functioned above all as proof of their family's illustrious heritage, given that they preserved it as part of a collection of trade-related materials including flags, licenses, letters, and genealogical works. The Kadoya did so not only in order to document their family history. This collection may also have served as an evidentiary arsenal in court battles over their ongoing domestic commercial shipping operation, which enjoyed tax exemptions granted with vermilion seals.[2]

Notes

1. Matsumoto Dadō, *Annan-ki* (1807), reproduced in "Jingū Chōkokan shozō jūyō bunkazai *Annan-ki* no shōkai," by Matsumoto Yoshihiro, *Mie Chūkyō Daigaku chiiki shakai kenkyū shohō* 22 (2010): 63.

2. Kado Kiyoshi, "Annan bōeki-ka Kadoya Shichirōbē kankei shiryō no gaiyō," *Mie Chūkyō Daigaku chiiki shakai kenkyū shohō* 21 (2009): 217; Kado Kiyoshi, *Mie-ken shi shiryō-hen kinsei 4* (Tsu-Shi: Mie-ken, 1993), 57–58, docs. 1–2.

Suggested Readings

Campbell, Tony. "Portolan Charts from the Late Thirteenth Century to 1500." In *History of Cartography*, vol. 1, *Cartography in Prehistoric, Ancient and Medieval Europe*, edited by J. B. Harley and David Woodward, 371–463. Chicago: University of Chicago Press, 1987.

Li Tana. *Nguyễn Cochinchina: Southern Vietnam in the Seventeenth and Eighteenth Centuries.* Ithaca: Cornell University Press, 1998.

Nakamura Hiroshi. "The Japanese Portolanos of Portuguese Origin of the XVIth and XVIIth centuries." *Imago Mundi* 18 (1964): 24–44.

Shapinsky, Peter D. "Polyvocal Portolans: Nautical Charts and Hybrid Maritime Cultures in Early Modern East Asia." *Early Modern Japan* 14 (2006): 4–26.

Waters, D. W. *The Art of Navigation in England in Elizabethan and Early Stuart Times.* New Haven: Yale University Press, 1958.

3 Elusive Islands of Silver: Japan in the Early European Geographic Imagination

OKA Mihoko 岡美穂子

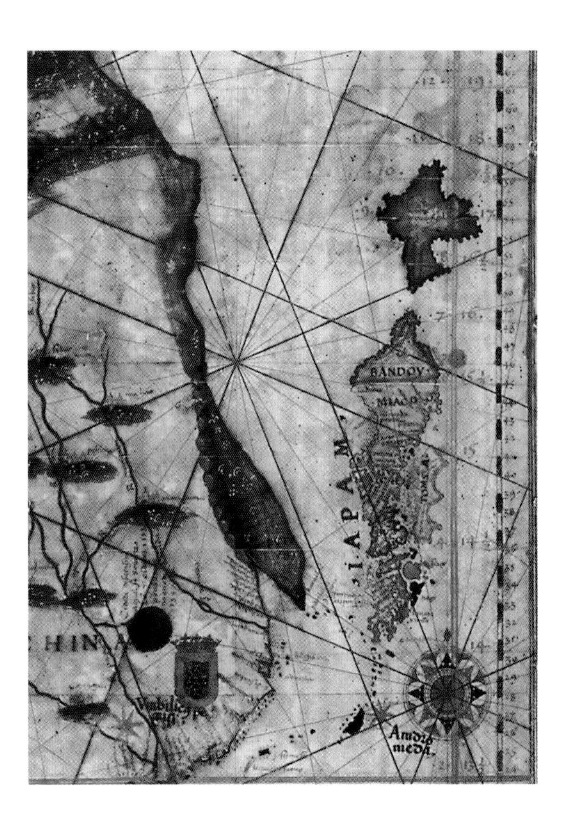

FIGURE 3.1 "General Chart of the Globe" [*Carta General do Orbe*] by Bartolomeu Velho, 1561; detail showing Japan ("Iapam"). Courtesy of Museo Galileo, Florence, Italy.

Geographic understanding of the world advanced greatly in the fifteenth and sixteenth centuries through the overseas exploration promoted by the two Iberian powers of the age, Portugal and Spain. Europeans started recognizing Japan as Marco Polo's Zipangu around the 1550s, as Portugal's commercial network expanded into the China seas. Zipangu was initially believed to be an island abundant in gold. But the first Jesuit who arrived in Japan, Francisco Xavier, reported otherwise. Based on information from Portuguese traders who sailed in the waters around Japan, Xavier associated the country with a different precious metal. Thereafter, Japan became famous among European seafarers as an island of silver.

Soon after the conquest of Malacca in 1509, the Portuguese started navigating the China seas with the guidance of Chinese traders or on board Chinese junks. After encountering people from the Ryukyu kingdom at Malacca, the Portuguese began exploring the Ryukyu Islands in 1517. A map of the world by an anonymous Portuguese mapmaker, conserved in the Biblioteca Vallicelliana in Rome, shows a part of East Asia as *LEQVIOS*, which includes the small island of *Iapam* as part of the archipelago of Ryukyu. The map is thought to have been made between 1545 and 1550, suggesting that the Portuguese knew more about the Ryukyu Islands in that period than about Japan. Japanese historian Murai Shōsuke has pointed out that Japan was considered part of the Ryukyu archipelago until the 1550s, when more accurate information concerning Japan was acquired by European (in this case Portuguese) mapmakers.

How did information on Japan become abundant enough in Europe to influence cartographers in the 1550s? The key players were Portuguese merchants, who started trading in Japan in the 1540s. The first empirical report on Japan—including an account of its natural features, products, people, and religion—was written by Portuguese trader Jorge Alvarez at the request of Francisco Xavier in 1547 at Malacca. When Alvarez returned to Malacca, he brought with him a Japanese man called Anjiro who gave the Jesuits additional information. According to a report by Nicolao Lancilotto, an Italian Jesuit present at the time, Anjiro provided information regarding Japan's abundant silver mines and gold mines and the production of copper, lead, tin, iron, mercury, and sulfur. Such knowledge, gained by Jesuits from Portuguese traders, helped bring about a new level of geographic understanding among European mapmakers.

Jesuits were not only keen on collecting secondhand information about Japan's geography and natural environment, however. They also conducted their own investigations after arriving in Japan. For example, after a battle between Japanese and Portuguese traders at the port of Hirado in northern Kyushu in 1561, it became necessary to locate a new trading site in Kyushu. It was the Jesuits who worked to find a port suitable for the entrance of large Portuguese vessels by sounding the depth of the water and measuring the surrounding natural features. In addition, before Nagasaki was confirmed as the new trading port, the Jesuits landed and walked on shore to investigate potential trading sites around the island of Kyushu. These explorations seem to have contributed to later maps of Japan.

In the 1560s, Portuguese depictions of Japan suddenly became much more detailed than those of previous decades. Consider the 1561 "General Chart of the Globe" (*Carta General do Orbe*) (fig. 3.1) by the outstanding cartographer, geographer, and cosmographer Bartolomeu Velho ("velho" is not a surname but a nickname meaning "old"). Bartolomeu's chart, conserved in the Accademia di Belle Arti (Florence), clearly (if crudely) identifies three islands—Honshu, Kyushu, and Shikoku—in addition to a large island described as a place abundant in gold and silver, which can possibly be identified as Ezo (Hokkaido).

If Portuguese cartography in Bartolomeu's day was known as the best in the world, it was due in no small part to the kingdom's religious tolerance. The founder of the leading school of mapmaking in Portugal has been identified as a Jew from Majorca named Jacome. Similarly, the scientific skills and know-how related to navigation—including cosmography, cartography, and shipbuilding—had all advanced under the influence of Jewish or converso (converted Christians, mainly from Sephardic Judaism) families. Many of these families had fled to Portugal after being expelled from the kingdoms of Aragon and Castile when the Spanish Inquisition broke out at the end of the fifteenth century. Reputable Portuguese mapmakers, like the Homem family, were known for being converso. In the 1550s, however, Portugal's religious tolerance came to a swift end. After the death of King John III, the Inquisition came to be harshly enforced in Portugal under the regent Catherine, a sister of Charles V of the Holy Roman Empire.

Like so many other Portuguese cartographers of his day, Bartolomeu Velho was known for his Jewish ancestry. Persecuted by the Inquisition in Portugal, he eventually fled to Nantes, a French port that served as a sanctuary for refugees of the Iberian Inquisition after Henry II of France offered protection to expelled Sephardic conversos from Portugal. By offering sanctuary to this Jewish cartographer—Bartolomeu would live in Nantes until his death in 1568—the French king was able to obtain confidential geographic knowledge.

Apart from Bartolomeu's 1561 "General Chart of the Globe," the mapmaker left his final work, "Figure of the Celestial Bodies" (*Figura dos Corpos Celestes*), conserved in the Bibliothèque Nationale de France (Paris). It includes an illustration of the Ptolemaic geocentric system with the globe at its center. This globe features a map of the world based on the latest geographic information Bartolomeu possessed. Here we can see the presence of a big island north of Honshu, confirming that Bartolomeu's knowledge of the Japanese archipelago in 1561 included Ezo.

Strikingly, this information does not seem to have been adopted by other Portuguese mapmakers. Among

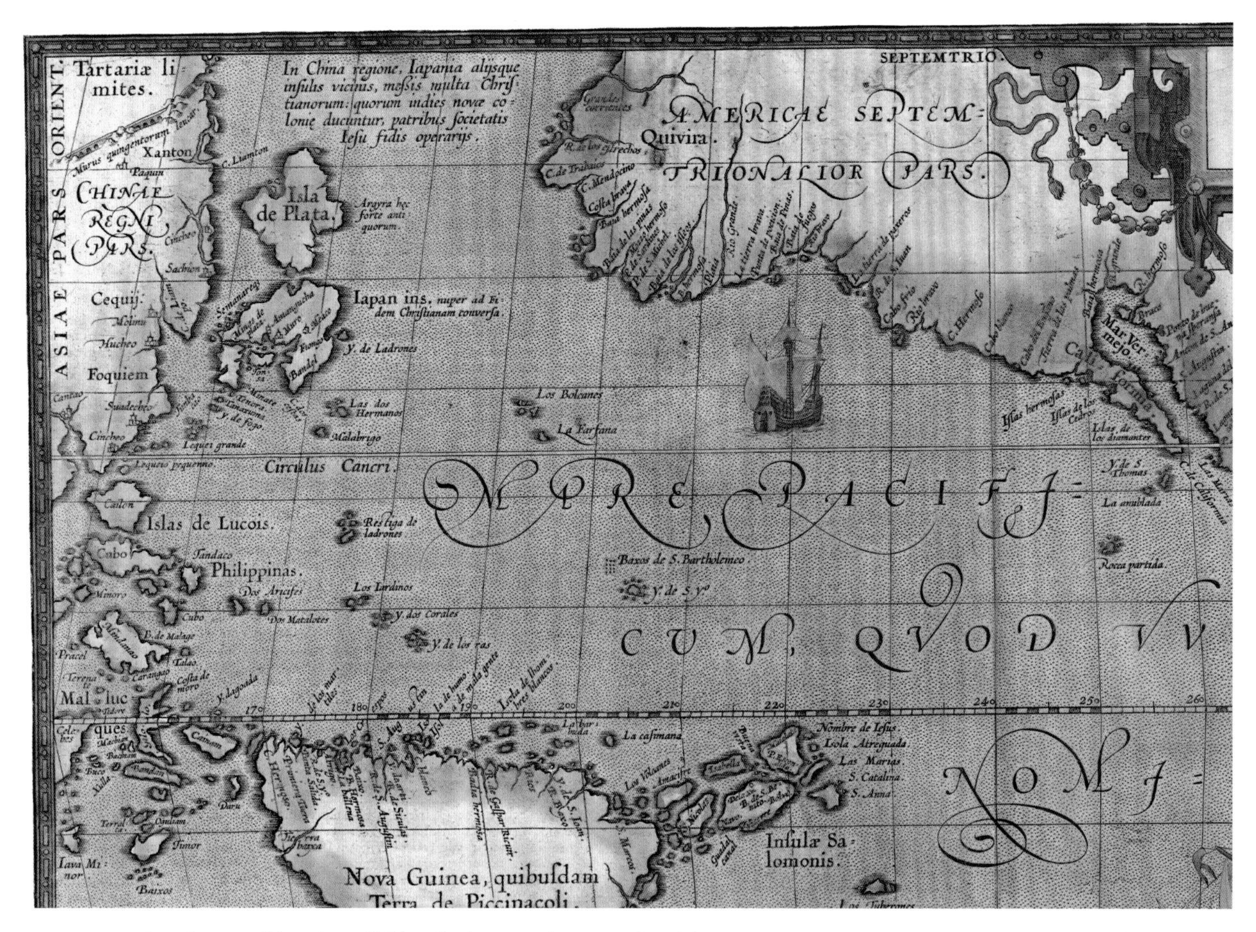

FIGURE 3.2 "Pacific Ocean" [*Maris Pacifici*] by Abraham Ortelius, 1589; detail showing Japan. Manuscript, 33 × 48 cm. Courtesy of the National Library of Australia.

the many world maps and atlases made during the sixteenth century, the only surviving map that seems to have a trace of Bartolomeu's influence is Abraham Ortelius's "Pacific Ocean" (*Maris Pacifici*), published in Antwerp in 1589 (fig. 3.2). In Ortelius's map, the island labeled Ezo is identified as an "Island of Silver" (Isla de Plata). The same island in Bartolomeu's "General Chart of the Globe" is accompanied by the description "This island has much gold and silver" (Nesta ilha ha muito ouro /plata). Interestingly, in other maps of Asia published by Ortelius before 1589, no island likely to be Ezo appears.

As to the mystery of why Bartolomeu's knowledge was not transmitted to Portuguese cartographers, the answer seems related to the circumstances of his life. None of Bartolomeu's maps remained in Portugal after he fled the country; because of that, the famous "General Chart of the Globe" (1561) and "Figure of the Celestial Bodies" (1568) could well have been unknown to Portuguese cartographers. Although the "General Chart of the Globe" has a dedication to D. Sebastian, it did not remain in Lisbon, but in Medici-ruled Florence. Considering that world maps were treated as state secrets at the time, it is possible the map was completed during his stay in Nantes under Henry II, who was married to Catherine de' Medici.

While the main silver mine in Japan first appeared in Bartolomeu's 1561 map, some cartographers thereafter included similar information. In this case, we may suppose that the mine in question was already well known among the Portuguese. This mine can be identified as Omori (Iwami) silver mine for two reasons: first, it was described as being close to Yamaguchi, and second, Iwami was the richest silver mine in Japan during the sixteenth century. Nonetheless, European traders believed that there were other islands rich in gold and silver in the Pacific Ocean near the archipelago of Japan. Ezo was believed to be such an island, as we have seen in the map by Bartolomeu.

The drive to discover the island's abundant gold and silver reserves prompted several expeditions by European explorers. The first attempt was undertaken by Sebastian Vizcaino, an ambassador sent by the viceroy of New Spain to thank the Tokugawa shogunate for aiding the shipwrecked governor of the Philippines, Rodrigo de Vivero, in 1609. In addition to his ambassadorial role, Vizcaino was charged with discovering Ezo's rich gold and silver reserves. The attempt failed, and the adventure almost ended in another shipwreck. Attempts were also made by several other dream-fueled European explorers. The Dutch East India Company (VOC) twice sent expeditions, in 1639 and again in 1643, to the Pacific Ocean with the intention of discovering the island. As a result of these expeditions, Europeans' knowledge of the geography of Ezo became much more detailed, although none could realize their dreams of discovering its fabled gold and silver.

Suggested Readings

Cortesão, Armando, and Luis de Albuquerque. *History of Portuguese Cartography*. 2 vols. Lisbon: Junta de Investigações do Ultramar, 1969–71.

Lach, Donald F. *Asia in the Making of Europe*. Vol. 2, *A Century of Wonder*. Chicago: University of Chicago Press, 1970.

Murai Shōsuke. *Nihon chūsei kyōkai shiron* [Collective historical studies on the boundary of medieval Japan]. Tokyo: Iwanami Shoten, 2013.

Schwade, Arcadio. "Japan in the Portuguese Cartography of the 16th Century." In *Transactions of the International Conference of Eastern Studies*, no. 8 (Tokyo, 1963).

4 Mapping the Margins of Japan

Ronald P. Toby

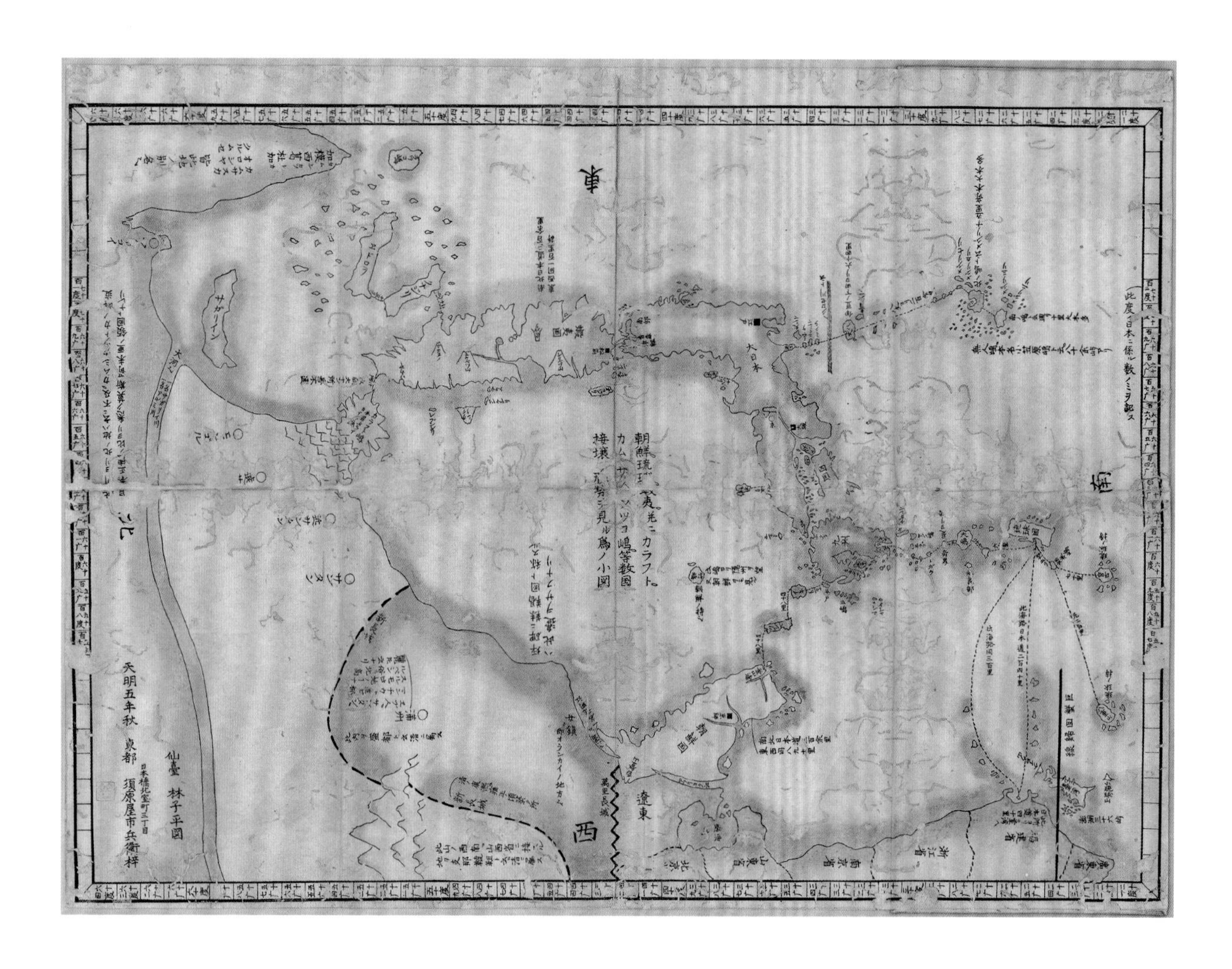

FIGURE 4.1 "A Small Map for the Purpose of Seeing the Shapes of the Numerous Bordering Countries of Korea, Ryukyu, and Ezo, as well as Sakhalin, Kamchatka, Sea-Otter Island, etc." [Chōsen Ryūkyū Ezo narabini Karafuto Kamusasuka Rakkojima nado sūkoku setsujō no keisei o miru tame no shōzu 朝鮮琉球蝦夷并ニカラフトカムサスカラツコ嶋等数国接壌ノ形勢ヲ見ル為ノ小図] by Hayashi Shihei 林子平, 1786. Published in Edo by Suharaya Ichibē 須原屋市兵衛. Woodblock print, 51 × 72.3 cm. Courtesy of the University of British Columbia Library, George H. Beans Collection. Oriented with East at the top. Dotted lines across maritime routes indicate shipping lanes and distances between ports.

A notable feature of domestically produced maps of Japan drawn or published in the seventeenth and eighteenth centuries is their seeming lack of concern with demarcating national boundaries.[1] Japan, after all, was—and is today—an archipelago separated from the nearest continental landfall by the Korea Strait (at two hundred kilometers, about six times the span of the English Channel at its narrowest point).[2] Although Japanese maps were quite clear about *internal* boundaries, neither official nor commercial maps made during these centuries specified exactly where "Japan" ended and someplace else began.

Indeed, while the shogunate's maps made it clear that the "sixty-six provinces and two [offshore] islands" were integral Japanese territory, they were less definite with respect to the outer bounds of Japanese territory: Ezo (now Hokkaido) and the Ryukyu Islands (including present-day Okinawa Prefecture). Neither had historically been part of the Japanese polity; Ryukyu remained a nominally independent kingdom until the 1870s, and most of Ezo remained beyond the pale of shogunal control. For most of the Tokugawa period, a Japanese daimyo held an outpost only at the southern tip of Ezo, from which he and his retainers traded with the indigenous Ainu people.[3]

The shogunate compiled four official maps of Japan before 1717, each based on different conceptions of the territory of "Japan." The first of these, compiled ca. 1639 (surviving in a hand-painted copy dated 1653), depicts a country comprising only the sixty-six-plus-two provinces—the main islands of Honshu, Shikoku, and Kyushu, the "Two Islands" of Iki and Tsushima, and dozens of small offshore islands—floating in the sea. No lands beyond the seas impinge on the vision of the map, so no boundaries are necessary; national boundaries, uncontested from without, were not an issue from within. A second effort, compiled around 1670, added a perfunctory rendition of Ezo in the north, while the third map included the Ryukyu Islands and noted the southern coast of Korea, where the island of Tsushima stationed men to trade. But the fourth map in this official series reverted to the coverage of its 1670 predecessor, excluding both the Ryukyu island chain and the Korean coast.

Commercially published maps, on the other hand, starting from the mid-seventeenth century, acknowledged non-Japanese countries "somewhere beyond the sea," frequently indicating them in the corners or margins. Typical was Ishikawa Ryūsen's colorful "Map of the Seas, Mountains, and Lands of Japan" (see fig. 7.1 below), which showed small bits of the foreign territories of Chōsen (Korea), Ezo, and Ryukyu bleeding into the map at the corners: acknowledging their existence while clearly marking their "foreign" character. Precisely where one transitioned from "Japan" to foreign territory, however, remained an open question.

This relaxed attitude toward defining Japan's borders could be sustained only so long as no unwelcome foreign powers threatened Japanese security. In the 1770s and 1780s, the rapidly expanding ambit of Russian activity presented just such a challenge. The mounting sense of foreign crisis only accelerated in succeeding decades as British, French, and American vessels—merchantmen, whalers, and warships—appeared more and more frequently off Japanese shores, with ever-more-insistent demands for access to Japan. These threats evoked a felt need to define the boundaries of "Japan," especially in the northeast, a need to which cartographers and strategists outside government were the first to respond.

The first cartographic attempt to specify the limits of Japanese territory, *Illustrated Complete Survey of the Three Countries* (*Sangoku tsūran zusetu*), was drafted in 1785 by Hayashi Shihei (1738–93), a samurai and scholar from the northeastern domain of Sendai, and printed the following year by one of Edo's leading publishers. Ironically, the book was banned in 1792 on the charge that the text advanced "strange and unorthodox opinions" (kikai isetsu), while its "maps of Japan and foreign countries are contrary to the geography." Shihei was put under house arrest in Sendai, where he died the following year. Unsold copies of his book were confiscated, the publisher fined, and the printing blocks destroyed; consequently, of the surviving copies, manuscript copies outnumber the printed version by about two to one.[4]

Hayashi's larger objective was to overcome what he saw as the major shortcoming of the countless published Japan maps available in the marketplace: their failure to place Japan in the geographic context of Northeast Asia. "Maps of our country only picture the ports of the countries across the seas on all sides [of Japan], but not the entire shape of those countries." But in order to succeed, he had to establish boundaries, something he accomplished in a set of maps including a small master map (fig. 4.1) showing Japan and the three neighbors which most concerned him: Korea, Ryukyu, and Ezo. To these, he added a fourth category as well: the "over eighty uninhabited islands sometimes called the Ogasawara Islands."

In his master map Hayashi's overriding concern was to clarify Japan's boundaries with its neighbors. He accomplished this, in the first instance, through color coding, most starkly in the boundary between the parts of the great northern island of Ezo that he regarded as foreign, which he called "The Country of Ezo" (Ezo-koku), and the relatively small area at the southern tip of the island where the Japanese daimyo house of Matsumae was based. The latter was, in Hayashi's reading, Japan's only land border with a foreign country (fig. 4.2).

In the southwest, likewise, Hayashi deployed color coding to mark Tokara as the limit of Japanese territory, adding a caption specifying Kikai-ga-shima as the point where Ryukyu begins: "Beyond here, Ryukyu territory" (Kore yori Ryūkyū-chi) (fig. 4.3). This accords with the

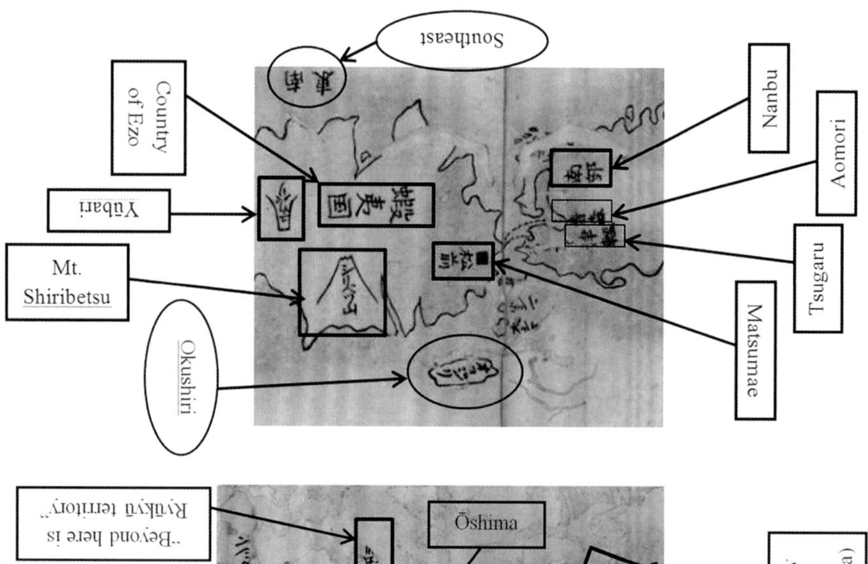

FIGURE 4.2 Detail of figure 4.1, showing northern Honshu and southern Ezo (Hokkaido). Hayashi uses color-coding to distinguish *Wajin-chi*, the area of Japanese residence around Matsumae in southern Ezo (shown here in green), from the much larger *Ezo-chi* inhabited by the indigenous Ainu (orange), which Hayashi explicitly regarded as foreign, non-Japanese territory. English labels added by author.

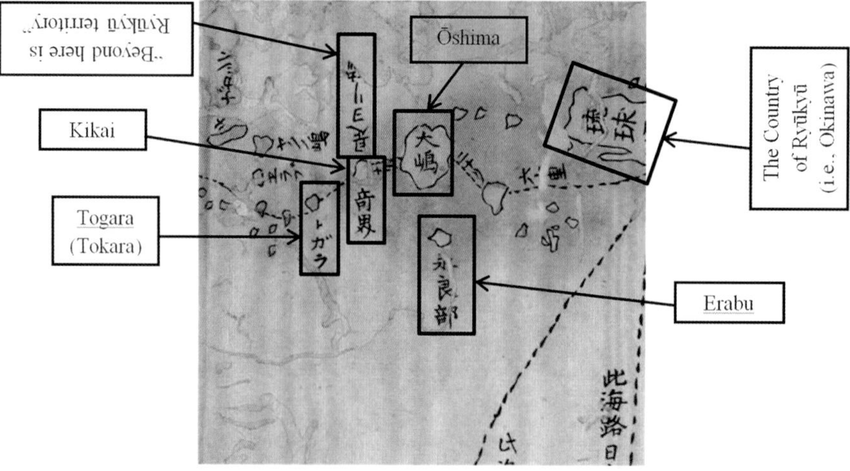

FIGURE 4.3 Detail of figure 4.1, showing part of Ryukyu. Here too, Hayashi uses color-coding to show the limits of Japanese territory, adding a caption specifying that at Kikai Island one crosses the outer boundary of Japanese territory and enters into the "Land of Ryukyu." English labels added by author.

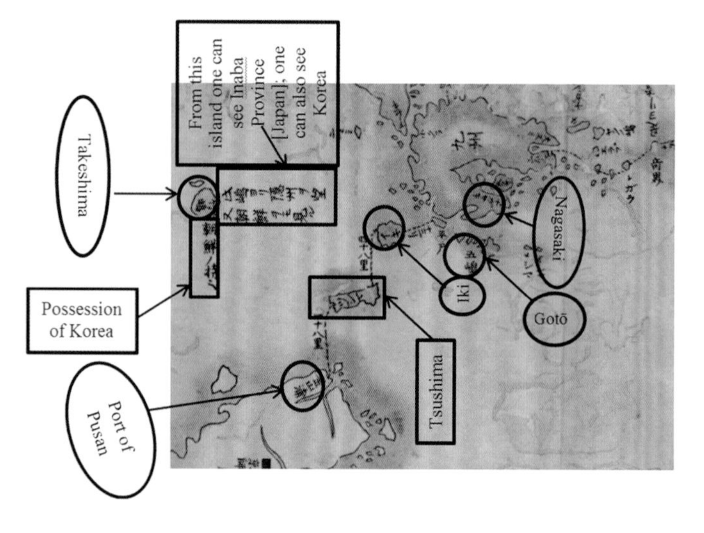

FIGURE 4.4 Detail of figure 4.1, showing Kyushu (green), the Korean peninsula (yellow), and various islands between. English labels added by author.

boundary on the 1696 provincial maps of Satsuma and Ryukyu, in which the Tokara Islands are the southern limit of Satsuma, and Kikai-ga-shima and Ōshima the northernmost of the Ryukyus. Hayashi's treatment of Korea (fig. 4.4) is equally unequivocal: territory he regards as Korean (including two small islands labeled "Takeshima") is shaded yellow, in clear contrast to the green shading for Japanese territory.[5]

As evidenced in the way they were handled in shogunal maps, Ezo and the Ryukyus were particularly ambiguous, if for different reasons. Neither was ever counted as one of the sixty-six provinces, whether in gazetteers or on maps. Although a daimyo house, the Matsumae, was based at the southern tip of Ezo, only about 2 percent of the island was designated as Wajin-chi (Japanese people's land); the remainder was Ezo-chi (Ainu land).[6] As the playwright Chikamatsu Monzaemon observed, if somewhat tongue in cheek, "Who knows where that country is? Is it Japanese territory or Chinese?"[7] The territorial identity of Ryukyu—in Japanese cognition, at least—would remain indeterminate until the 1870s, when it was formally integrated into the Japanese state.[8]

Although Hayashi's work was suppressed, his map was but one among many responses, official and unofficial, to the rising sense of foreign threat—from Russia in the north, initially, but increasingly in the south and along the Pacific coast as well—that gave urgency to defining the boundaries of Japan. The Tokugawa dispatched a series of expeditions, beginning in 1784, to explore Ezo and beyond, and eventually took over direct management of Ezo (see chap. 32). During the third expedition (1798), sent "to ascertain and regulate the border with foreign countries," the shogunal official Kondō Jūzō erected a marker on the island of Iturup, proclaiming it "Etorofu, in Great Japan" (Dai-Nihon Etorofu).[9]

Two years later, Inō Tadataka obtained bakufu authorization to map Japan's northern boundary; over the next two decades Inō and his team would survey the entire Japanese coastline (see chap. 29 below).[10] The resultant "Inō maps" are celebrated as the most precise maps of Japan until after the Meiji Restoration, and not without reason.[11] But his maps ignore almost everything except the coastline, revealing that Inō's primary goal, like Hayashi Shihei's, was not to map the country's internal geography, but to define the outer margins of Japan.

Notes

1. In this context I use "national" to denote large territorial units such as Japan, Korea, China, or Russia.

2. The straits are punctuated by the "two islands" of Iki and Tsushima, counted as provinces of Japan. The shortest distance from Tsushima to the Korean coast is only about 50 km.

3. The Jesuit Jeronymo de Angelis reported that when he visited Ezo in 1618, the daimyo told him that "Matsumae is not Japan[ese territory]." Quoted in Kikuchi Isao, *Ainu to Matsumae no seiji bunka-shi: kyōkai to minzoku* (Tokyo: Azekura Shobō, 2013), 20.

4. My calculation, based on extant print and manuscript copies listed in the online *Nihon kotenseki sōgō mokuroku* and *Kōnitsuki Ōshū shozai Nihon kosho sōgō mokuroku bēsu*, supplemented with information on North American collections. Peter F. Kornicki, "Manuscript, Not Print: Scribal Culture in the Edo Period," *Journal of Japanese Studies* 32, no. 1 (2006): 23–52, notes the proliferation of MS copies of Hayashi's work. Hayashi's other major work, *Kaikoku heidan* (Military talks for a maritime country), met the same fate. For an account of the case and the texts of the verdicts, see Taira Shigemichi, *Hayashi Shihei: Sono hito to shisō* (Sendai: Hōbunkan, 1977), 247–50.

5. It is unclear whether Hayashi's "Takeshima" corresponds to the disputed islets today known in Japanese as Takeshima and in Korean as Tokdo: http://www.dokdo-takeshima.com/1785-japanese-map-by-hiyoshi-shihei.html.

6. On the Wajin-chi/Ezo-chi distinction, and the gradual expansion of the Wajin-chi, 1550–1800, see Hirakawa Arata, *Kaikoku e no michi* (Tokyo: Shōgakukan, 2008), 60–61; David L. Howell, *Capitalism from Within: Economy, Society, and the State in a Japanese Fishery* (Berkeley: University of California Press, 1995), xv (map), 191, n. 2.

7. Chikamatsu Monzaemon, *Taishokukan* (1711), in *Chikamatsu Monzaemon zenshū*, vol. 6 (Tokyo: Shun'yōdō, 1922–24), chap. 8, 281–334.

8. The Ryukyu kingdom was declared a part of Kagoshima Prefecture in 1871, a domain (*han*) in 1872, and a Japanese prefecture in 1879; the king was transferred to Tokyo and integrated into the modern Japanese aristocracy as a marquis (*kōshaku*).

9. Murao Motonaga, ed., "Kondō Morishige jiseki-kō," in *Kondō Seisai zenshū*, vol. 1 (Tokyo: Kokusho Kankōkai, 1905), 6; Hirakawa Arata, *Kaikoku e no michi* (Tokyo: Shōgakukan, 2008), 67.

10. Inō (1745–1818), a retired sake merchant from Sawara (in today's Chiba Prefecture), has become something of a culture hero in recent years. A global search for "Inō Tadataka" in the National Diet Library yielded 1,874 books and articles, ninety of them just since 2010.

11. For an 1827 MS copy of Inō's map of Ezo, http://dl.ndl.go.jp/info:ndljp/pid/1286165.

Suggested Readings

Batten, Bruce L. *To the Ends of Japan: Premodern Frontiers, Boundaries, and Interactions.* Honolulu: University of Hawai'i Press, 2003.

Harrison, John A. *Japan's Northern Frontier: A Preliminary Study in Colonization and Expansion, with Special Reference to the Relations of Japan and Russia.* Gainesville: University of Florida Press, 1953.

Walker, Brett L. *The Conquest of Ainu Lands: Ecology and Culture in Japanese Expansion, 1590–1800.* Berkeley: University of California Press, 2001.

5 The Creators and Historical Context of the Oldest Maps of the Ryukyu Kingdom

WATANABE Miki 渡辺美季

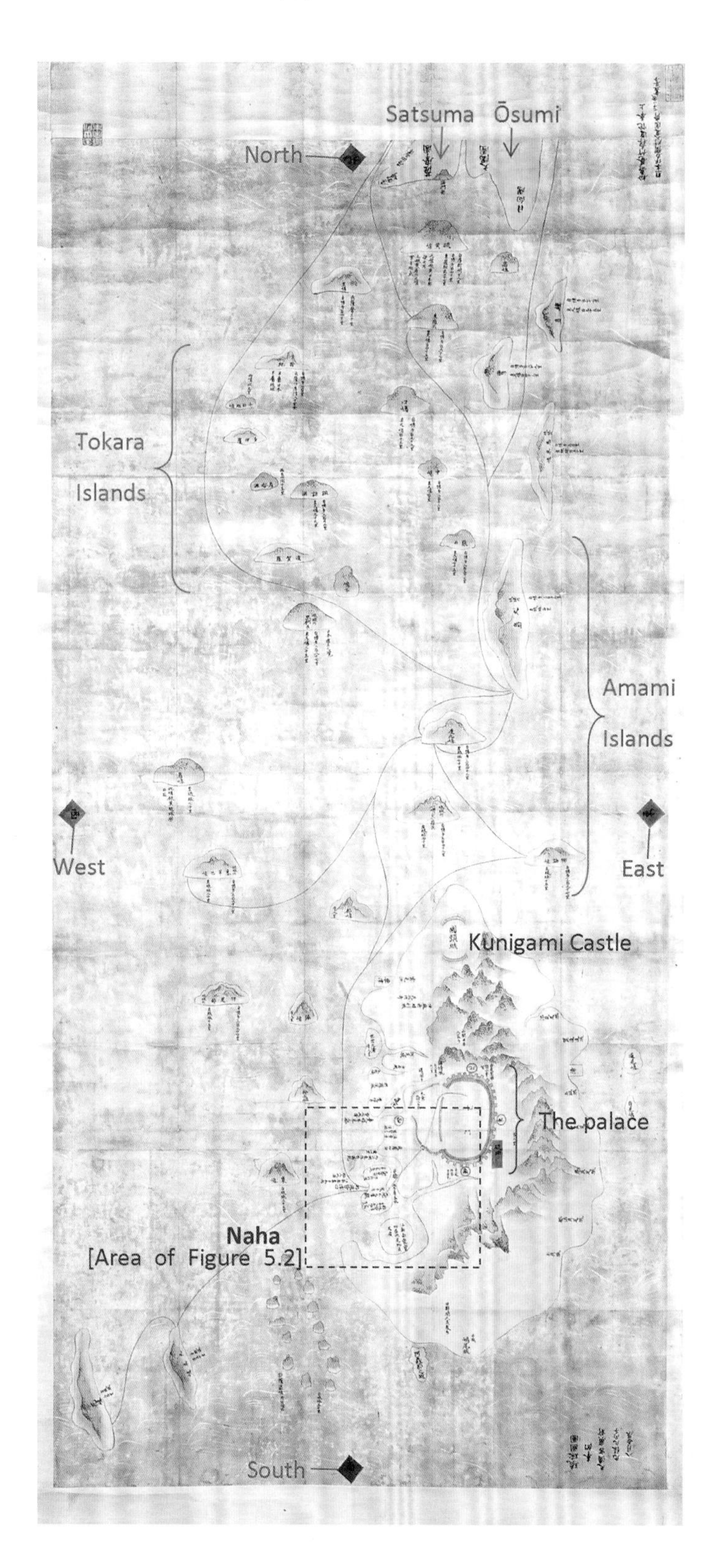

FIGURE 5.1 "Map of the Ryukyu Kingdom" [Ryūkyū-koku zu 琉球国図], seventeenth-century copy of a lost fifteenth-century original. Manuscript, approximately 88 × 176 cm. Courtesy of the Okinawa Prefecture Museum. English labels added by author.

The islands known today as Okinawa Prefecture were once the site of the Ryukyu kingdom.[1] After its foundation sometime around the twelfth century, the Ryukyu kingdom went on to form a tributary relationship with Ming China during the fourteenth century and eventually became subject to Japan at the beginning of the 1600s. The Ryukyu kingdom disappeared completely in 1879 when it was forcibly absorbed into Japan proper.

During the fifteenth century, however, this seemingly frail kingdom served as a major hub of international maritime trade. A map detailing the Ryukyu kingdom at the peak of its prosperity came to the attention of researchers in 2004 and has been much discussed since.[2] This map will be referred to below as map I (figs. 5.1 and 5.2). One reason this map has so fascinated the academic community is that it describes the kingdom in greater detail than the oldest map of the Ryukyu kingdom known to exist (produced in 1471), which we will call map II (fig. 5.3). It seems likely that map I was created using a lost third map (map III) as the basis for its cartography.

Map I begins at the top with the Satsuma and Ōsumi areas of southwest Kyushu and goes on to depict the Tokara and Amami island chains leading down to the Ryukyu kingdom itself. Roads and sea-lanes are depicted in red. Of these, one from the north and another from the south both lead to Naha on the main island of Okinawa. One of the main characteristics of map I is the detailed information it provides regarding Naha, which was both the center of the Ryukyu kingdom's trade activities and its largest port.

A detail of this area (fig. 5.2) shows Naha drawn as a small hourglass-shaped island near the center. Where the two sea-lanes previously mentioned connect with Naha (see point *a*), a note on the map reads: "Mouth of Naha Port. Ships from China (Kōnan), Southeast Asia (Nanban), and Japan enter here." Comments written directly on the island include "Naha Port—Houses of Japanese and Ryukyu people located here" (*b*) and "Kume Village—Houses of Chinese people located here" (*d*). A shrine in which Japanese gods were worshipped is also labeled on the map (*e*). It is quite clear that non-Ryukyu

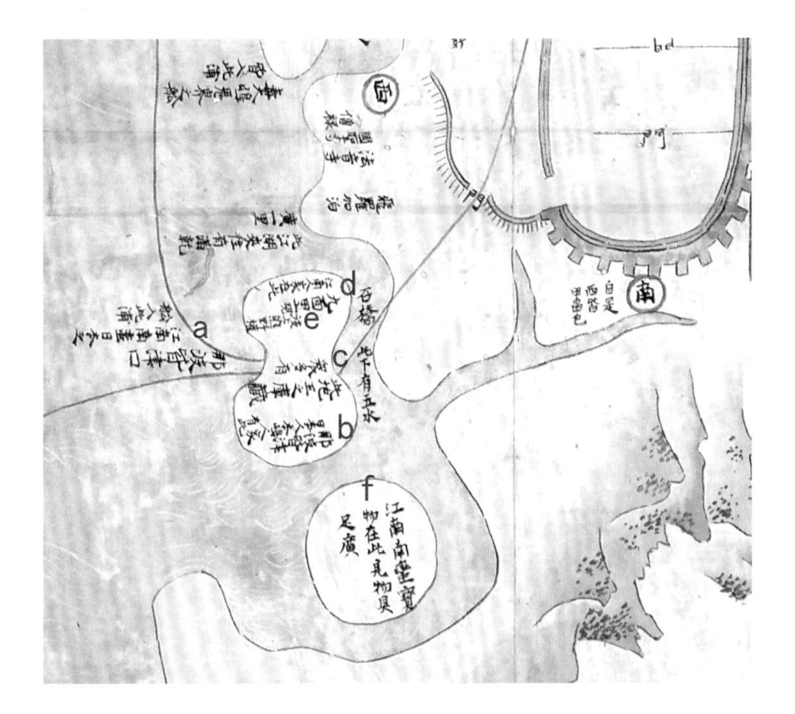

FIGURE 5.2 Detail of figure 5.1, showing Naha. Labels added by author.

people lived fulltime on the island. It also seems that while the Chinese population was concentrated, Japanese and Ryukyu homes were intermingled.

In the center of the island containing Naha (*b*), the map reads, "Many of the king's storehouses are located here." Another small island is similarly labeled "Chinese and Southeast Asian treasures kept here." The map also notes a stone bridge leading from the port area to Shuri, the king's palace. In general, map I presents a view of Naha as a bustling and cosmopolitan port under royal control.

A quick perusal of map I beyond the vicinity of Naha reveals the existence of sixteen castles, several rural settlements, two more ports, and four Buddhist temples. Several smaller islands are also named; significantly, their distances are given from major ports in Kyushu (Hakata or Satsuma). Map I's particular use of the terms "from Hakata" and "to Ryukyu" emphasize that Hakata was considered the point of departure and Ryukyu the destination. This makes it likely that many of those who used map I were Hakata merchants.

These details correlate well with the fact that Ryukyu served as the nodal point for trade between East and Southeast Asia during the fifteenth century. Ming China only permitted its merchants to trade with tributary nations, and Chinese citizens were forbidden by a general ban on maritime activities to go to sea. Under these circumstances, the Ryukyu kingdom capitalized on its tributary status to deliver large quantities of Chinese goods to Southeast Asian countries like Siam (Thailand), Malacca, and Java as well as to Japan and Korea. Ryukyu could also carry the products of these nations back to China.

Yet the Ryukyu prosperity illustrated in map I was not to last long. The sixteenth century saw a rapid weakening of the Ming position and large-scale entry into the international trade scene by private Chinese merchants, who displaced Ryukyu traders first in Korea and later in Southeast Asia. In addition to these troubles, the Ryukyu kingdom was unable to fend off invasion by Japan in 1609 and found itself subject thereafter to both the Ming (and subsequently the Qing) dynasty of China and the Tokugawa shogunate.

Careful analysis shows that map I describes the Ryukyu kingdom at its peak in the late fifteenth to sixteenth century. However, map I is not in fact a product of Ryukyu's golden age of trade, but was produced in Japan toward the end of the seventeenth century. This anachronistic map was made in 1696 by Takemori Dōetsu, a samurai from Fukuoka domain—including Hakata

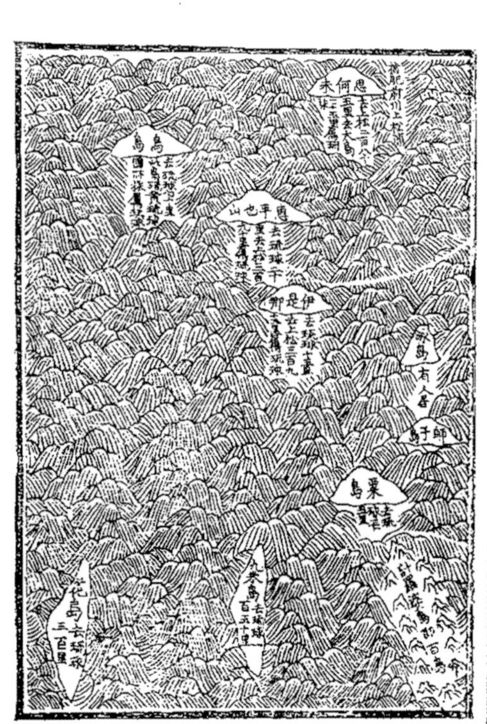

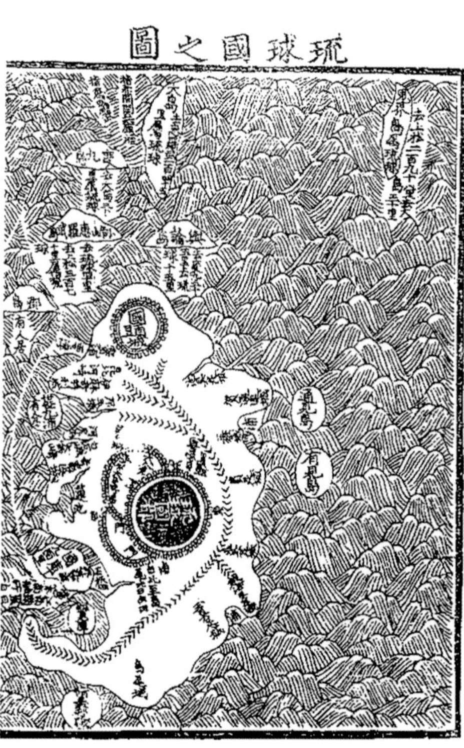

FIGURE 5.3 "Map of the Ryukyu Kingdom" [Yugung-guk chi t'o 琉球国之図], from *Records of Lands to the East of the Sea* [Haedong chegukki 海東諸国記], by Suk-chu Sin 申叔舟, 1471. Woodblock print. Courtesy of the Historiographical Institute, University of Tokyo.

town—and donated to the famous Dazaifu Tenman-gū Shrine. For many years it was buried and forgotten in a museum storehouse. Once the map was unearthed, however, scholars were fascinated to find that it contains a richness of historical detail surpassing even that of map II—an actual product of the golden age of Ryukyu trade.

Map II was compiled by order of the Korean king in 1471. Named "Map of the Ryukyu Kingdom," map II was part of a larger atlas, called the *Haedong chegukki,* focusing on the islands to the east of Korea (including Japan).[3] Recognized as the oldest extant map of the Ryukyu kingdom, map II was seen as the greatest source of geographic detail regarding Ryukyu until the recent discovery of map I.

A comparison of the Naha area in each map quickly reveals the differences. For example, point *a* in the above detail is described on map II as simply "Mouth of the port. Chinese, Southeast Asian, and Japanese ships arrive here." Points *b, c, d,* and *f* are variously labeled as "Naha Port," "national warehouse," "Kume Village," and "treasury." There is no comment on map II corresponding to point *e* on figure 5.2. As this small sample reveals, map I provides much richer detail about Ryukyu than does the older map II. It therefore seems highly unlikely that map I could have been produced as a copy of map II. Yet scholars recognize undeniable links between the two maps. How can this relationship be interpreted?

One solution proposes the existence of a lost third map (map III) that served as the basis for construction of both map I and map II. Although no copy now exists, there is mention in the historical record of a "Map Showing the Distances between Hakata, Satsuma, and Ryukyu" presented to the Korean government by the Hakata monk-merchant Dōan in 1453. A merchant who traded heavily with Ryukyu, Dōan also served as the Ryukyu king's emissary on three trips to Korea. Records show that one reason he took map III was to impress on the Koreans just how difficult the journey was between Hakata (or Korea) and Ryukyu. Analysis suggests that map I bears a closer resemblance to map III (and may even be a direct copy of it), although map II also seems to be based on map III.

Map III was probably still extant in the area of Hakata during the end of the seventeenth century and would have been available to Takemori Dōetsu, who presumably found some chance to make a copy of it. While no one could have foreseen such a development, it may be that the key to interpreting what was thought to be the oldest map of Ryukyu in existence (map II) is a later copy (map I) of a yet earlier original (map III).

What sort of person was Takemori Dōetsu?[4] Born in approximately 1625, Takemori was a vassal to the daimyo of Fukuoka. In his youth, he spent twenty years studying medicine and Confucianism in Kyoto, returning in midlife to Fukuoka to serve the daimyo. Takemori held no particularly high rank, nor did he accomplish much of note. It seems he lived a very anonymous life as a samurai. Yet the more we learn about him, the more intriguing Takemori becomes from a cartographic perspective. This little-known samurai donated at least a dozen maps to the shrine in Dazaifu, including images of Ming China, Korea, Ezo (Hokkaido), and Kyoto. Due to the fact that only his signature appears on the printed maps and his is the only name mentioned in the shrine's donation list, it must be assumed that Takemori was solely responsible for the production and donation of those maps.

This leads us in turn to ask why a samurai like Takemori would have spent time collecting, copying, and donating maps to the Tenman-gū Shrine. Although nothing directly is known on this score, it is highly likely that Takemori was influenced by a cultural network of samurai-scholars centered around Kaibara Ekiken that was active in Fukuoka domain during his lifetime. Kaibara Ekiken was an intellectual polymath who found considerable fame in his own time.[5] We know that Takemori was a member of his circle because his name appears in the diaries and writings of both Kaibara and one of his disciples. One characteristic of this group was their focus on actively buying, lending, and copying books—and maps.[6] While it is unknown exactly how Takemori came into possession of his personal map collection, we can assume that he was assisted by the Kaibara network.

In this way, the early maps of Ryukyu open up doors of inquiry beyond the scope of the maps themselves. By studying them, we can gain insight into the way maps were made, preserved, copied, collected, and circulated among scholar elites in early modern Japan. In the process, we can catch a glimpse of the people involved in all of these endeavors. When considering any map, it is well worth keeping our minds open to such people, and researching them as well as the contents of the map itself.

(Translated by Bret Fisk)

Notes

1. More accurately, modern Okinawa Prefecture consists of the southern half of the Ryukyu archipelago.

2. For pertinent research see Asato Susumu, "Dazaifu jinja kyūzō Ryūkyū kokuzu ni miru jūgoseiki no Ryūkyū Ōkoku," *Urasoe shiritsu toshokan kiyō* 15 (2004): 73–90; Uezato Takashi, Fukase Kōichirō, and Watanabe Miki, "Okinawa kenritsu hakubutsukan zō Ryūkyū kokuzu: Sono shiryōteki kachi to *Haedong chegukki* to no kanrensei ni tsuite," *Komonjo kenkyū* 60 (2005): 24–45; Saeki Kōji, "Kaitōshokokuki no Nihon Ryūkyū zu to Ryūkyū kokuzu," *Kyūshū shigaku* 144 (2006): 80–91.

3. Shin Sukchu, *Haedong chegukki* (*Kaitōshokokuki: Chōsenjin no mita chūsei no Nihon to Ryūkyū*), ed. Tanaka Takeo (Tokyo: Iwanami Shoten, 1991).

4. The following is based on Watanabe Miki, "Takemori Dōetsu to chizu hōnō: Sekai zu, Hizen Nagasaki zu no shōkai o chūshin ni," *Kyūshū shigaku* 146 (2006): 1–47.

5. Nishinihon Bunka Kyōkai, ed., *Fukuoka ken shi, tsūshihen, Fukuokahan, bunka (jō)* (Fukuoka: Nishinihon Bunka Kyōkai, 1993), 46.

6. Tadashi Inoue, *Kaibara Ekiken* (Tokyo: Yoshikawa Kōbunkan, 1963).

6 The Introduction of Dutch Surveying Instruments in Japan

SATOH Ken'ichi 佐藤賢一

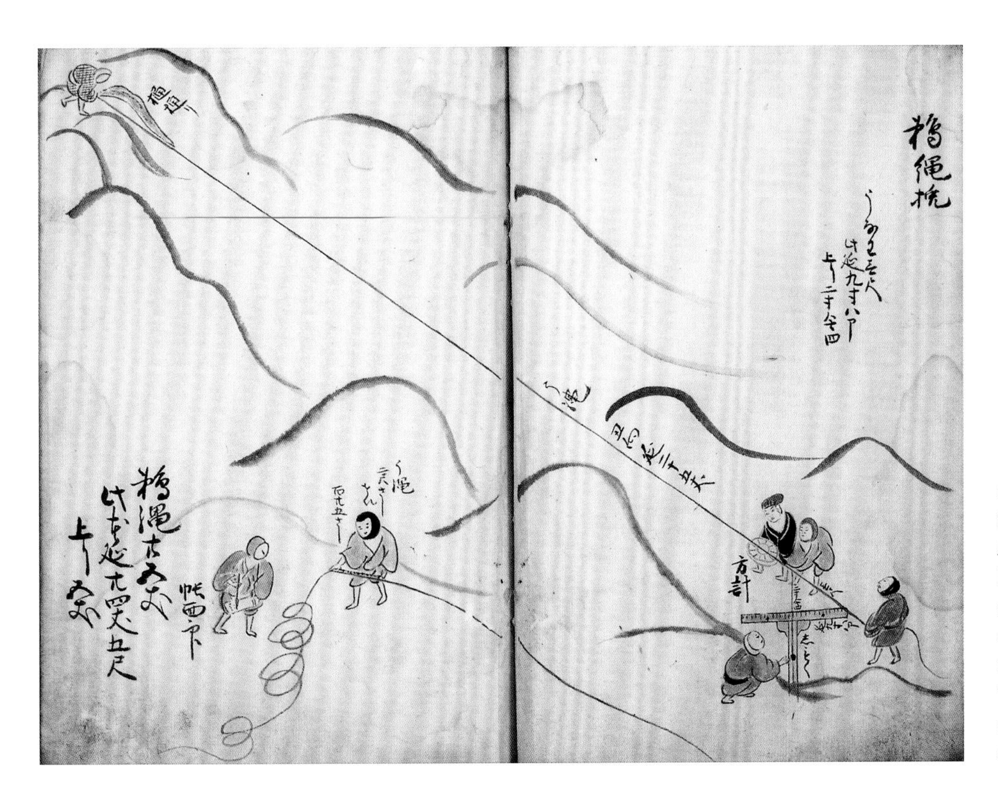

FIGURE 6.1 Illustration from *Pictorial Record of Mountains* [Sanchū ken-bunzu 山中見聞], manuscript, late 1500s. Courtesy of the University of Electro-Communications (Denki Tsūshin Daigaku). Depiction of slope angle survey during mine excavation.

Although closely related, the history of cartography and that of surveying are rarely discussed together. We talk about the map as an artifact, a historical document, or sometimes as a tool for reconstructing past geography. Meanwhile, we often approach surveying from the perspective of technological history. Yet surveying is part of the mapping process; by rights, the history of the map and the history of surveying should be considered together. Unfortunately, whereas finished maps are likely to be preserved as historical material and kept for future generations, surveying techniques were rarely documented prior to the modern age, and surveying instruments often did not survive. For this reason, it is extremely difficult to paint a comprehensive history of maps and surveying.

In the case of Japan, we know almost nothing about the practical side of surveying prior to the Edo period. Although some private estate drawings and maps from ancient times are extant, we do not know how they were produced. Maps of small areas must have been drawn freehand while the mapmaker was looking at the terrain, while domain maps were presumably drawn with recourse to memory and imagination. In the seventeenth century this situation began to change. The traders and Jesuit missionaries from Western Europe who came to Japan in the late 1500s introduced a variety of novel technologies, including previously unknown techniques for mine excavation, navigation, and gunnery. We can imagine that European surveying techniques were also brought into Japan at this time. While some of the maps introduced from Europe during this period survive, however, there is almost no surviving documentation to indicate what surveying techniques were actually used in the field (fig. 6.1 being a rare exception). In the case of mine excavation, the accuracy of surveying at the time can only be inferred from measuring the remains of the tunnels.

What we do know is that up until the early 1600s, the Society of Jesus operated schools in Kyushu where Japanese students learned about Christian teachings as well as basic natural science. Even in later ages, the study of astronomy (beginning with Aristotle's cosmology) remained widespread in Japan as a legacy of these Jesuit schools. Interestingly, the concept of a spherical earth was brought into Japan at this time and gained a certain degree of understanding among the Japanese.

The maps brought over from Western Europe during these decades, especially world maps, remained in Japan and exerted considerable influence over Japanese people's sense of the world. The nautical charts discussed by Peter Shapinsky also attracted considerable attention. But the 1602 world map created for the Chinese by Matteo Ricci arguably played the biggest role in educating the Japanese about the size of the world and the distribution of its major landmasses. Popular books after the seventeenth century often contained crude images of the world modeled after Ricci's map (fig. 6.2). We can also confirm that some Japanese produced world globes in the late seventeenth century.

Before long, the trading system centered on Spain and Portugal would disintegrate due to policy changes implemented by the Tokugawa shogunate. All Iberians were banished from Japan in the 1630s, when the Dutch East India Company (VOC) was given a monopoly on European trade with Japan. Subsequently, information about mapping and surveying practices in Europe would be brought into the country by the VOC. At the time, the practice of translation from European languages into Japanese was just beginning, and it was extremely rare to document imported techniques or to translate books. Considering the rudimentary nature of European lan-

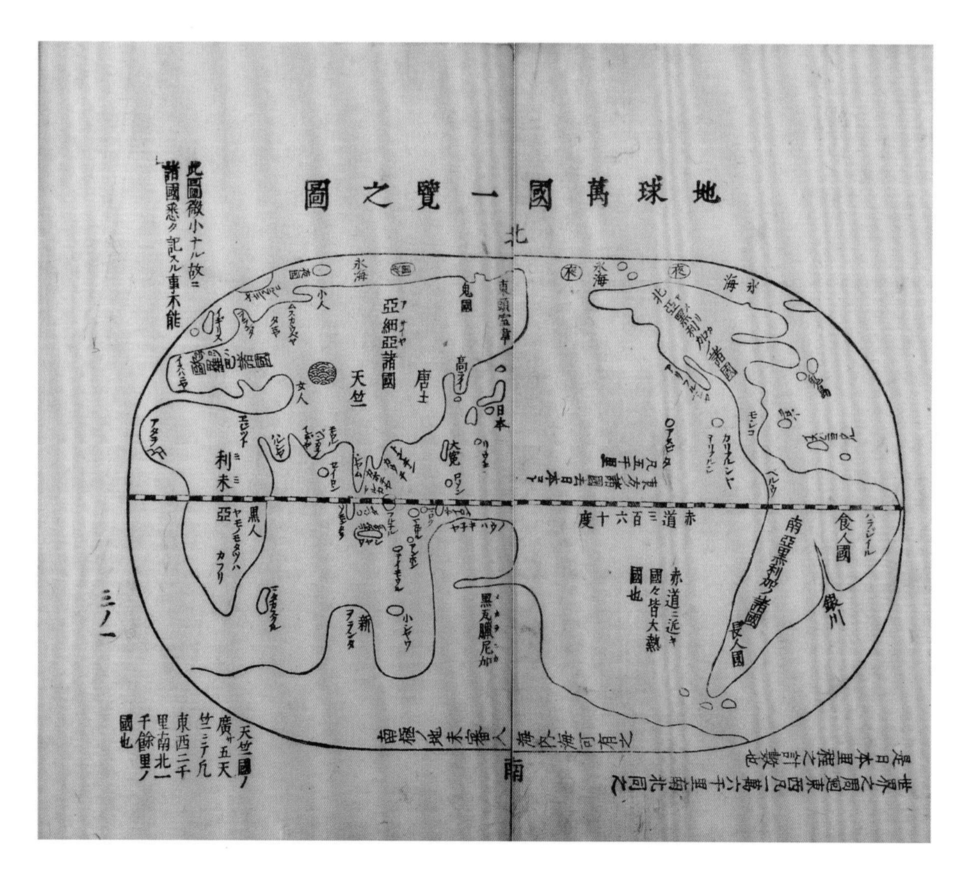

FIGURE 6.2 "Map of the Myriad Countries in the World at a Glance" [Chikyū bankoku ichiran no zu [地球萬國一覧之圖], by Nishikawa Joken 西川 如見. Illustration from *Japanese Trade with the Chinese and Other Countries, Enlarged* [Zōho Ka-i tsūshō kō 増補華夷通商考], 1708, vol. 3, 4–5. Courtesy of the University of Electro-Communications (Denki Tsūshin Daigaku). This is a crudely simplified version of the world map drawn by Matteo Ricci for the Ming court in 1602.

guage ability among the Japanese of the day, it seems likely that techniques from abroad diffused slowly as people learned by watching them in practice. It is unsurprising that surviving documents merely summarize imported techniques to the extent that people understood them. The surveying practices introduced by the Dutch were no exception.

A significant breakthrough came when the Tokugawa shogunate commissioned the production of *kuniezu* (provincial map series covering all of Japan). As discussed in chapter 10, this massive effort brought various innovations to mapping practice in Japan and led to the diffusion of mathematical surveying techniques. The official who was initially put in charge of the project requested the VOC to provide geometry lessons to Japanese officials during the first round of *kuniezu* production in the 1640s. Historical materials from the Dutch archives indicate that the lessons included discussion of maps.

Although it has long been known from Japanese archival fragments that European techniques were used in producing the *kuniezu*, we can see a clearer picture by cross-referencing historical materials from the VOC archives. In the past, scholars assumed that surveying techniques were imported through Portugal or Spain; we now know for certain that the Dutch were involved as well. Because of its national scope, the shogunate's mapping project in the mid-seventeenth century spread the latest surveying methods virtually throughout Japan. Numerous field surveyors and mathematicians were employed nationwide, spreading Dutch cartographic techniques across the country. But what specific techniques did the Tokugawa shogunate adopt? The journal kept by the head of the trading house of the VOC contains several clues.

A journal entry for December 20, 1647, for example, states that Masashige Inoue—the official who was charged with overseeing *kuniezu* production—invited "an assistant who is well-versed in astronomy and geometry" to teach his subordinates and "draw the territorial borders on the map of seventeen states belonging to the Dutch Republic." A similar lesson was conducted a few days later. In addition, Inoue showed great interest in examining the surveying instruments owned by the Dutch merchants. A journal entry of September 27, 1647, mentions that Inoue asked a member of the Dutch delegation to demonstrate the use of a cross-staff. Later, the cross-staff became a common surveying instrument (called *garātobōko* in Japanese, after the Dutch *graetboogh*). These episodes involving Inoue were probably not the only opportunities Japanese had to learn surveying techniques from the Dutch; there must have been other undocumented instances where the Japanese and Dutch came into contact. We do not know, however, who the instructor may have been or what manuals, if any, were used.

Provincial maps were commissioned by the shogunate not once, but four times during the seventeenth century. By the fourth round, in the 1690s, systematization of surveying techniques had considerably advanced and experienced officials had begun to synthesize the knowledge accumulated up to that point. At this time, several distinct Japanese surveying schools emerged, and textbooks were compiled. Many of the surveying instruments depicted in these early Japanese manuals are quite similar to the ones used in the Dutch Republic and other regions of Europe. For example, in addition to the cross-staff, there are plane tables and instruments called *kikugenki* that are identical to the European graphometer. Although the exact path by which they were brought into the country is not known, the instructions for using these instruments to produce large-area maps had already been set down and systematized by the 1690s.

In Europe, surveying techniques were advancing steadily during the sixteenth and early seventeenth century, particularly in the Dutch Republic. Professional surveyors were emerging in various places, and basic manuals were being published. By contrast, the men driving surveying techniques in Japan at the time were essentially amateurs: low-level samurai who happened to be adept at mathematics. Most belonged to an academy of one sort or another.

Three Japanese surveyors launched schools in the seventeenth century, all advocating methods that employed the traverse survey and plane table (notwithstanding minor differences in the types of instruments used). The leading school was the Shimizu Academy, headed by Shimizu Sadanori. Shimizu was a retainer of the Tsugaru domain, located in the northernmost part of Honshu, where his maps and landscape sketches can be found today at the Hirosaki City Library. A second influential school was founded by Kibe Shiroemon, who served the powerful Matsudaira daimyo. Kibe produced a coastal map of Echigo as well as a map of the Tone River watershed near Edo. Since their surveying techniques are much the same, Kibe is presumed to share an instructional lineage with Shimizu. The third noteworthy surveyor of this era was Matsumiya Kanzan, a specialist in military tactics. Matsumiya provided a comprehensive description of surveying techniques in his book *Techniques of Protraction* (*Bundo yojutsu*), which described in detail the equipment imported from the Dutch Republic as well as the method of *kuniezu* production. Matsumiya's teacher had in turn been taught surveying techniques as part of courses in gunnery run by the Dutch in the 1650s.

In short, a number of skilled Japanese surveyors took charge of mapping in the field starting in the late seventeenth century. Their work preceded by a full one hundred years the precise coastal surveys of Inō Tadataka (discussed in chap. 29). Over the course of the eighteenth century, the basic surveying techniques deployed in making the *kuniezu* would become common throughout Japan, being eventually passed down to Inō himself. While the national map of Japan produced by Inō and his team is deservedly famous, we should not overlook the fact that local surveyors throughout the archipelago had long since laid the foundation for its production. Inō's map—like the *kuniezu* before it—was a product of surveying techniques introduced from abroad in the seventeenth century.

Suggested Readings

Brown, Philip C. "The Mismeasure of Land: Land Surveying in the Tokugawa Period." *Monumenta Nipponica* 42 (2) (1987): 115–55.

Ravina, Mark. "Wasan and the Physics That Wasn't: Mathematics in the Tokugawa Period." *Monumenta Nipponica* 48 (2) (1993): 205–24.

7 The European Career of Ishikawa Ryūsen's Map of Japan

Marcia YONEMOTO

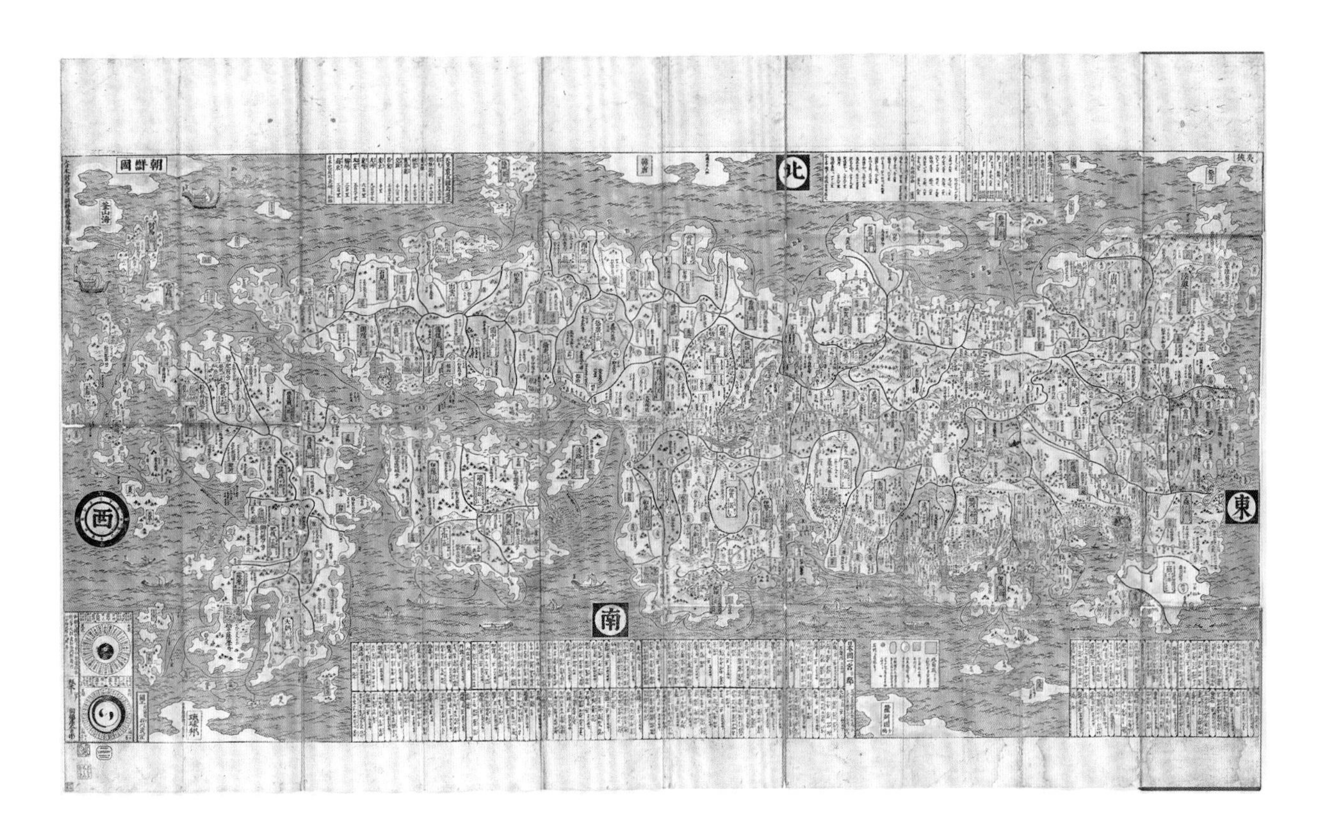

FIGURE 7.1 "Map of the Seas, Mountains, and Lands of Japan" [Nihon kaisan chōriku zu 日本海山潮陸図], by Ishikawa Ryūsen 石川流宣, published in Edo by Sagamiya Tahē 相模屋太兵衛, 1694. Woodblock print, hand colored, 81 × 171 cm. Courtesy of the C. V. Starr East Asian Library, University of California, Berkeley.

Ishikawa Ryūsen was a writer and a woodblock-print artist who lived and worked in Edo between approximately 1680 and 1720. Although he is not widely known today, he was quite prolific in his time, authoring and illustrating a variety of texts, ranging from comic fiction to scholarly treatises on geography, arithmetic, and architecture. In the late 1680s, Ishikawa began making single-sheet woodblock-print maps of Japan and its main cities. One of his best-known works is entitled "Map of the Seas, Mountains, and Lands of Japan," first published in 1689 and reprinted frequently thereafter (fig. 7.1). This map was taken to Europe in the late seventeenth or early eighteenth century, where, after its revision and publication by a prominent Dutch scholar, it became one of the most influential cartographic images of Japan. Through this process of translation, Ishikawa's map of Japan acquired all the symbolic trappings characteristic of early modern European representations of empire, endowed with meanings wholly unintended and unanticipated by the Edo mapmaker himself.

It is ironic that Ishikawa's maps should have become so important in Europe, for Ishikawa himself was far from a cartographic expert. Although educated and highly literate, he had no formal scientific training and did not conduct any actual surveying. Instead he adapted various existing sources, both narrative and visual, to make his maps. Accuracy not being his primary concern, Ishikawa sought to blend old cartographic and visual conventions with new information that would be of interest to the late seventeenth-century viewer. For example, his maps emphasize the sixty-six provinces of Japan, ancient geographic units that were first established during the classical period (ca. seventh century) and were deeply embedded in the early modern spatial consciousness. However, while the province evoked an archaic image of imperial unity, by the Tokugawa period it was not the province but the daimyo's domain that was the relevant political entity. Ishikawa's maps cleverly integrate these old and new geographic units by clearly depicting the provinces—which are outlined in black and, in most printed versions, visually distinguished through the use of color—while also adding rectangular labels listing the names of daimyo within each province and the productivity of their lands. Tables at the borders of the map list the general conditions and productivity of land in each province and traveling distances between towns on the main trunk roads. Such information was useful to the merchants, pilgrims, and other travelers who took to the roads in increasing numbers as the eighteenth century approached. Although Ishikawa's map of Japan focuses primarily on the archipelago itself, it also depicts other countries in the East Asian region with which Japan had frequent diplomatic and trade relations in the early modern period. The Korean Peninsula, the Ryukyu Islands (site of present-day Okinawa Prefecture), and the island of Ezo (now Hokkaido) appear at the edges of the map, hovering, respectively, at the top left, lower left, and top right of the printed frame.

While Japanese mapmakers were envisioning their country as an integrated and independent spatial and political realm, by contrast, their European counterparts began to conceive of the archipelago as an outpost of Western empire. Following the so-called "expulsion edicts" of the Tokugawa shogunate in the 1630s, which barred Spanish and Portuguese missionaries and traders from the country, the only Europeans left in Japan were the Protestant, nonproselytizing Dutch traders. Much of the information exchanged between Japanese and Europeans after that date was therefore mediated by the Dutch. Considering this close and exclusive connection, it is not surprising that the Dutch scholar Adrien Reland produced one of the earliest and most influential cartographic images of Japan in the eighteenth century. Reland, a scholar of "Oriental" languages and religions at the University of Utrecht, created a large-format map of Japan entitled *Imperium Japonicum*, which was published in Utrecht by Wilhelm Brodelet (fig. 7.2). Even a cursory inspection reveals that this map is all but identical to Ishikawa's "Map of the Seas, Mountains, and Lands of Japan." Reland evidently obtained a copy of Ishikawa's map through one of his students, who in turn had acquired it through personal connections from a prominent East India Company official who had served in Nagasaki. The shape and orientation of the three main islands of the archipelago in the Reland map are identical to those in the Ishikawa map, as are the representations of Ezo, the tip of the Korean Peninsula, and the Ryukyu Islands. Gone from Reland's map, however, are the tables listing traveling distances and the information about land productivity of the domains. In their place is an elaborate colophon dedicating the image to Reland's patron, the abbot Jean-Paul Bignon (1662–1743)—religious lecturer to the French king, royal librarian, and scholar of the Japanese and Chinese languages and history. Angels hold aloft the bishop's miter, cross, and crown above the colophon, which acknowledges the borrowed "Japanese map" that Reland used as a model. At the lower left is an illustration depicting a group of monks in the process of inscribing a dedicatory poem to Bignon on a large stone tablet. Rendered in flawed schoolbook Latin in the style of Ovid's elegies, the poem seems at first to be a tribute to Japan, but the reader soon realizes that this is not the case. Instead, the poem is a paean to Dutch imperial might, which brings under its sway far-flung locales like Japan. In translation, the poem reads:

> Behold the Queen of the Sea, you who first be-spy
> The horses of the Sun as they rise from the rose-tinted sea,
> And you who broadly span twice thirty shores,
> Thus, you must be viewed as abundant locales to be given to our people.
> Here we, the Dutch, have founded our settlement. Could I believe
> That at last there can be any place untouched by our countrymen?[1]

FIGURE 7.2 "The Empire of Japan" [*Imperium Japonicum*], by Adrien Reland, 1715. Copperplate engraving, 48.5 × 60 cm. Utrecht: Wilhelm Brodelet, Private Collection, London, UK.

By adding the colophon and illustrations, Reland imagined a new and superior status for the Dutch in Japan—ignoring the political reality that the Dutch were able to maintain a limited presence in Japan only at the behest of the shogunal government. He emended Ishikawa's map by removing all references to its original context, adding labels and annotations in Latin, and embellishing the map with the cross and the crown, symbols of church and state. Further, through his flowery poem, Reland made an explicit if exaggerated statement about the power and authority of the Dutch over Japan. In this way, Reland did more than merely transcribe—he transformed the "seas, mountains, and lands" of Japan into part of a vast, imagined European-controlled empire in Asia.

Needless to say, the Dutch never colonized Japan, and it is debatable whether Reland knew exactly what kind of government pertained there in the eighteenth century. Before Englebert Kaempfer's *History of Japan* appeared in 1727, many European observers confused the shogun with the emperor, failing to realize that Edo and Kyoto hosted two separate but mutually dependent seats of government.[2] For its part, Kaempfer's *History* also contained a map of Japan modeled directly on another of Ishikawa Ryūsen's works, the "Outline Map of Our Realm" of 1687. Like Reland's derivation, Kaempfer's was subsequently reproduced and adapted in European atlases as late as the nineteenth century. Seen from this perspective, Ishikawa's map of Japan had a stunningly successful career, encompassing some 120 years and crossing two continents and multiple national boundaries. Yet throughout this process, Ishikawa himself remained anonymous—indispensable to the European mapmaker but unseen and largely uncredited. And the further European mapmakers got from the original Japanese text, the more the map of Japan became the vehicle for European imaginings of this remote world as an extension of their own. One could argue that Ishikawa Ryūsen's map escaped its author, careering off on its own independent and uncontrollable trajectory.

The influence of Ishikawa's maps on European cartography waned in the early 1800s following the arrival of Japanese maps based on updated surveying techniques. Having acquired a set of Inō Tadataka's new maps of Japan's coastline while serving in Nagasaki during the 1820s, the German doctor Philipp Franz von Siebold upon his return to Europe published the most accurate maps of Japan produced in the West up to that time. Siebold's maps were subsequently used by the British navy, and later by the American naval expedition headed by Commodore Matthew C. Perry that officially "opened" Japan to the West in 1853. Once Siebold published his maps, those of Reland and others slipped into antiquarian obscurity and the career of Ishikawa's map came to a close, pushed into retirement as the need for military, diplomatic, and trade relations with Japan cast Euro-American dreams of Japanese empire in a new and pragmatic light.

Notes

1. I would like to thank Lionel Jensen of the University of Notre Dame for his translation of and commentary on this poem.

2. Due to the politics of patronage and the bad luck that beset its author, the English version of Kaempfer's *History of Japan* was the first version of the text to be published.

Suggested Readings

Bodart-Bailey, Beatrice. *Kaempfer's Japan: Tokugawa Culture Observed.* Honolulu: University of Hawai'i Press, 1999.

Hubbard, Jason. *Japoniæ Insulæ, The Mapping of Japan: Historical Introduction and Cartobibliography of European Printed Maps of Japan to 1800.* Houten, The Netherlands: Hes & De Graaf, 2012.

Schmeisser, Jörg. "Changing the Image: The Drawings and Prints in Kaempfer's *History of Japan*." In *The Furthest Goal: Englebert Kaempfer's Encounter with Tokugawa Japan*, ed. Beatrice Bodart-Bailey and Derek Massarella, 132–51. Folkestone: Japan Library, 1995.

Walter, Lutz, ed. *Japan: A Cartographic Vision; European Printed Maps from the Early 16th to the 19th Century.* Munich: Prestel, 1994.

Wigen, Kären. *A Malleable Map: Geographies of Restoration in Nineteenth-Century Japan.* Berkeley: University of California Press, 2012.

Yonemoto, Marcia. "Envisioning Japan in Eighteenth-Century Europe: The International Career of a Cartographic Image." *Intellectual History Newsletter* 22 (2000): 17–35.

———. *Mapping Early Modern Japan: Space, Place, and Culture in the Tokugawa Period (1603–1868).* Berkeley: University of California Press, 2003.

8 A New Map of Japan and Its Acceptance in Europe

Matsui Yoko 松井洋子

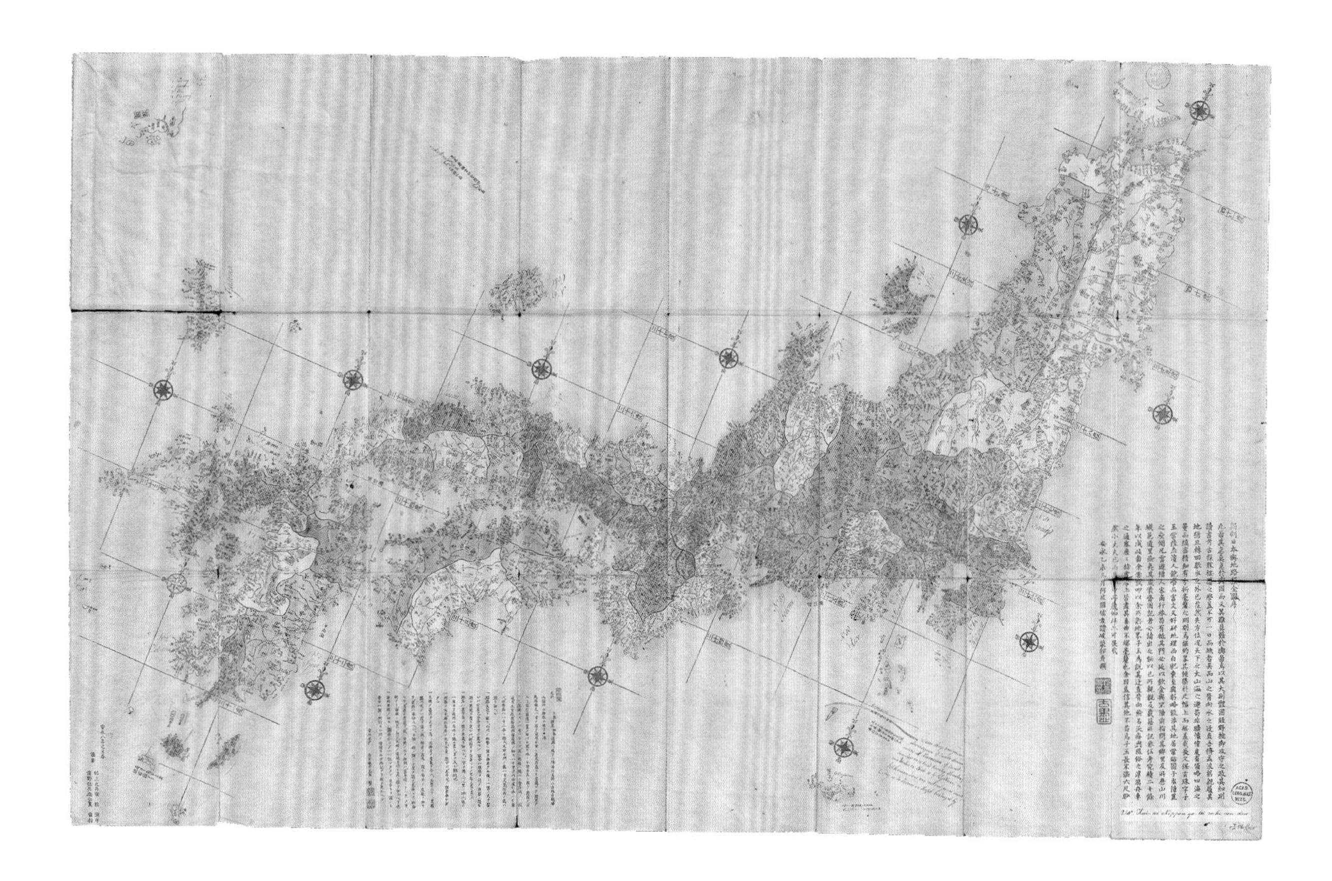

FIGURE 8.1 "Revised Route Map of Japan" [Kaisei Nihon yochi rotei zenzu 改正日本輿地路程全図] by Nagakubo Sekisui 長久保赤水, 1779. Woodblock print, hand colored, 83 × 134 cm. Courtesy of the Leiden University Library, Leiden, the Netherlands (Serrurier. 220a).

Toward the end of the seventeenth century, colorful maps based on the work of Ishikawa Ryūsen—the cartographer whose famous map of Japan is discussed in the preceding essay—gained considerable popularity in Japan. Over the years, numerous variations were published in the Edo area with additional information about the eastern part of Japan. In addition to their decorative function, the Ryūsen-style maps also had practical purposes, offering ample information for travelers and showing basic political features that allowed the general public to form a vivid image of their country.

These popular Ryūsen-type maps were not displaced until a century later, when Nagakubo Sekisui first published his "Revised Route Map of Japan" (fig. 8.1). This 1779 map, widely known as the *Sekisui-zu,* represented a landmark in Japanese cartography: it was the first printed map to put Japan on a grid.[1] Its horizontal lines express degrees of latitude, while its vertical lines, although not meridians in a strict sense, mark off distance in an east-west direction.

This linear grid was not the only striking innovation of Nagakubo Sekisui's design. A second novelty of the *Sekisui-zu* was its adherence to a fixed scale, specified in the legend as one *sun* to ten *ri* (approximately 1:1,296,000). A third feature was its unprecedented spatial accuracy. Although the map was not based on an actual land survey (being compiled from existing maps and documents), its outline shape of Japan as well as its internal administrative boundaries were drawn with a high degree of precision. Fourthly, decorative elements were deemphasized, and travel and political information was stripped out altogether. Last but not least, the map was saturated with more than four thousand place-names—approximately five times as many as in the Ryūsen-type maps. The legend indicates that the compiler took pains to place each toponym in the right spot. Taken together, these features took Japanese cartography a big step closer to the modern Western standard.

What do we know about the cartographer who made this great leap forward? Nagakubo Sekisui (1717–1801) was born to a wealthy peasant family in the domain of Mito. Sekisui was both well educated and well traveled. Acquiring a broad Confucian education in his youth, he distinguished himself among local intellectuals throughout his long life in Mito. His geographic interest led him to northern and southwestern Japan. During one of his trips, he spent time in Nagasaki, the site of the Dutch trade and the home of *rangaku*, or the study of Western science. In 1777, at the age of sixty-one, he was appointed lecturer for the lord of Mito and invited to Edo.

In a varied career that included writing books on astronomy and itineraries, submitting political opinions, compiling geographic sourcebooks, and more, Sekisui's most remarkable accomplishments were in the realm of cartography. In addition to the gridded map shown here, he also designed and published a world map and compiled historical atlases of the Chinese empire. To be sure, Sekisui was not a scientist in the modern sense. In many ways, he was a typical intellectual of his time, operating in a culture that placed a high value on accuracy and rationality. Because of his travel experiences and his membership in a variety of cultural networks, he was able to create precise, practical, and up-to-date maps that attracted the attention of discerning readers who were not satisfied with popular products like the Ryūsen print.

Sekisui's map was revised and republished five times in five decades (starting in 1791, and continuing after Sekisui's death with posthumous printings in 1811, 1833, 1840, and 1844). A large number of imitations and pirated editions were also made. Consequently, quite a few maps of this type survive—not only in Japan but also in other countries. Although few foreigners came to the Japanese archipelago during the late eighteenth and the early nineteenth century, those who did acquired the *Sekisui-zu* as the latest map of Japan. In fact, the map presented here was evidently the first one to be exported to Europe. This particular copy was included in the collection of Philipp Franz von Siebold, which was eventually donated to the library of Leiden University.[2] In his book on Japan, Von Siebold explained that he had obtained a copy of the new map of Japan that had originally belonged to Hollander Isaac Titsingh (1745–1812).[3] Titsingh had spent three and half years in Japan, where he served three separate terms as the chief factor of the Dutch East India Company in Nagasaki between 1779 and 1784. During his stay, he became determined to compile a complete description of Japan and collected a variety of books and other materials for that purpose. He asked some interpreters to translate them privately. (The interpreters of Dutch language in Nagasaki were all Japanese officials; having daily contact with the Dutch residents, these interpreters sometimes acted as information brokers. Titsingh developed personal relationships with them through teaching language.) After leaving Japan, he served for seven years as director of the Dutch trade in Bengal (1785–92) before ending his career as special envoy to the imperial court of Beijing (1794–95).

On January 2, 1792, Titsingh wrote to his brother, "I also include in the same bundle of papers a new map of Japan, made by order of the daimyo of Moets and given to me as a present; the place-names of the different provinces are numbered and to be found in the explanation. . . . The numbering of these place-names has cost me a lot of work." Although he confused Mito with Mutsu (Moets), the "new map" he referred to here can be clearly identified as the first edition of Sekisui's "Revised Route Map of Japan," reproduced here as figure 8.1. With a preface by a famous Confucian scholar, dated 1775, this woodblock print map was carved by Murakami Kyūbē in Osaka and published by Asano Yahē in 1779. From the place-names and coastlines it can be identified as the very first version of Sekisui's map, despite the fact that it lacks the seal of the publisher. It is colored by hand, and the combination of the eight colors is different from that in other extant prints.

The most remarkable feature of this particular copy, however, is the insertion of numerous handwritten toponyms and numbers in red and black ink. The names of all the provinces are noted in Roman letters, with associated numbers in red. In some cases, local place-names have also been added, especially around the bay of Nagasaki, where the Dutch trading post was situated. Furthermore, every place-name in each province was numbered sequentially in black ink. The legend shows the symbols for castle towns, sights, temples, shrines and so on, but the numbers are assigned without reference to those distinctions. Isaac Titsingh himself inserted them, as noted in his letter. The preface, legend, and other Japanese text on the map were not translated into Dutch except for a few short explanatory notes.

After his return to Europe in 1796, Titsingh was eager to realize his great dream to publish a comprehensive work on Japan simultaneously in Dutch, English, and French editions. Due to the unfavorable political and economic circumstances in the midst of the Napoleonic Wars, however, he eventually had to settle for depositing his manuscript in Dutch with the Royal Netherlands Academy of Sciences in 1811, hoping that times would soon change for the better.[4] A twenty-two-page list embedded in the manuscript, entitled "Explanation of the great map of Japan, of its 67 provinces and its islands," contains more than 4,200 place-names written out in Roman letters. The numbers added to each place-name correspond to the numbering on the Sekisui map reproduced here. This enormous index is what Titsingh meant when he mentioned an "explanation" in his letter; the map and the list, now separated from each other, were originally produced as a set. Although the Japanese interpreters must have helped him transliterate many Japanese place-names during his stay, Titsingh had to finish the list by himself. If the arduous labor of compiling and Romanizing over four thousand Japanese place-names seemed worth Titsingh's time, it was because he intended to grasp Japan in its totality through this combination of map and gazetteer.

Von Siebold, who consulted Titsingh's work, appreciated Sekisui's map as a veritable geographic dictionary and continued to consult it even after he had obtained various newer maps including some accurate survey drawings. In other words, while Sekisui's map upon first appearing was quite revolutionary for its spatial accuracy, for its fixed scale, and for placing Japan on a grid, to European Japanologists the greatest benefit of the *Sekisui-zu* was the knowledge of the country manifested in the map's multitude of place-names.

Notes

1. Regarding the *Sekisui-zu*, see Baba Akira, "Chizu no shoshigaku—Nagakubo Sekisui 'Kaisei Nihon yochi rotei zenzu' no baai," in *Chizu to ezu no seiji bunka shi*, ed. Kuroda Hideo et al. (Tokyo Daigaku Shuppankai, 2001), 383–430; Kinda Akihiro and Uesugi Kazuhiro, *Nihon chizu-shi* (Tokyo: Yoshikawa Kōbunkan, 2012), 203–13; Marcia Yonemoto, *Mapping Early Modern Japan: Space, Place, and Culture in the Tokugawa Period (1603–1868)* (Berkeley: University of California Press, 2003), 35–43.

2. The map is held under inventory number Serrurier 220a. The label attached to it confirms that it was "Titsingh's Original." Philipp Franz von Siebold mentioned this map in *Nippon/Archiv zur Beschreibung von Japan*, ed. F. M. Trautz (Berlin: Wasmuth, 1930–31), vol. 2, 76–77 and 253–54.

3. Regarding Titsingh, see Frank Lequin, *Isaac Titsingh (1745–1812): Een passie voor Japan/leven en werk van de grondlegger van de Europese Japanologie* (Alphen aan den Rijn: Canaletto, 2002); *De particuliere Correspondentie van Isaac Titsingh 1783–1812*, ed. Frank Lequin (Alphen aan den Rijn: Canaletto, 2009).

4. The manuscript entitled *Beschryving van Japan* is deposited in the National Library of the Netherlands as KA (Koninklijke Academie Collection) 147. See Matsui Yōko and Frank Lequin, "Titsingu korekushon no Nagakubo Sekisui Kaisei Nihon yochi rotei zenzu," *Report of the Center for the Study of Visual Sources, an Affiliate of the Historiographical Institute* 45 (2009): 4–11; Frank Lequin, *Varia Titsinghiana: Addenda & corrigenda* (Leiden: Titsingh Instituut, 2013), 123–26.

Suggested Readings

Boxer, Charles R. *Jan Compagnie in Japan, 1600–1850: An Essay on the Cultural Artistic and Scientific Influence Exercised by the Hollanders in Japan from the Seventeenth to the Nineteenth Centuries.* The Hague: Martinus Nijhoff, 1950.

Yonemoto, Marcia. *Mapping Early Modern Japan: Space, Place, and Culture in the Tokugawa Period (1603–1868).* Berkeley: University of California Press, 2003.

9 The Arms and Legs of the Realm

Constantine N. Vaporis

FIGURE 9.1 "Proportional Linear Maps of the Major Turnpikes" [Gokaidō sono hoka bunken mitori nobe ezu 五海道其外分間見取延絵図], Edo period, detail showing Kami Suwa Station on the Kōshū Highway [Kōshū Kaidō Kami Suwa eki 甲州街道上諏訪駅]. Manuscript (scroll), 60.3 × 958 cm. Courtesy of the Tokyo National Museum.

Early in the nineteenth century, the shogun's magistrate of road affairs created a series of maps known as the "Proportional Linear Maps of the Major Turnpikes." Belonging to the last of the five major categories of Edo-period maps (which depicted, respectively, the whole country, an individual province, a region, a city, or a road), the Turnpike series emerged from a lengthy survey of road conditions. Shogunal officials began the survey in 1800 and carried out another in 1843 to compare the changes in the intervening years, particularly population figures and the number of houses at each post station. The late date of production of these maps indicates that even with the long years of peace that defined the Tokugawa era, the shogunate never lost its strategic concerns in administering the highway network. The continuing historical importance of the series is underscored by their official designation as Important

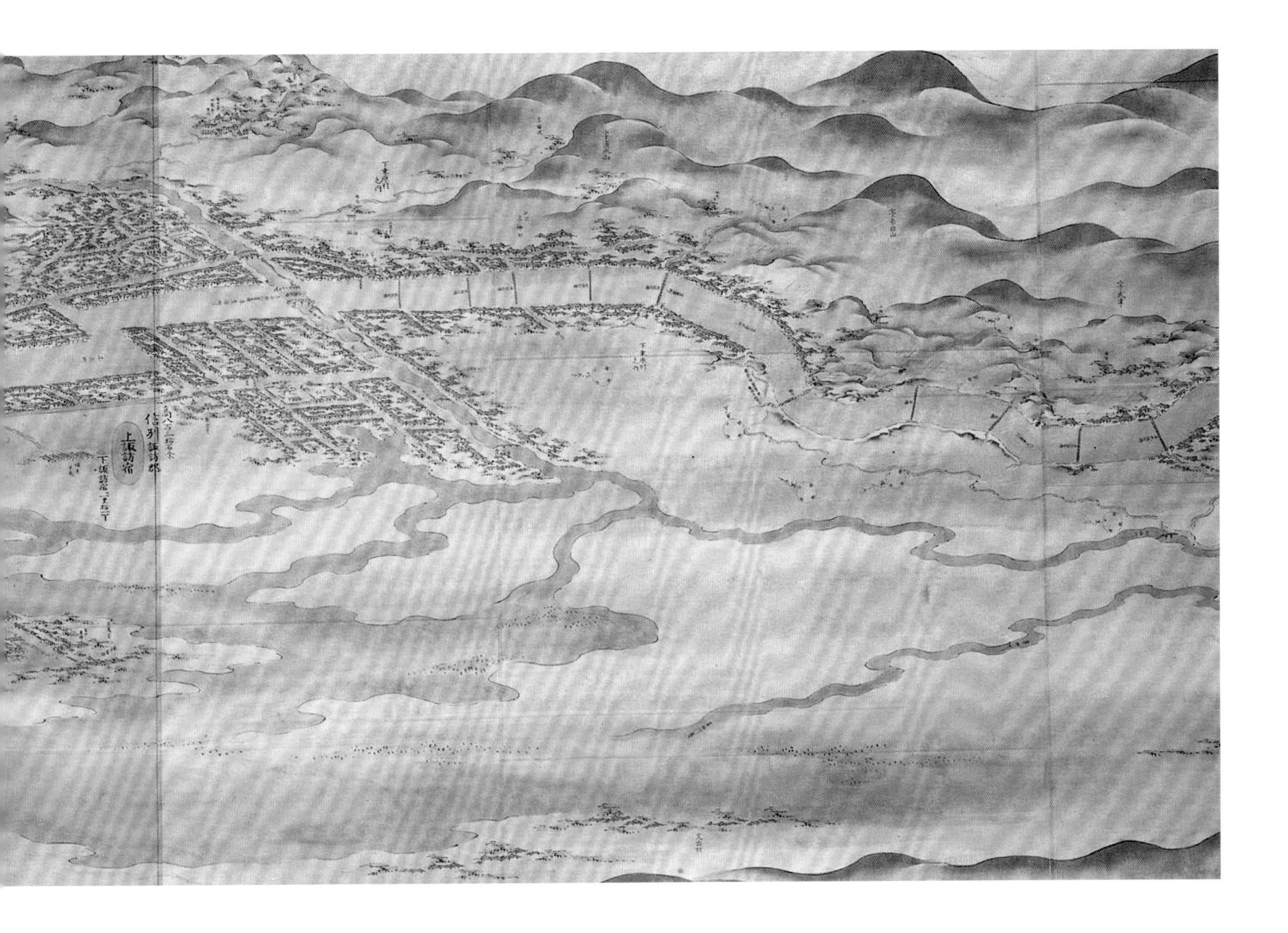

Cultural Property by the modern Japanese government.

These route maps give us a late Tokugawa-era snapshot of roads that, while initially designed for military and political control, gradually became the site of voluminous traffic, drawing millions of ordinary travelers and pilgrims. We see very little evidence here of that lively bustle, and only a restrained view of the significant urbanization in the post stations that line the road. To some degree, this reflected the shogunate's anachronistic view of Japan as an agrarian society—a view that continued to remain a part of official discourse despite the high level of urbanism and commercial activity that were attributes of Tokugawa life.

The "Proportional Linear Maps of the Major Turnpikes" are hand-drawn manuscripts, of which only three copies were made. All remained in Tokugawa hands, with one kept in Edo Castle and the other two held in the office of the magistrate of road affairs. Each was produced in horizontal scroll form, roughly sixty centimeters in width. Drawn to a scale of 1:1800, their hallmark is accurate topographical and infrastructural detail based on direct measurement.

The official purpose of these maps is evidenced by the care with which they were produced. Despite the massive scale of the project—104 scrolls—the maps were made in a refined manner, with hand-painted additions in color, by an anonymous team of cartographers and artists. Not intended for popular dissemination, the maps, rather than emphasizing the practical aspects of traveling, focus on the physical infrastructure and environment of the roads. They were meant to be comprehensive, including every shrine, temple, river, and bridge. Their perspective and details reflect the political centrality of the shogunate, highlighting its strategic and administrative concerns.

The Tokugawa regime created its centralized highway network early in the seventeenth century out of political and strategic needs. Seeking to stabilize a society fractured by more than a century of civil war, the new government took command of the country's main highways for the passage of its martial forces, the movement of its officials, and the conveyance of communications. The importance of the highway network to the Tokugawa state is reflected in the comments of one of its officials, Tanaka Kyūgu, who referred to the roadways as the "arms and legs of the realm."[1] A central aim of Tokugawa policy was to prevent the disruption of communications by local powers and to keep traffic flowing. It achieved this by nationalizing the trunk lines of the system, abolishing private barriers and tolls, and prohibiting the disruption of transport services.

Placing the turnpike system's central hub in the administrative center of Edo also reflected the centrality of the Tokugawa government. Four of the five main highways radiated directly outward from the shogun's capital, while the fifth branched off from one of the other four. Eight additional branch roads completed the network. The two images included here show, respectively, the Kōshū Road (fig. 9.1), which passed through several mountainous provinces in central Japan, and the Tōkaidō (fig. 9.2), the famous coastal road that linked Edo with Kyoto.[2]

Both maps contain a remarkable level of detail, noting the structure of bridges (stone, wood, or earth), the position of mileage markers, and the location of post-station offices and inns for daimyo. Careful attention was paid to the right-angle turns in the road, particularly in the approaches to castle towns such as Kami-Suwa (fig. 9.1), which doubled as a post station. The insertion of ninety-degree turns in the road at the approaches to these settlements reflected the defensive mentality of the Tokugawa, for the progress of potentially hostile forces could be most effectively stopped at these choke points.

The maps also contain information on religious edifices (shrines and temples), the population of the post stations, and surrounding topography. Even farmers' houses were not overlooked. In extending their depiction of territory far beyond the road itself, these official maps exceeded the function of typical road maps. Since all the roads in the turnpike network were Tokugawa territory, regardless of the domain through which they passed, this extension of the gaze far beyond the borders

FIGURE 9.2 "Proportional Linear Maps of the Major Turnpikes" [Gokaidō sono hoka bunken mitori nobe ezu 五海道其外分間見取延絵図], Edo period, detail showing checkpoint at Hakone by the edge of Lake Ashi on the Tōkaidō Highway. Manuscript (scroll), 60.1 × 254.6 cm. Courtesy of the Tokyo National Museum.

of the road implied Japan's unity under Tokugawa rule. Furthermore, by depicting the road in a totalized view from above, and empty of all inhabitants, the cartographers implied Tokugawa authority over the entire political space that was Japan.

While drawn in a standardized fashion, these maps clearly exaggerate certain features. The coastal road, for example, is drawn to look much wider than it actually was, accenting its symbolic importance.[3] Similarly, castles belonging to the Tokugawa family were drawn larger than on other maps, while details within the castle complex were obscured under the conventional cloud technique.

The section of the map shown in figure 9.2 centers on Hakone, one of the fifty-seven post stations on the Tōkaidō. The name Hakone is distinguished by the oval cartouche that surrounds it, while a notation to the left informs the viewer of the distance to the next station. All post stations fulfilled the various functions of rest stop, transport center, communications node, and recreation area. Hakone had the additional task to serve as one of fifty-three inspection stations established by the shogunate on the five major turnpikes. Wherever possible, these checkpoints were located in strategic areas at which traffic could be easily controlled. The stretch of road containing the inspection station is surrounded by fences and gates, and the entire enclosure is overlooked by a guard tower on the hillock above it.

The initial military purpose of checkpoints like Hakone was to prevent a surprise attack by the last holdouts of the coalition that had opposed the Tokugawa in 1600. Once peace was established, that mission expanded to include monitoring traffic, inspecting guns, and enforcing the institution of alternate attendance. Under the requirements of that system, which was formalized in 1635, daimyo wives and children were required to remain in Edo as hostages when the lord returned to his domain.

This stretch of road around Hakone is, like the entire series, drawn with great concern for accuracy, including the exact number of households, the names of stone bridges, the location of public notice boards, and the names of temples and shrines that faced the road (as well as some more distant ones). Within the inspection station compound, the locations of a public notice board, a well, and a target range for muskets are clearly marked. Topographical detail is not missed either; the names of the mountains on both sides of the lake behind the post station are noted. Only the iconic landmark of Mount Fuji, in the distant background and distinguished by a lack of color, required no labeling.

In short, the "Proportional Linear Maps of the Major Turnpikes" reveal an official view of Tokugawa Japan's highway network: a record that was utilized by its officials for administration, but that could also have been useful in the event of war. Reflecting the administrative and martial concerns of their creators, these maps provide a remarkable visual tour of the central arteries that coursed through the terrain of Tokugawa Japan.

Notes

1. Tanaka Kyūgu, "Minkan shōyo," as quoted in Constantine Nomikos Vaporis, *Breaking Barriers: Travel and the State in Early Modern Japan* (Cambridge, MA: Council on East Asian Studies, Harvard University, 1994), 17.

2. On the turnpike network, see Vaporis, *Breaking Barriers*, chap. 1.

3. On road widths, see Vaporis, *Breaking Barriers*, 36.

Suggested Readings

Kodama Kōta, ed. *Gokaidō bunken nobe ezu*. 103 vols. Tokyo: Tōkyō Bijutsusha, 1977–85.

Traganou, Jilly. *The Tōkaidō Road: Traveling and Representation in Edo and Meiji Japan*. London: Routledge, 2004.

Vaporis, Constantine Nomikos. *Breaking Barriers: Travel and the State in Early Modern Japan*. Cambridge, MA: Council on East Asian Studies, Harvard University, 1994.

10 Visualizing the Political World through Provincial Maps

Sugimoto Fumiko 杉本史子

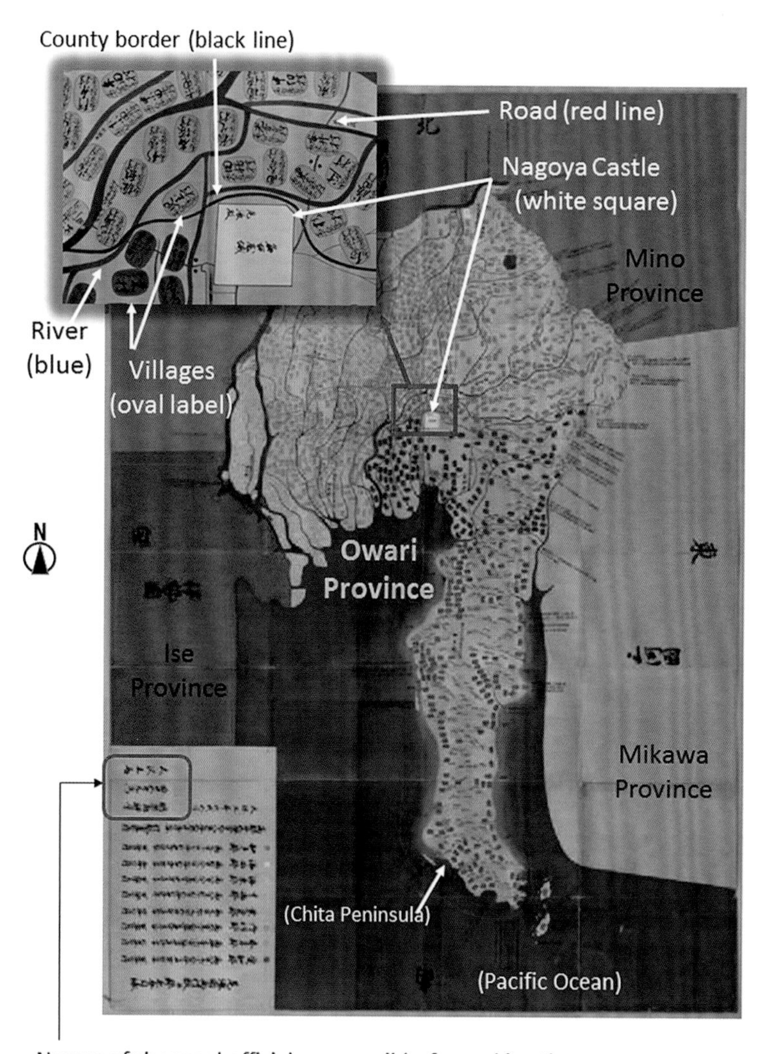

FIGURE 10.1 "Owari Provincial Map" [Owari kuniezu 尾張国絵図], 1835. Manuscript, 295 × 419 cm. Courtesy of National Archives of Japan. Detail (upper left) shows the area around Nagoya Castle. English labels added by author.

In 1823, a young man named Philipp Franz von Siebold traveled from Europe all the way to Japan, a remote island country off the eastern coast of the Eurasian continent. This gentleman had been born in 1796 in the German-speaking town of Würzburg, located within the waning Holy Roman Empire. He made the journey in order to work as a doctor in the Dutch East India Company's branch office in Nagasaki. From the perspective of Europeans at the time, Japan had been discovered by the Portuguese in the sixteenth century and had only just come to be explored by the Russians and the Dutch. It was still thought of as an unknown country.[1] Once established in Nagasaki, the young man collected all manner of plant specimens and books about Japanese ships, language, history, and law, which he then sent off to the Netherlands. He encountered truly surprising maps too, in particular a series of large handmade maps composed of multiple sheets of paper drawn to phenomenally large scales. The young man reproduced twenty of these maps at a reduced scale and sent them back home—an astonishing feat if we consider the difficulty of accessing such politically sensitive documents.

Called *kuniezu* in Japanese, the giant handmade provincial maps that so impressed the German doctor had been made for the Tokugawa shogun by local lords throughout the archipelago (fig. 10.1).[2] The standard set by the shogunate called for roads to be indicated at a scale of 1:21,600 in today's terms. (At the time, units of measurement were not uniform across Japan, so the actual scale varied from map to map.) Even the smallest *kuniezu* measured over nine feet on a side, and it took several people working carefully to unfold a single map without damaging it (see fig. 57.1, at the end of this volume). In any case, successive shoguns ordered the construction of provincial maps at regular intervals between the seventeenth and nineteenth centuries.

The young man who encountered *kuniezu* in Nagasaki offered high praise for these impressive maps while pointing out what he considered to be their unfortunate use of oval labels for villages and square labels for cities; he thought it would have been better to let those spaces appear naturalistically. Such elements may have seemed strange for Europeans and therefore defective, but they were anything but strange within premodern Asia, since they constituted basic conventions of mapmaking in this context. Geometric labels for castles and villages appear in all the *kuniezu*. The names of local lords and their castles appear within white squares. Oval labels indicate villages with land tax obligations and display both the name of the village and its productive yield measured in rice. Even deserted villages were included for the sake of maintaining consensual agreement between the shogun and the lords. In other words, *kuniezu* were designed to depict the world under shogunal and domainal rule rather than the physical characteristics of topography.

The province, or *kuni*, was an administrative unit used since ancient times, when Japan was subdivided into sixty-odd regions. Originally, the emperor dispatched representatives to govern each province. However, by the Tokugawa era, the emperor no longer exercised such power, the governance of provinces having fallen under the control of local lords, all of whom were vassals ultimately answerable to the shogun. These samurai-ruled territories did not always coincide with provinces; lands belonging to one powerful lord might exceed the boundaries of a particular province, while in other cases one province might be divided up among a number of lesser lords. However, conceiving of Japan as a configuration of sixty-eight cohesive provinces was a powerful ideal that persisted into the early modern period. *Kuniezu* made that ideal visible.

Kuniezu reveal the strategy of using colors and shapes as a way to indicate the organization of an administrative system. This is made manifest both in the aforementioned use of geometric labels and the application of colors. Red, blue, black, and white constitute the four primary colors, as customarily expressed in the Japanese language.[3] In the maps, red is used for roads, blue for waterways, and black for the boundaries of county-level districts (*gun*). Pure white squares mark castles. Compound colors (other than the four primary colors) designate mountains, vegetation, religious structures, lookout facilities, and villages. Black distinctly marks borders between the districts that make up a single province. Villages are colored according to the district in which they are located (fig. 10.1, detail inset). Almost all *kuniezu* dating from the eighteenth century follow this color scheme.

The deployment of a symbolic system that made such extensive use of color must have made a very strong impression on those who viewed *kuniezu*. Earlier maps had usually been rendered simply in black ink, only occasionally making use of two or three colors. Featuring such innovations, this newly standardized and differentiated geographic language ended up becoming a commonly shared idiom among those in power. For when those responsible for making *kuniezu* (for the most part, daimyo) submitted the maps to the shogun, they made duplicate copies for safekeeping in their own libraries. In this way, maps of identical content existed within the shogunate and in each province.

The spatial language of *kuniezu* provided a precise grammar for each province: (1) lines for spatial demarcation between districts; (2) a system of colors and geometric labels for administrative and military structures; and (3) place-names and indications of productivity for villages, which served as a basis for legal understanding among shogun and lords. The practical nature of this grammar established a form of geographic representation with much utility for governing—so much so that it was extended to other kinds of maps as well.

For example, a map of falconry grounds within the province of Owari (figs. 10.2 and 10.3) utilized the same grammar found in the *kuniezu* submitted by the lord of Owari to the shogun. As in many other parts of the world, hunting with falcons in Japan was a powerful symbol of a ruler's control over his territory. This particular map closely resembles the Owari *kuniezu* submitted by a lord to the shogun. As it happens, this falconry map was a work in progress, and no finished version exists. (There are several working labels attached to this copy, as can be seen clearly in fig. 10.3.) Nonetheless, it shows quite clearly what kind of map its makers intended to create. The image depicts the entire province of Owari, indicating the castle as a square and the villages as ovals. Since it was made for a local lord, however, rather than the shogun, the falconry map uses a more glamorous gold for the lord's castle, while a silver line encompasses an area extending around the castle. This bold use of different colors to indicate political distinctions is very much in the spirit of *kuniezu*, but the falconry map uses color to emphasize hunting grounds allocated by the lord to his senior vassals, rather than to demarcate administrative districts.

The effect of this particular color-coding scheme was to vividly display the hierarchical power structure within Owari. Each of the allocated falconry grounds was given a different color: red, brown, or ochre. By contrast, the hunting grounds used by the lord did not require any special color. This made for a clear division between those who bestowed colors and those upon whom colors were bestowed. In effect, color classification of this nature functioned as a symbol for land control by the daimyo. It also served as a reminder that even the highest-ranking retainers were recipients of privilege and could not lay claim to the ubiquitous land of their

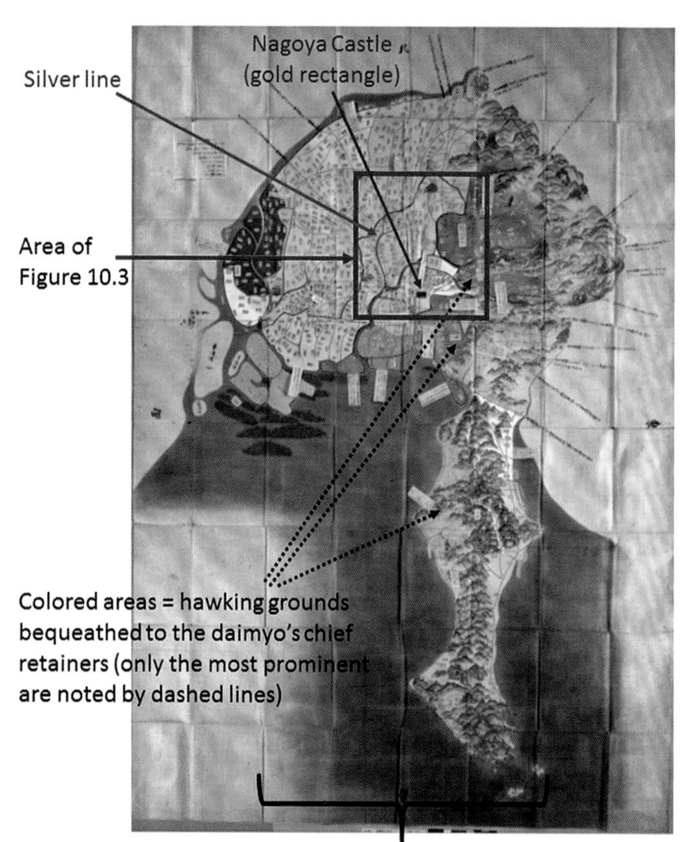

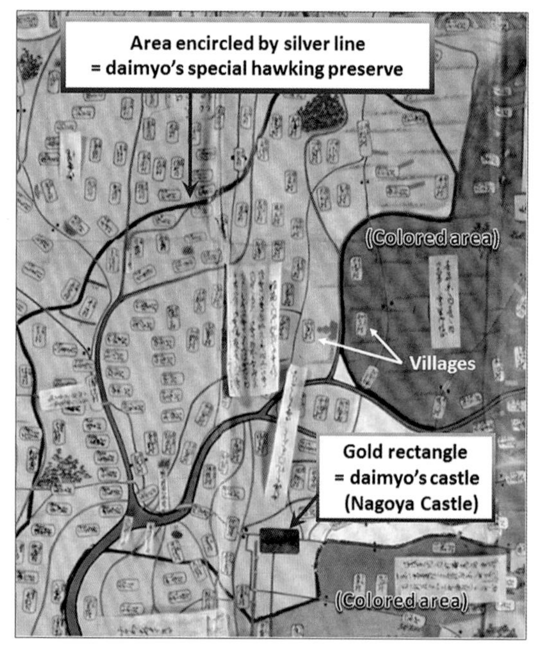

FIGURE 10.2 "Falconry Map" [Otakaba no ezu 御鷹場之絵図], date unknown. Manuscript, 404 × 282 cm. Courtesy of Tokugawa Bijutsukan. English labels added by author.

FIGURE 10.3 Detail of figure 10.2, showing the area around Nagoya Castle. English labels added by author.

lord. Moreover, the silver line around the castle frames a hunting area with particular importance for the lord. Like the brilliant gold used for the castle itself, silver expressed the radiant authority of the lord. And this radiance stands out all the more since it illuminates the areas that mark the full extent of the lord's authority.

The young man in Nagasaki who sent the *kuniezu* back to the Netherlands returned to Europe in 1830. In 1832, he began work on the publication of a large book called simply *Japan*, which was to be printed in German, Dutch, French, and Russian. Siebold was proud of having brought "Japan" back with him in order to show Europe. At the same time, however, the Japanese observed and gazed upon Europeans as well. Woodblock portraits of the Dutch East India Company, where Siebold worked, were sold as souvenirs that captured the peculiar customs of the exotic foreigners in Nagasaki.

Siebold's *Japan* was published in 1835, the same year that the Tokugawa shogunate compiled the final *kuniezu*. By this time, Japan was becoming a society where private networks for commercial and cultural interaction were flourishing. Financial instruments such as securities linked merchants in far-flung cities, while the circulation of poetry and painting penetrated into villages and towns across the land. Private individuals observed and wrote about the political activities of the shogun and lords; some even satirized the authorities in print.[4] Many had started to question the notion of a world ruled by samurai. In other words, by the time the *kuniezu* were introduced to Europe, they conveyed an outdated image of what was in fact a dynamic, fast-changing society.

(Translated by Robert Goree)

Notes

This essay draws on the results of a research project concerning the relationship between Siebold's maps and other provincial maps in the Edo era, under principal investigator Onodera Atsushi (JSPS KAKENHI Grant Number 1920130).

1. Miyazaki Katsunori, "Shiiboruto Nippon no furansugo han," *Hakubutsukan kenkyū hōkoku* 6 (2008): 1–32.

2. The shogunate produced this map in 1835 as one of eighty-three maps of territories under its jurisdiction, based on information it solicited from various *kuni*. The orientation of the English translations in the image of the map reproduced here follows the irregular orientation of the Japanese text in the original map. Parenthetical English explanations have been added to aid understanding and are not translations of the original Japanese. The *kuni* of Owari appears at the center of the map. This *kuni* corresponds to the western half of present-day Aichi Prefecture. The Chita Peninsula juts out in a southerly direction into the Pacific Ocean.

3. Ōno Susumu, *Nihon no iro* (Tokyo: Asahi Sensho, 1979).

4. Sugimoto Fumiko, "Ikoku iiki to nichijō sekai," in *Kinseiteki shakai no seijutsu*, ed. Arano Yasunori, Ishii Masatoshi, and Murai Shōsuke (Tokyo: Yoshikawa Kōbunkan, 2010), 273–90; Sugimoto Fumiko, "Chōkan fūkei no naka no shōgun," in *Boshin sensō no shikagaku*, ed. Hakoishi Hiroshi (Tokyo: Bensei Shuppan, 2013), 343–80.

Suggested Readings

Ikegami, Eiko. *Bonds of Civility: Aesthetic Networks and the Political Origins of Japanese Culture*. Cambridge: Cambridge University Press, 2005.

Kouwenhoven, Arlett, and Matthi Forrer. *Siebold and Japan: His Life and Work*. Leiden: Hotei, 2000.

Siebold, Philipp Franz von. *Manners and Customs of the Japanese*. Rutland: Tuttle, 1973.

11 Fixing Sacred Borders: Villagers, Monks, and Their Two Sovereign Masters

SUGIMOTO Fumiko 杉本史子

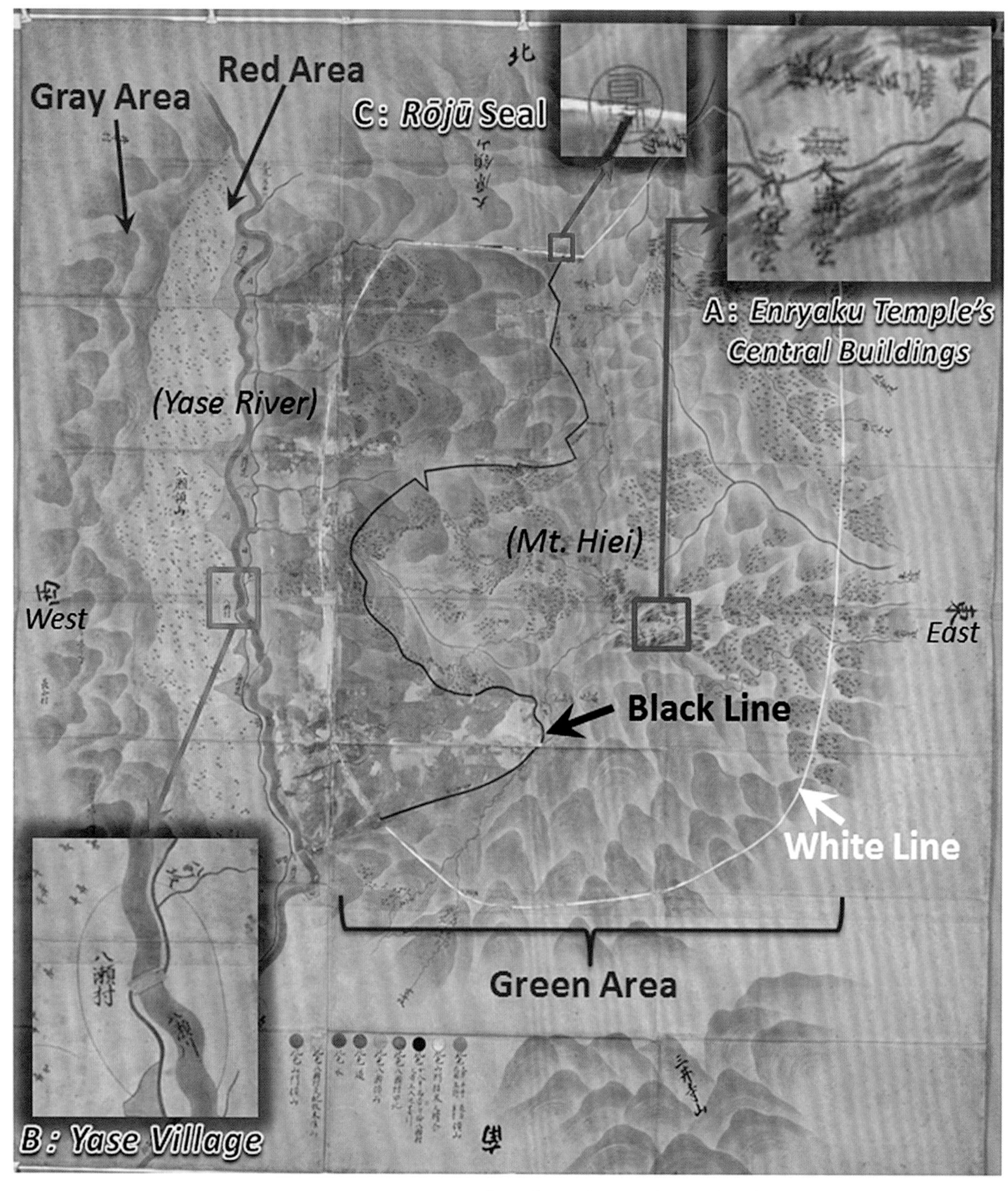

FIGURE 11.1 "Sacred Boundaries Set for Enryaku-ji by the Shogunal Government" [Sanmon kekkai saikyo ezu 山門結界裁許絵図], 1708. Manuscript, 144 × 127 cm. Yase-no-dōji Foundation, Rekishi shiryōkan, Kyoto, Japan. English labels added by author.

For more than a century of civil war starting in the second half of the fifteenth century, people across the Japanese archipelago took up arms to defend their rights. With the termination of widespread warfare under the unified political settlement of early modern times in the late sixteenth century, however, it became illegal to resort to military force in order to safeguard claims. Japan was becoming a society where governing bodies and legal courts solved conflicts. It was the first time in Japanese history that cooperative villages could legally lodge complaints with shogunal courts.

This essay introduces a border map produced during one such dispute (fig. 11.1).[1] On one side was Mount Hiei's massive Enryaku-ji temple complex; on the other side, the people of Yasemura, a village located at the base of the mountain. In this society of stark social stratification by birth, the very possibility that an official public map could be used to resolve a border dispute between two such socially disparate groups marks the situation as early modern.

The purpose of the map was to show the relationship between two boundary lines, indicated in black and white, and the areas occupied by the disputants: Enryaku-ji in green and Yasemura in red. The green area covers most of the map and reveals Mount Hiei in its towering position to the northeast of the Kyoto basin, the seat of imperial power. The buildings belonging to the temple are scattered across the mountain (*A*). By contrast, a much smaller red area to the left indicates the settlements, fields, and hills of the village (*B*). The gray areas on the edges of the map indicate terrain with no connection to the disputed territory.

The two border lines contrast in both color and shape. By and large, the white line describes a continuous elegant oval with no link to the peaks depicted. It is an abstract demarcation imposed on the mountain. Following the physical contours of the terrain itself, by contrast, is the jagged black line, along which are conspicuously indicated actual elements of the topography. Unlike the white line, the black line is marked with seals stamped by several *rōjū*, the highest-ranking officials in the Tokugawa administration (*C*). In order to appreciate the significance of these boundary lines, we need to know something about both Enryaku-ji and the two kinds of sovereign powers that existed simultaneously in the Japanese archipelago when the map was made.

The temple complex of Enryaku-ji traces its origins to the eighth century, when a priest turned his back on the world in order to practice Buddhist austerities in a hut located high atop Mount Hiei. For this and later priests at Enryaku-ji, the mountain was not just a natural environment but a sacred space for praying while moving throughout the peaks. Devout emperors and members of the nobility from Kyoto congregated within Enryaku-ji over time, and the temple evolved into a socially elite institution, amassing enormous power as it accumulated vast tracts of land in various parts of the archipelago. The temple complex even developed military authority of its own, supporting a powerful army of warrior monks.

In land-poor Yasemura, meanwhile, villagers augmented their livelihood by making charcoal and other products from trees harvested up and down the slopes of Mount Hiei. The women of Yasemura played a key role in this work. During the medieval period, the men in the village also worked as servants at the temple. The uniforms and hairstyles they wore in this capacity marked them as immature adults, and so they came to be known as "temple lads." To this day, people living in the area refer to themselves as "temple lads of Yase" (*Yase-no-dōji*).

The villagers made sure that the authorities recognized their right to profit from these activities. In particular, they had long been authorized to carry the palanquins of high-ranking courtiers who wished to ascend to Enryaku-ji. The "temple lads of Yase" boasted the memory of having shouldered Emperor Godaigo's palanquin up Mount Hiei, and, in the fifteenth century, they escorted the Ashikaga shogun up the mountain.[2] This meant that they had influence over the success of politically significant processions involving members of the imperial court. The proof of this exceptional duty was a special tax-exempt status.

As explained in the introduction to part I, samurai in the fifteenth and sixteenth centuries fought for hegemony over the regional domains they had built up over time throughout the archipelago. But these hostilities did not bring ruin to the imperial court. To be sure, the emperor and his direct subordinates were allowed to reside only in the palace and immediate environs of Kyoto. By contrast, the shogun regarded all members of the samurai class as his own vassals, using this prerogative to exert control, albeit indirectly, throughout the archipelago. Even so, the title of shogun was conferred by the emperor, and the shogun used the authority of the emperor to support his own claims to power.

This duality between the emperor and the shogun forms an important backdrop to the map. Under the unifiers of Japan in the sixteenth and seventeenth centuries, Enryaku-ji was destroyed and rebuilt, this time as a subordinate institution designed to serve the interests of samurai rule. Before this time, Enryaku-ji in all probability restricted the movements of the villagers from Yasemura on Mount Hiei. Under the Tokugawa, however, the monks no longer had such policing authority, since they like everyone else in the new political settlement were prohibited from using force to defend their rights.

The map reproduced here was made by the shogunal government at the request of Kōben Hosshin'nō (a son of Emperor Gosai), who had been appointed head of Enryaku-ji. Kōben Hosshin'nō wished to clarify the sacred borders of the temple lands under the new order of Tokugawa rule. The shogunate distributed identical copies of the map to both the temple and the village; the copy examined here belonged to Yasemura. As mentioned at the outset, it would have been unimaginable before the early modern period for these two parties, which were so sharply separated by status, to receive the same official legal document. Moreover, it was of great significance that the "temple lads of Yase" attempted to defend their interests by leveraging documents issued by the two different sovereigns: claiming tax-exempt status with documents issued to them by successive emperors while appealing to business permits that had been issued to them by several generations of Kyoto administrators under Tokugawa rule. In their wisdom, the villagers exploited the circumstances of plural sovereignty to advance their own interests.

The resulting map shows two boundaries authorized by the shogunate. The white line encompassing Enryaku-ji represents a barrier separating the sacred from the profane. Women, oxen, and horses were not allowed to pass beyond this line.[3] While the white line confirmed the worldview of the priests, the black line marked a property boundary determined by the shogunal government. Across the black line, no villagers could cross, regardless of gender.

This decision was a major blow to Yasemura, since it severely restricted where the villagers could collect wood on Mount Hiei. The villagers worked frantically to overturn the verdict by legal and political means. They submitted petitions over and over again to shogunal officials in Kyoto and Edo. They also appealed directly to the shogun through the auspices of the noble Konoe family, which had historically acted as the village's guarantor. At the time, the shogun's legal wife was from the Konoe family and might have been able to secure a hearing for the petitioners. After two years, although the villagers failed to change the decision, they succeeded in securing shogunal recognition of their inherited annual tax exemption.

The system of dual sovereignty that lay behind this map experienced tremendous upheaval during later centuries. The last Tokugawa shogun relinquished his authority in 1867, and an emperor wielding real power took his place. But in 1945, after the Japanese empire's defeat in World War II, the emperor turned into a "symbol of Japan" under the new constitution, once again assuming a position with no governing authority. During the same span of time, Enryaku-ji and Yasemura likewise transformed themselves. In 1994, the temple compound of Enryaku-ji was designated a Unesco World Heritage Site on account of its twelve hundred years of history and tradition. Today many people make pilgrimages to this "mother of Japanese Buddhism." As for the people of Yasemura, they have written a history of their village and preserved their historical documents with great care.[4] In 2010, the archive held by Yasemura was designated an Important National Property in Japan. In 2012, the Museum of Kyoto celebrated this national distinction by making publicly available for the first time Yasemura's precious historical treasures, including this border map.

(Translated by Robert Goree)

Notes

1. This map consists of one large paper sheet measuring 144 ×127 cm. Owing to its size, the map was spread out and viewed on tatami mats, rather than held in the hands. Costly pigments customarily used for painting, such as verdigris, derived from minerals, are applied liberally to the surface. Most of the map is based on surveys of the actual topography, with faithful depictions of each mountain peak. Green, red, and gray are used to divide the space into sections: green for Enryaku-ji, red for Yasemura. The gray areas on the northern, southern, and western perimeter of the map were not relevant to the dispute between Enryaku-ji and Yasemura and are accordingly represented with such an unassuming color as gray. White and black lines mark boundaries, which, together with the use of the three other colors, makes for a vivid representation of complex power relations. The names and seals belonging to four *rōjū* (the highest-ranking post within the shogunal administration) indicate those who validated the map. The seals of the *rōjū* correspond to those displayed above the black lines on the front side of the map. Three sections of the map have been enlarged: detail map A shows the three main buildings linked to the sprawling complex of Enryaku-ji; detail map B shows the oval label that marks Yasemura's location along the river; detail map C shows the *rōjū* seals validating the boundary line.

2. Nishiyama Gō, "Koshi o kaku Yase dōji," in *Yase dōji tennō to satobito*, ed. Kyoto Bunka Hakubutsukan (Kyoto: Kyoto Bunka Hakubutsukan, 2012).

3. Take Kakuchō, "Hieizan no kekkai ni tsuite," *Nihon bukkyō gakkai nenpō* 59 (1994): 91.

4. Uno Hideo, *Yase no dōji rekishi to bunka* (Kyoto: Shibunkaku Shuppan, 2007).

Suggested Readings

Inose Naoki. *Tennō no kagebōshi*. Tokyo: Chūkō Bunko, 2012.

Sugimoto Fumiko. "Justice in Early Modern Japan Revisited: Society and Justice." In *A Multilateral Comparative Study on Documents from the Ninth to the Nineteenth Centuries* (annual report of the National Institutes for the Humanities), trans. Sugahara Miu, 286–99. 2012.

12 Self-Portrait of a Village

Komeie Taisaku 米家泰作

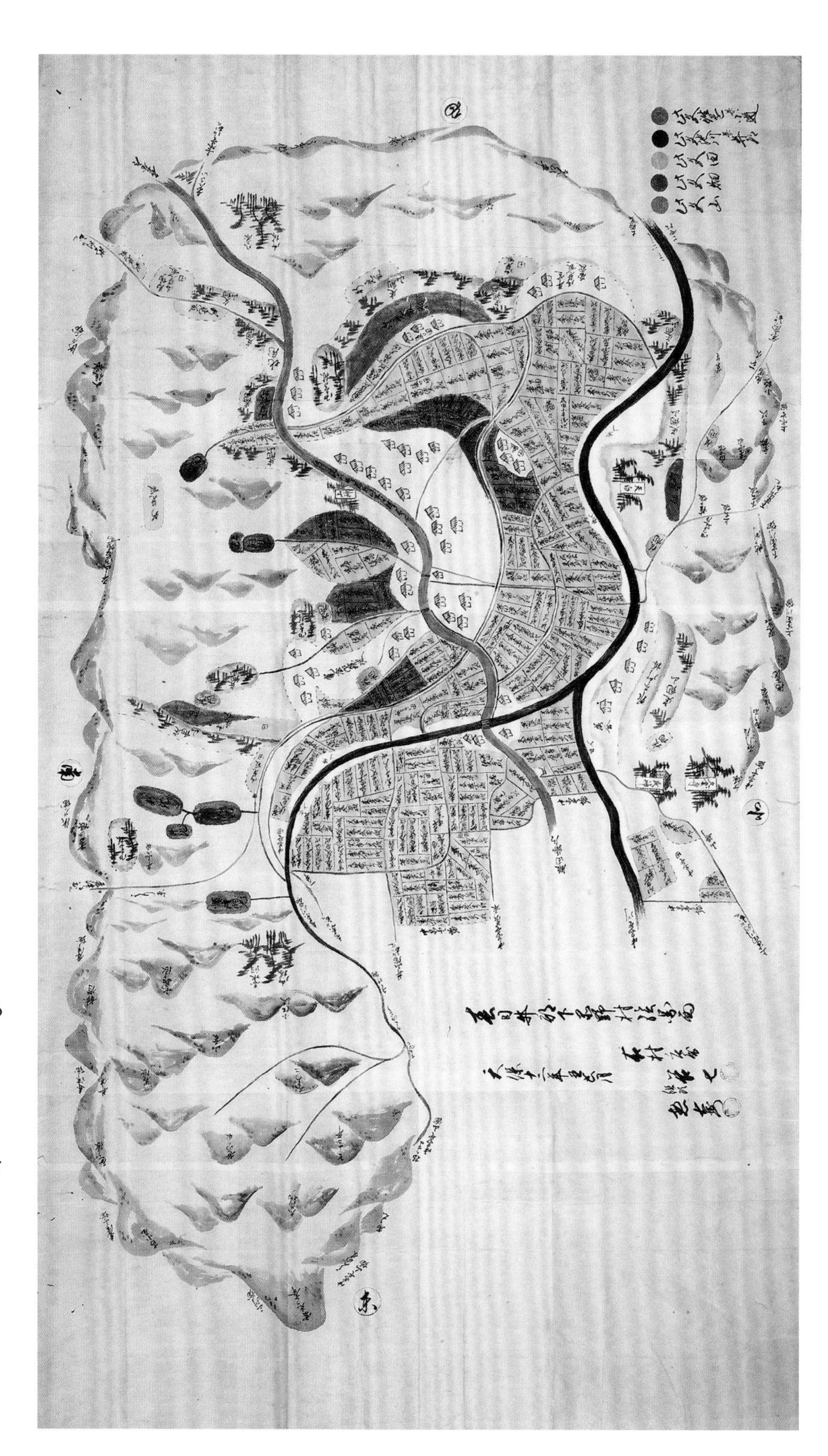

FIGURE 12.1 "Map of Shimo-Shinano Village, Kasugai County" [Kasugai gun Shimo-Shinano mura ezumen 春日井郡下品野村絵図面], by Zenshichi 善七 and Sōemon 惣右衛門, 1841. Manuscript, 61 × 107 cm. Courtesy of Tokugawa Institute for the History of Forestry, Tokyo, Japan.

The *mura-ezu*, or pictorial map (*ezu*) of a village (*mura*), was one of the most common types of maps produced in early modern Japan. Feudal lords frequently commanded village leaders to submit images of their fields and woodlands, drawn in a standardized manner. In general, village maps were produced not by professional experts or painters but by literate peasants who had the ability to draw pictorial maps. Administrative officers in the samurai class used these maps as a basic resource to understand the geographies and economies of their regions. Most *mura-ezu* provide only an unsophisticated overview of the rural landscape without recourse to scientific surveying. Yet their pictorial illustrations—showing settlements, agricultural fields, and commons—are indispensable sources for contemporary scholars seeking to reconstruct past landscapes.

Many pictorial village maps are now held in the official libraries or archives of the prefectural governments, which inherited documents from the early modern daimyo. Additionally, some local communities have kept duplicates of pictorial maps in their own archives. For example, eighteen maps have been preserved in the Ido District of Yoshino County, Nara Prefecture. Sixteen of these are copies of village maps originally submitted to the lords at various periods; one is a prototype map of a fictitious village, and one is a specific map of disputed land with a neighboring village.[1]

There remain some questions about *mura-ezu* in spite of their ubiquity in local historical archives. In particular, who introduced the conventions of village mapping to feudal domains across Japan, and when did this occur? The uniform style of the surviving documents—including the size of the paper on which they are drawn (generally a rectangle with a length of one to two meters), their color scheme (generally blue streams, red roads, yellow fields, black houses, and green forests), and other features (such as legends and directions)—suggests that *mura-ezu* were not invented independently in each domain but emerged as part of a more widespread effort. To date, however, no order commissioning village maps by the unified authority of either the Toyotomi or the Tokugawa governments has been discovered. Instead, scholars imagine that *mura-ezu* came into being through a more indirect process: one in which officials who had grown familiar with other kinds of maps (including the *kuniezu* discussed in chap. 10) came to appreciate the utility of cartography at the local level as well.

According to Kimura Tōichirō, a pioneer of research in this area, four main occasions warranted the creation of *mura-ezu*: (a) land registration, (b) a change of lords or magistrates, (c) land disputes, and (d) petitions for construction or reclamation.[2] We can assume many feudal lords found the *mura-ezu* to be useful and adopted this manner of mapping, especially for the first two situations noted above. A nationwide system of land registration had been authorized by Toyotomi Hideyoshi at the onset of the early modern era in the late sixteenth century, and new lords or magistrates were routinely provided with village maps by their predecessors as basic administrative documents. It is noteworthy that several daimyo sought to complete more accurate maps of all villages under their rule.[3] In those cases, village mapping did not rely on villagers' drawings but on more scientific land surveys that were part of the lords' official administrative projects.

The case of Owari Domain illustrates the systematic collection and use of *mura-ezu* by a feudal lord. This sizable domain was administered by a branch family of the Tokugawa, who administered all of Owari Province (today's western Aichi Prefecture) as well as some lands in neighboring provinces. Village maps produced in Owari survive today in two archives that inherited historical documents from the domain: the Aichi Prefectural Library and the Tokugawa Institute for the History of Forestry.[4] The former has at least 102 copies of village maps and a series of atlases called *Son'yu zenzu*, comprising copies of the maps; the latter has 2,751 sheets of *mura-ezu*. These maps were mainly the results of three mapping projects, commissioned in 1791, 1841, and 1844.[5] While the first and the last sets were used to compile official geographies of the domain, the fact that copies of the first set of maps (made in 1791) were also bound into the atlases mentioned above—and retained at the office of rural management of Owari Domain—suggests that they were useful for rural administration. Additionally, the orders issued to each village in the second project, in 1841, indicated not only that local place-names and vassal lands should be drawn in detail, but also that any changes to the landscape since 1791 should be noted. This shows that *mura-ezu*, at least in the case of Owari Domain, had become a basic resource for governing the territory by the early nineteenth century.

Figure 12.1 shows a typical *mura-ezu* from the second Owari mapping project, of 1841.[6] It does not have a fixed scale of the kind required for scientific accuracy; different areas are represented at various scales, and the shape of the village as a whole has been distorted to fit the rectangular sheet. The central settlement (houses) and arable area (yellow paddy fields and brown dry fields) are exaggerated in size, while the surrounding forests (green hills) are foreshortened. Nonetheless, this pictorial overview would have been very informative to the domain's officers. Small written notes over the arable areas indicate the names of nine vassals who had rights to the tax income from various fields. Many vassals in Owari Domain enjoyed tax rights in multiple villages, so the arrangement and allocation of their lands were important tasks for rural administration. Other elements were also important for understanding the geography of the village. A heavy red line represents a local highway between Owari and Shinano Provinces, and a notation along the road indicates a relay point for commodities—a feature that provided villagers many benefits. Blue circles (ponds) and lines (canals) show the irrigation network that supported rice agriculture, while green is used to depict the forests that provided essential fodder, fuel, and fertilizer. White areas along both sides of the rivers represent former fields, destroyed by flooding, which were now exempt from taxation. Experienced rural officers could read from the *mura-ezu* the quality and quantity of water, fields, and forests, which in turn determined the agricultural economy of the village.

The case of mapping in Owari Domain attests not only to the utility of village maps for rural administration but also to rural residents' own cartographic abilities or *map literacy*. Given the high rate of literacy in early modern Japan, we should not be astonished at the fact that almost all villages could produce self-portraits of their *mura* without calling on the services of a professional expert or artist. According to the records of Okoshi Village in Owari Domain, representatives of the village produced four maps within just three weeks after the *mura-ezu* project of 1841 was ordered.[7] Three of these were submitted to the lord, while one was kept for the village's records, partly as a reference for the next occasion of mapping. We can surmise that the skill required to draft maps was not necessarily beyond the abilities of those who were sufficiently literate to serve as official representatives of villages. Although they were not scientific mapmakers, they had the ability to represent the landscapes in which they had spent their lives. This foundation of cartographic literacy would play an important role in the early Meiji state, which called on all local municipalities to produce cadastral maps after 1873.[8] The making of Japan's modern cadastral maps started not as a new type of scientific land survey but as a revision or elaboration of *mura-ezu*, supported by the effort, experience, and cartographic literacy that Japanese villagers had accumulated over many generations.

Notes

1. Komeie Taisaku, *Chū-kinsei sanson no keikan to kōzō* (Tokyo: Azekura Shobō, 2002).

2. Kimura Tōichirō, *Mura-ezu no rekishi-chirigaku* (Tokyo: Nihon Gakujutsu Tsūshinsha, 1979).

3. For example, Tokushima (Awa) Domain produced *bunken mura-ezu* and other *bunken* regional maps in the early nineteenth century. In this case, *bunken* means accurate scale of reduction.

4. The Aichi Prefectural Library provides high resolution images of *mura-ezu* on its website (Japanese only): http://www.aichi-pref-library.jp/gazou/search/. Tokugawa Institute for the History of Forestry provides a list of old maps on its website (Japanese only): http://www.tokugawa.or.jp/institute/.

5. Kawamura Hirotada, *Kinsei ezu to sokuryōjutsu* (Tokyo: Kokon Shoin, 1992).

6. *Map of Shimo-Shinano Village, Kasugai County* [*Kasugai Gun Shimo-Shinano Mura ezumen*]. The village is in today's Shinano-chō, Seto City, Aichi Prefecture. Tokugawa Institute for the History of Forestry: map 108-1. Another version of the same map (map 108-2) is reproduced in Setoshi-shi Hensan Iinkai, ed., *Setoshi-shi: Shiryō-hen 1* (Seto: Seto City, 1985).

7. Document 2879 in Ichinomiya-shi, ed., *Shinpen Ichinomiyashi-shi: Shiryō-hen 8* (Ichinomiya: Ichinomiya City, 1968).

8. Satō Jinjirō, *Meiji-ki sakusei no chisekizu* (Tokyo: Kokon Shoin, 1986).

PART II

Public Places, Sacred Spaces

Introduction to Part II

Kären Wigen

By the mid-1600s, after centuries of debilitating warfare, a stable regime had taken hold in the Japanese archipelago. Tokugawa Ieyasu's victory in 1600 allowed him to command submission from the majority of the great lords; capturing the coveted title of "barbarian-quelling generalissimo" (*sei-i taishōgun*, or more simply *shōgun*) in 1604—and then maneuvering his son into the same prestigious position after his own early retirement—Ieyasu successfully translated military might into a lasting dynasty. The power of his house was cemented with a second victory over the remaining Toyotomi loyalists in 1615. After putting down a last-ditch rebellion in the far south some twenty years later, the Tokugawa would face no major challenges to their rule for the next two centuries. Exhausted by decades of warfare, the remaining daimyo sheathed their swords and turned to the business of governing their domains.

In making Japan one of the rare war-free zones in the early modern world, the Tokugawa unwittingly set the stage for an explosion of cartography. Part I of this volume has already explored the burgeoning mapmaking enterprise of the new state. But even as official mapmaking was spurred by the administrative demands of the Pax Tokugawa, commercial mapmakers, too, were beginning to reap the benefits of peace. By the eighteenth century, publishers were catering to a broad public appetite for maps. Key to the flourishing of commercial cartography were rapid advances in three related areas: urbanization, transportation, and literacy.

First and foremost was the rise of cities. In East Asia as in Europe, cities were the cradle of the publishing industry, and the Edo era saw Japan's first nationwide wave of urban growth. Ordering all samurai off the land and into castle towns set off a tremendous construction boom, culminating in the massive expansion of the Tokugawa's own headquarters on Edo Bay. By the end of the seventeenth century, Edo was home to nearly a million people, followed closely by Osaka and Kyoto with more than three hundred thousand souls apiece. Outside these three great cities, the countryside supported more than two hundred medium-sized castle towns, with a dense network of market towns and post stations between them. By 1750, fully a fifth of Japanese lived in towns of three thousand or more.[1]

Closely related to urban growth was an upswing in travel. During the centuries before 1600, banditry and extortionate toll taking had conspired to make Japan's roadways inconvenient and unsafe. Under the Pax Tokugawa, by contrast, a national turnpike network centered on Edo was cleared, patrolled, and maintained. To be sure, the government-run post-station system was designed primarily to serve officials. But in a culture where business and pilgrimage alike supplied respectable reasons for extended trips away from home, commoners soon began to take to the roads as well. It did not take long for enterprising merchants to see opportunities for profit in catering to ordinary travelers. From ferries and hostels to bars, shops, and brothels, a full range of establishments sprang up along the turnpikes of Tokugawa Japan. The resulting culture of mobility, in turn, created a burgeoning market for guidebooks and maps.

A third way the peace dividend benefited cartographers was by rewarding education. The widening cash economy of the Edo era made basic literacy and numeracy a sound investment, even as it put money in more farm families' pockets to pay for their children's schooling. By the late eighteenth century, young people were gathering in classrooms throughout Japan, with temple schools and private academies penetrating even remote mountain villages. The resulting expansion of the literate populace again contributed to expanding the market for printed matter of all kinds, including maps.

Taken together, these three developments—a wave of city building, an upswing in travel, and a steady growth in readership—encouraged Japanese publishers to print and market a wide range of geographic and cartographic material. While some maps were embedded in guidebooks or geography readers, the typical commercial map was a single-sheet affair. Designed by an amateur or professional artist, transferred to a woodblock by artisanal carvers, and printed on sturdy mulberry-bark paper in relatively small batches (usually a few hundred copies at a time), these maps survive in the thousands in contemporary collections. While some were simple utilitarian objects issued in black and white, others could be quite elaborate, embellished in beautiful colors and clearly meant for display. Ambitious print shops competed for the services of skilled artists, whose repertoire might also include landscape prints, human portraits, and bird's-eye views. The resulting maps reveal a striking degree of stylistic and thematic latitude, despite the existence of official censorship. Some of these maps have already been featured in part I (e.g., in chaps. 7 and 8). But the essays that follow here showcase many more examples, both typical and singular, from this expansive cartographic corpus. Our selection highlights three representative subjects: the city, the sacred, and the itinerary.

We lead off with what was perhaps the most popular genre of the later Tokugawa era, the individual city view. Kyoto in particular was well represented, and no wonder; its treasure trove of temples and historical sites appealed to both visitors and armchair travelers in the Tokugawa period, as it does today (see the chapters by Uesugi Kazuhiro). But publishers competed to offer up-to-date images of Edo, Osaka, and smaller towns as well. Most such maps had at their core an aerial view of the street grid, labeling each block, bridge, and canal in the most densely populated neighborhoods. This plan view of the urban core, in turn, was typically ringed by foreshortened perspective drawings of nearby harbors and hills, suggesting the way the surrounding topography opened out in all directions. Maps of this kind, featuring a prominent city and its setting through a combination of aerial and pictorial perspectives, became a mainstay of the later Edo woodblock print business (chapters by Tamai Tetsuo, Uesugi, Mary Elizabeth Berry, Paul Waley, and Ronald Toby).

If secular maps of cities constituted one popular subject for Japanese mapmakers, sacred geography was another. Commercial cartographers found a ready market for maps of the divine at a range of scales, from grand cosmic visions to single sacred sites. Reflecting the combinatory character of Edo-era religion, cartographies of the sacred might point to a variety of polestars. Some of the images discussed below locate Japan in a Buddhist world, centered on India and its mythical Mount Mehru (chapters by D. Max Moerman and Sayoko Sakakibara). Others disregard that framework entirely, focusing instead on indigenous landscape features associated with Shinto *kami* (chapters by Miyazaki Fumiko, Anne Walthall, and Henry D. Smith II). Likewise, we see a great variety of scale in this group of artifacts. Cartographers of the sacred might zoom in to depict individual rocks, or zoom out to apprehend the archipelago or the planet as a whole. Whatever their framework or focus, it is critical to note that religious maps of the Edo era never partook of a single, stable "tradition." On the contrary, the maps in this section attest not only to the pluralism of Tokugawa sacred cartography but to its dynamism over time.

The third and final major genre of popular cartography, taken up here under the heading of "Travelscapes," consists of maps and guidebooks that catered to people on the move. As with sacred cartography, so with itinerary maps: this genre, too, is notable for its diversity and dynamism. Route maps were also characterized by a playfulness and exuberance of design. It is perhaps fitting, then, to conclude part II with an eclectic sampler of route maps on a range of themes, with subtexts that range from the playful (chapter by Robert Goree) to the political (chapter by Kären Wigen and Sayoko Sakakibara). The last two chapters in this unit (by Nicolas Fiévé and Roderick Wilson) spill over the 1868 divide, demonstrating how readily the tropes of early modern travel cartography were adapted to the era of the railroad.

Tokugawa Japan has been called a "world within walls."[2] In a limited sense this is true; outside contact was carefully constrained and channeled. But seclusion was never total. Throughout the Tokugawa era, but especially after the lifting of the ban on European books in 1720, the door was deliberately held ajar, allowing information and ideas from abroad to filter in. Nowhere is this fresh air more vividly in evidence than in commercial cartography. The domain of map publishing was in some ways a vast field of experimentation during the eighteenth and nineteenth centuries, as Japanese cartographers played with different ways to incorporate new information and imported technologies into their designs. The result is a lively variety of projections, perspectives, idioms, inks, and worldviews. It is our hope that readers will enjoy the way the interdisciplinary essays in part II echo the plurality of their subject matter in the polyvocality of their approaches.

Notes

1. John Whitney Hall, "The Castle Town and Japan's Modern Urbanization," *Far Eastern Quarterly* 15 (1955): 37–56.

2. Donald Keene, *World within Walls* (New York: Columbia University Press, 1999).

13 Characteristics of Premodern Urban Space

TAMAI Tetsuo 玉井哲雄

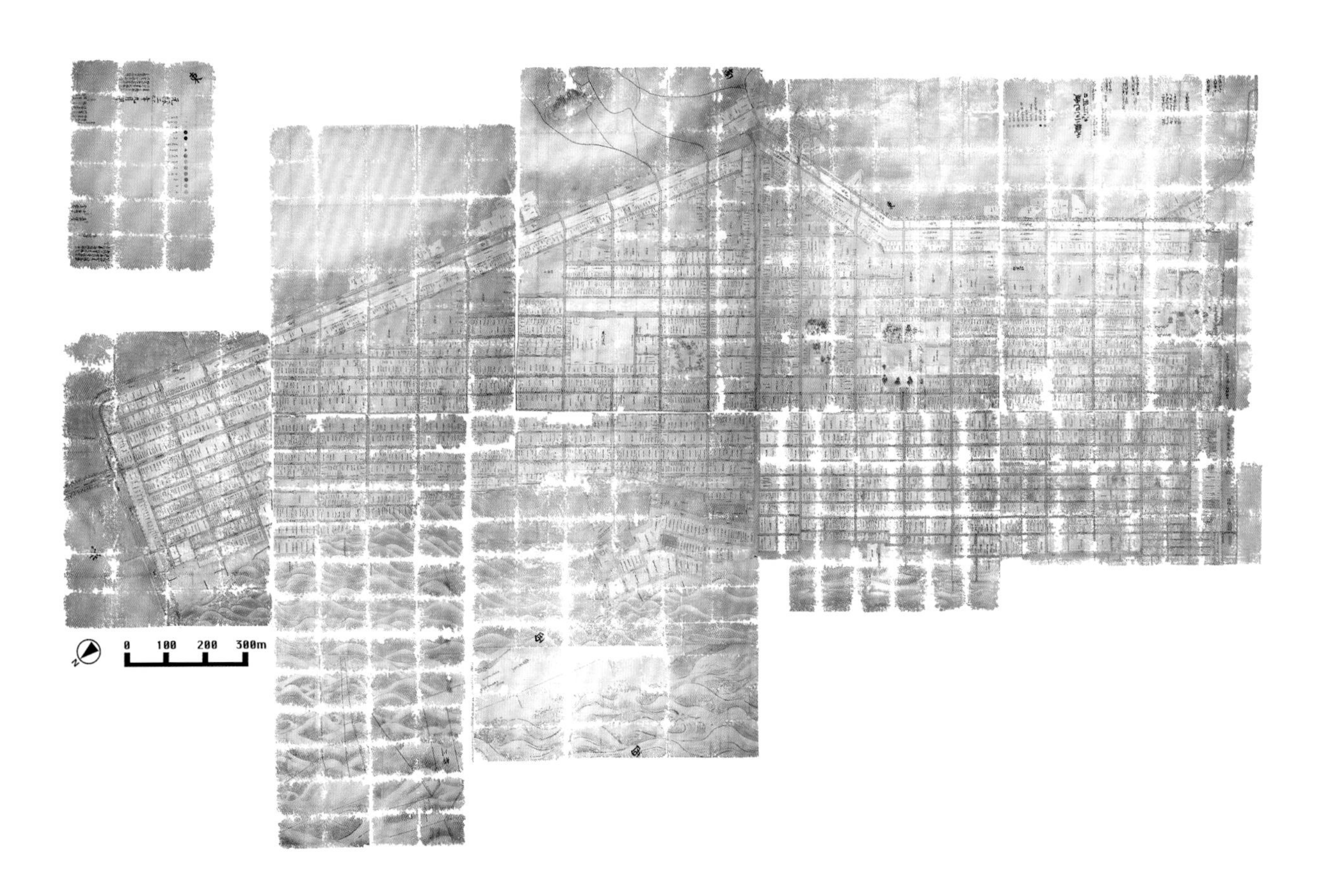

FIGURE 13.1 "Great Map of Sakai from the Second Year of the Genroku Era" [Genroku ni nen Sakai ōezu 元禄二年国堺大絵図], 1689. Manuscript in nine sheets, approx. 500 × 1,100 cm. Courtesy of National Museum of Japanese History.

The bustling port of Sakai lies south of Osaka at the eastern end of the Inland Sea. Literally meaning "border," the city's name is derived from its position straddling the ancient provinces of Settsu and Izumi. During the late medieval period, Sakai prospered as the easternmost node of an international trade route connecting Osaka Bay with the Inland Sea and leading out past the Kanmon Straits as far as the East China Sea. The economic power acquired through foreign trade allowed Sakai to stand its ground during the fifteenth century against the contending daimyo of the Warring States period.

Along with Kyoto and Hakata, Sakai is one of a handful of Japanese medieval cities that went on to prosper in the same location during the Tokugawa era as well. For this reason, Sakai offers a valuable case study of the changing allocation of urban space over time. "Allocation of space" here refers to the layout of streets, the distribution of city blocks created by these streets, and the particular way in which such city blocks were subdivided into residential lots. The current essay will consider these characteristics of Tokugawa urban space with reference to a map of Sakai from the Genroku era (1688–1704) (fig. 13.1).

According to Jesuit missionaries who visited Sakai in the late 1500s, a century before this map was made, medieval Sakai was designed with an eye to defense. The city at that time was surrounded by a circular moat, and each of its districts had a gate facing the moat that was closed at night. In addition, the moat could be deepened and turrets erected whenever conflict broke out. Run by its powerful merchant families, Sakai has often been compared to the communes or free cities of medieval Europe.

Sakai's medieval prosperity came to an end in 1569 at the hands of the warlord Oda Nobunaga. Its moat, the symbol of its status as a free city, was filled in by Toyotomi Hideyoshi in 1586. Not only was the shape of the city drastically altered, but the leading merchants were also forced to relocate elsewhere. Finally, with the defeat of the Toyotomi clan during the Summer Siege of Osaka, neighboring Sakai was destroyed by fire in 1615. The city was subsequently rebuilt under the direct control of the Tokugawa shogunate. During the early seventeenth century, Sakai briefly reclaimed some of its former prosperity by serving as a base for chartered trade with Southeast Asia. All such trade was brought to an end, however, in the mid-seventeenth century. At that point the economic center of the region shifted to Osaka, and Sakai effectively lost its function as a port by the eighteenth century.

Remnants of pre-Edo buildings and other relics have been discovered at most archaeological excavation sites throughout Sakai's city limits. Among layers of scorched earth from various fires, archeologists have found remains of storehouses and foundation stones as well as ceramics, tiles, and coins. The location and direction of street gutters indicate the extent to which the early modern allocation of urban space differed from that of both medieval and modern times. Traces of the original circular moat reveal that a new moat was dug during the Tokugawa era. Furthermore, the location of the medieval moat deep within the confines of the Edo-era moat shows how much smaller the medieval city was compared to the Sakai of the early modern era.

The map reproduced here clearly describes Sakai city at the end of the seventeenth century during the prosperous Genroku period. As can be seen in the accompanying image, the full map consists of nine very large sheets that together cover nearly thirty tatami mats (an area roughly 5 × 11 meters). The area of Sakai encircled by the main moat was divided vertically down the center (north to south) by a major national turnpike, the Kishū Road. This road in turn was crossed at a right angle by the Ōshūji Road, running east to west, effectively dividing the city into four quarters. A grid-like system of streets running parallel to the main axes subdivided each quarter into regular blocks of approximately 120 square meters each. Temples were lined up on the eastern edge of the city and ran its entire length. This aspect of its urban plan is one that Sakai shared with many other castle towns of the period, demonstrating the shogunate's influence on the city layout.

At a more detailed level as well, the allocation of space shown in this map is similar to that found in other castle towns of the Tokugawa period. As discussed in chapter 19, both sides of a street along a given block were considered to form a single neighborhood unit. The rectangular residential lots that reached back from these roads were identified on the Sakai map both by owner and by size. It can be inferred therefore that the map was originally used to establish property rights within the city—an issue of keen interest to tax collectors. The considerable length of time that this particular map was used for such purposes is attested by the many patches of paper added to it to update ownership information from the late seventeenth to the mid-eighteenth century.

Both the large scale of this map (1:3,250) and its high degree of precision are unrivaled by any other urban map from this early date in Japan. Its detailing of both ownership and plot measurements reveals that the process of dividing city blocks into *machiyashiki* (long, narrow wooden houses)—a common practice during the Tokugawa era—was already well underway by the Genroku era. The fact that such a map was produced at this time suggests that the late seventeenth century was a watershed in premodern conceptions of urban space in Japan.

It is instructive to compare this Genroku-era image of Sakai with contemporary depictions of Edo, which were more commonly produced. An Edo map known as the *Koken-ezu* is similar to this one in that it depicts individual lots and labels them by owner (or owner's agent) as well as specifying their dimensions. The *Koken-ezu* was compiled twice, in the early and mid-eighteenth century. Both of the resulting maps noted the sale prices of residential lots, or *koken-kin*, from which their name is taken.

While various contemporary sources confirm that the *Koken-ezu* were used in urban administration, no direct precursors for them have been found. It is known from shogunal edicts and other sources, however, that a map known as the *Genroku kensu ezu* did exist at one point. As its name suggests, this map was produced in the same period (the Genroku era) as the map of Sakai; it too was labeled with information regarding plot owners and sizes. Maps of other castle towns were also being produced for the first time during these years. The Genroku era can therefore be considered an important moment in the development and management of premodern urban space.

When the Genroku map is compared to the topography of modern Sakai, it is apparent that there have been drastic changes during the intervening years. Yet the framework of roads originally established within the circular moat is still very much in evidence, and we can see how the modern city's layout follows the basic plan of the medieval city. How can we understand this relationship between the "free city" of medieval Sakai and its subsequent incarnations?

In order to answer this question, it is helpful to compare Sakai with Kyoto and Hakata. In the case of Kyoto, the differences between the excavated (medieval) grid and aboveground (early modern) street layouts reveal a two-stage process: the filling in of an overall rectangular street plan during the postclassical era followed by the subdivision of the resulting city blocks via the construction of alleyways. This in turn laid the foundation for the different approaches taken to urban space allocation in the Tokugawa era (such as the direction in which roads were laid out) by Kyoto and Sakai. The same might be true for Hakata as well, but there are no maps of Hakata from this era of the same quality as the Genroku map of Sakai. This underscores just how important the cartographic record of Sakai is for understanding the general process of change experienced by cities transitioning from a medieval to an early modern form.

(Translated by Bret Fisk)

Suggested Readings

Morris, V. Dixon. "Sakai: From Shōen to Port City." In *Japan in the Muromachi Age*, edited by John Whitney Hall and Toyoda Takeshi, 145–58. Berkeley: University of California Press, 1977.

Tamai Tetsuo. "Genroku ninen Sakai ōezu." *Rekihaku* (2010): 160, http://www.rekihaku.ac.jp/outline/publication/rekihaku/160/witness.html.

14 Evolving Cartography of an Ancient Capital

UESUGI Kazuhiro 上杉和央

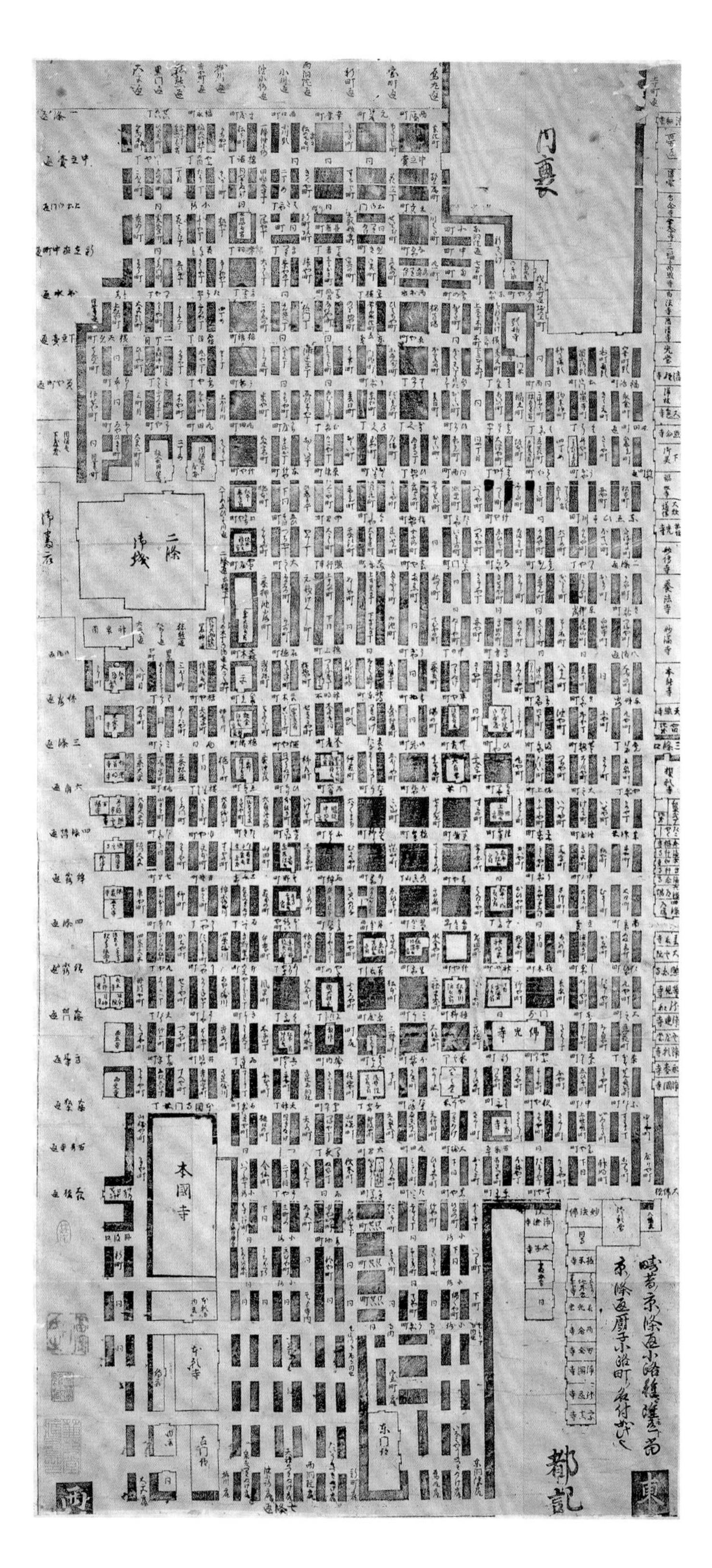

FIGURE 14.1 "Record of the Capital" [Miyako no ki 都記], ca. 1626. Woodblock print, 116 × 55 cm. Courtesy of Ōtsuka Collection, Kyoto University Library.

Kyoto's centrality as a political, economic, and cultural capital, established with the city's construction in 794, eroded during the political upheaval of the medieval period and all but collapsed once the Tokugawa shogunate established its administrative headquarters in Edo in the early seventeenth century. The city nevertheless weathered these changes by enduring as a cultural center, securing its place as Japan's most famous city, and becoming a leading urban tourist destination in early modern times. Throughout this storied history, Kyoto has perennially been a favorite subject for cartographers. This essay provides a glimpse of that popularity by way of an introductory overview of three prominent map styles of Kyoto: hand-drawn manuscript maps, commercially published maps, and bird's-eye views. The main features of these three modes are discussed in reference to their historical contexts in order to suggest some of the broad changes in how the city has been mapped.

Hand-painted maps, the first genre to emerge, were initially tied to urban planning and city management. The original city plan of Kyoto followed the rectilinear street patterns of Chinese capital cities. In all likelihood, the construction and administration of such an urban grid required maps, and indeed documentary evidence suggests the existence of such maps in ancient times, even though none survive today. The oldest extant maps of Kyoto are found in medieval copies of the *Engishiki*, a book concerning ancient law and customs. The maps themselves are thought to have been taken from a larger series of early maps dating from the middle of the twelfth century.[1] Interestingly, the purpose of these maps was not to represent the landscape of Kyoto as it existed at the time of their production, but rather to preserve a memory of its features from an earlier time. For example, one of the maps represents the western side of the old capital city, and another map represents the eastern side, which seems to suggest that they were not made for the purpose of city management, but rather as a record for confirming successive historical changes to the city itself.

So far as we can judge from existing maps of Kyoto, cartography for practical administration would have to wait until early modern times. The Tokugawa shogunate appointed the Nakai clan, who were master carpenters in Kyoto, to manage the entire realm's public works, including the production of city maps. As a result, the Nakai produced some highly detailed survey maps of Kyoto, including two entitled "Picture Map of the Capital" (*Rakuchū ezu*), made between 1630 and 1650.[2] These well-known maps are a boon to our understanding of bygone city management, since they indicate road widths, road frontages, and the depth of major houses lining the streets.[3] Moreover, these maps, which Kyoto's administrators most likely used as a basic tool for city management, include invaluable information about the inhabitants of these houses as well as land use more generally and place-names throughout the city.

A second major genre of Kyoto cartography was tied to the explosion of woodblock printing in the Edo period. Commercial publishing emerged in Japan during the seventeenth century as a phenomenon with tremendous cultural significance. Not only were works of fiction and treatises of all sorts published from that time onward, but many kinds of maps were printed too. The first business-minded publishers plied their trade in big cities such as Edo, Osaka, and Kyoto in order to meet the growing demand for city maps in particular. They made improvements to such maps in response to consumer needs throughout the early modern period. As a result of their efforts, the extant corpus of maps from the Edo era is remarkable for its quantity and variety.

The first published map of Kyoto from this period also happens to be the oldest extant published map of a Japanese city. Printed about 1626, "Record of the Capital" (*Miyako no ki*, fig. 14.1) reflects traditional conventions of cartographic representations of Kyoto as an ancient capital. In fact, its shape resembles that of the maps found in *Engishiki*.[4] In depicting Kyoto this way, the anonymous mapmaker of "Record of the Capital" clearly chose not to include areas of the city that would have been significant for his contemporaries, such as the city's older suburban areas, which constituted the core zone of economic activity in the inner area of the city by the sixteenth century, and famous tourism sites, which were of interest to travelers.

Maps of Kyoto published after "Record of the Capital" moved beyond conventional cartographic conceptions that had endured since ancient times, aiming instead to represent with greater fidelity the central area of Kyoto (*rakuchū*) and its suburbs (*rakugai*). During a time when leisure travel had become widely popular throughout the archipelago, with many people visiting Kyoto to see historical sites and other famous places, many publishers entered into the mapmaking fray and competed with each other in creating new styles of maps throughout the seventeenth century. Grasping the significance of the trend and investigating problems with previously published maps of Kyoto as well as the features of maps of other cities, Kyoto-based publisher Hayashi Yoshinaga bested the competition in 1686 when he published a new kind of large-sized map, titled "New Enlarged Kyoto Map" (*Shinsen zōho Kyō ōezu*), which contained practical information about everyday life and sightseeing in Kyoto.[5] Hayashi had the foresight to make the map easy to read and useful for travelers. He was also savvy in exaggerating the scale of certain buildings, such as Kiyomizu Temple (now a World Cultural Heritage Site), to which he had painters add attractive colors after printing in order to make such sites even more conspicuous. "New Enlarged Kyoto Map" and its many imitators can be thought of as convenient urban maps that also functioned as guidebooks.[6]

In the wake of Hayashi's successful innovation, consumers began asking publishers to make maps in a diversity of sizes. Some wanted larger maps to be spread out at home on tatami mats. Others requested small portable maps for sightseeing. Another key consideration for publishers was how to differentiate city maps from guidebooks, which grew in popularity during the eighteenth century. Takehara Kōbē is a representative ex-

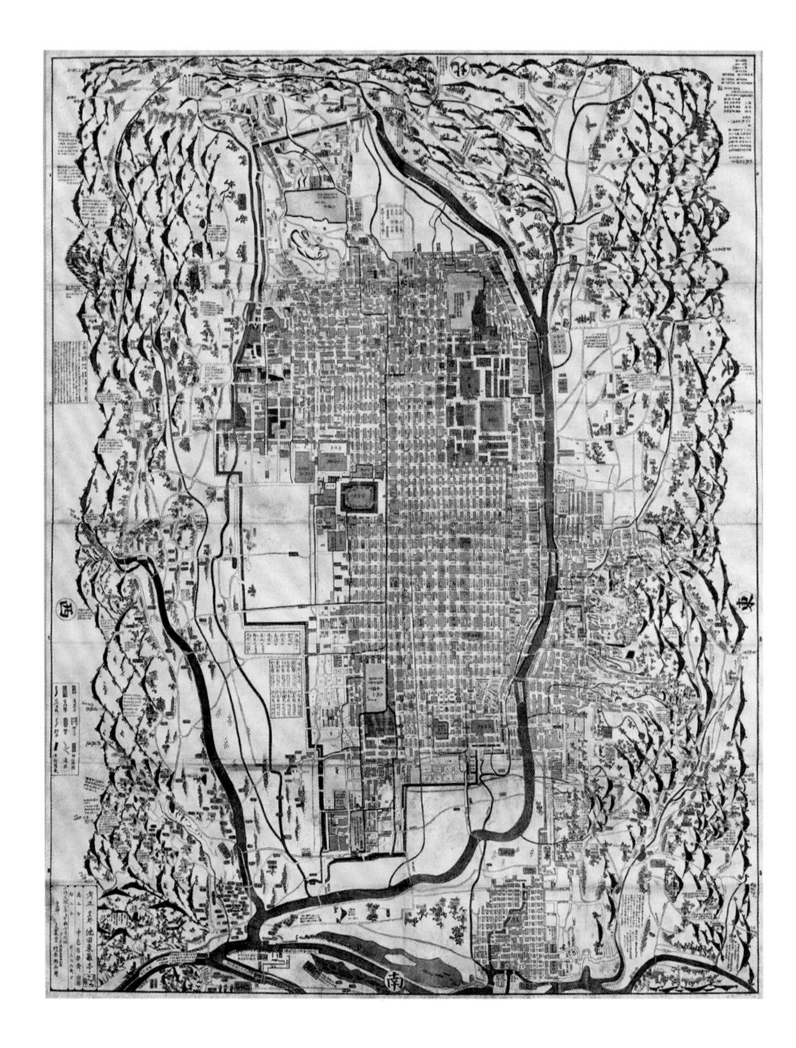

FIGURE 14.2 "Revised Detailed Kyoto City Map" [Kaisei Kyō machi oezu saiken taisei 改正京町御絵図細見大成], by Takehara Kōbē 竹原好兵衛, 1831. Woodblock print. Private collection.

ample of a publisher who responded to such market demands. This publishing house was founded in the second half of the eighteenth century but did not fully bloom as a mapmaker until 1831, when it came out with the "Revised Detailed Kyoto City Map" (fig. 14.2), the largest map of Kyoto ever published, at over 1.8 meters (nearly six feet) in length. According to Takehara's own account, the firm's objective had been to create "the greatest of all published maps of Kyoto." Whether or not the map achieved such superlative status is open to debate, but the fact that it was republished many times before the end of the early modern era suggests that it was indeed commercially successful. Perhaps this was owing to the detailed quality of the geographic information it contained, which arose in part from Takehara's diligent effort to resolve discrepancies vis-à-vis the physical world. Takehara also made maps of Kyoto in many different sizes, one of which was less than half a meter (1.6 feet) in length. The firm's entrepreneurial zeal found further expression in the leaflets it used to advertise its maps. Takehara judged itself to be a general purveyor of Kyoto maps; the historical results of this firm's innovative mapmaking attest to the legitimacy of such a claim.[7]

The third major idiom that cartographers used to depict Kyoto was the oblique perspective style, known in the West as the bird's-eye view. Landscapes depicted from an aerial view had existed in Japan since ancient times, but it was not until the sixteenth century that cities became a significant subject for the artists engaged in producing such views. Lavish paintings on folding screens, called "Scenes in and around the Capital" (*Rakuchū rakugai zu*), are one such format for portraying the city center and outskirts of Kyoto. The oldest extant example of the genre shows the city as it appeared between 1520 and 1540.[8]

"Scenes in and around the Capital" are noteworthy for their vivid depiction of people as well as landscapes in and around Kyoto. By contrast, woodblock prints published in the nineteenth century that adopt a bird's-eye view in order to depict Kyoto reveal a strong tendency to focus on the landscapes themselves.[9] For instance, "Panoramic Map of Kyoto" (*Karaku ichiran zu*), drawn by ukiyo-e painter Kō Kazan (1784–1837) and published in 1808, is an early example of such a bird's-eye view that emphasizes the physical landscape over its human inhabitants (fig. 14.3). Since the streets running from east to west across the city were not drawn straight or parallel to each other, the result is a technically inaccurate map. Nevertheless, the use of a bird's-eye view perspective allowed the artist to show the features of Kyoto's landscape with impressive precision, especially the differences between the inside and outside of the city proper,

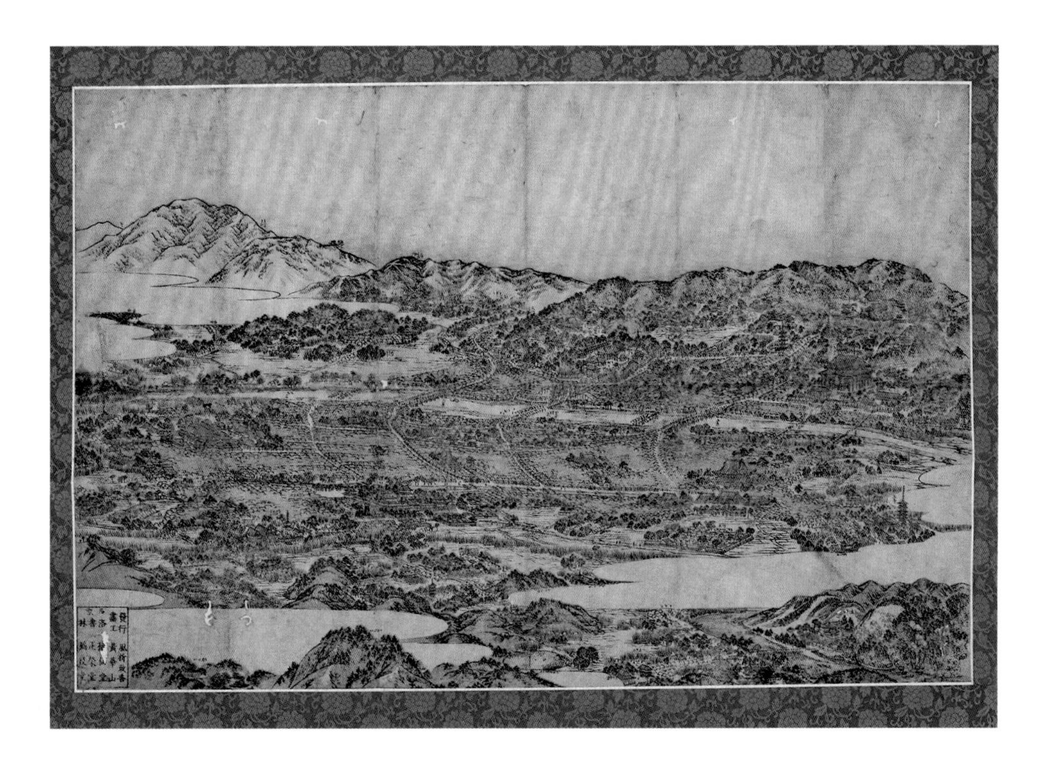

FIGURE 14.3 "Panoramic Map of Kyoto" [Karaku ichiranzu 花洛一覧図], by Kō Kazan 黄華山, 1808. Woodblock print, 41.6 × 64.2 cm. Courtesy of Waseda University Library.

as well as the contours of the skyline, which reveals temples, shrines, and castles rising above ordinary houses with tiled roofs.

Kazan's new view of the city—along with those of innovative painters like Maruyama Ōkyo and Kuwagata Keisai—seems to have stimulated the imagination of later ukiyo-e painters, several of whom used similar techniques to integrate considerable pictorial information into maps. Such images only grew in popularity toward the end of the Edo era, when the mounting influence of the emperor led to the reemergence of Kyoto as a political center. During this turbulent time, people yearned for up-to-date news about current events in Kyoto. Bird's-eye views of Kyoto produced at the time helped to satisfy such a desire with coded visual effects in the sky.[10]

Notes

1. Kinda Akihiro and Uesugi Kazuhiro, "Landscapes and Maps," in *A Landscape History of Japan*, ed. Kinda Akihiro (Kyoto: Kyoto University Press, 2010), 196.

2. The unabbreviated title is *Kan'ei go manji mae rakuchū ezu.*

3. A digital image of one such map from the 1640s is available on Kyoto University Library's website, http://edb.kulib.kyoto-u.ac.jp/exhibit/maps/map070/lime/map070.html.

4. Kinda and Uesugi, "Landscapes and Maps," 207.

5. A digital image of *Shinsen zōho Kyō ōezu* is provided on the National Diet Library's website.

6. Kinda and Uesugi, "Landscapes and Maps," 207.

7. Kinda Akihiro and Uesugi Kazuhiro, *Nihon chizu shi* (Tokyo: Yoshikawa Kōbunkan, 2012).

8. In the collection of the National Museum of Japanese History.

9. Kinda and Uesugi, "Landscapes and Maps," 213.

10. Sugimoto Fumiko, "Jiji to chōkanzu: Bakumatsu, aratana kūkan no tanjō to Gountei Sadahide," *Chiba kenshi kenkyū* 16 (2008): 45–67.

Suggested Readings

Kinda Akihiro and Uesugi Kazuhiro. "Landscapes and Maps." In *A Landscape History of Japan,* edited by Kinda Akihiro, 189–221. Kyoto: Kyoto University Press, 2010.

McKelway, Matthew. *Capitalscapes: Folding Screens and Political Imagination in Late Medieval Kyoto.* Honolulu: University of Hawai'i Press, 2006.

Smith, Henry D., II. "World without Walls: Kuwagata Keisai's Panoramic Vision of Japan." In *Japan and the World,* edited by Gail Lee Bernstein and Haruhiro Fukui, 3–19. London: Macmillan Press, 1988.

15 Historical Landscapes of Osaka

Uesugi Kazuhiro 上杉和央

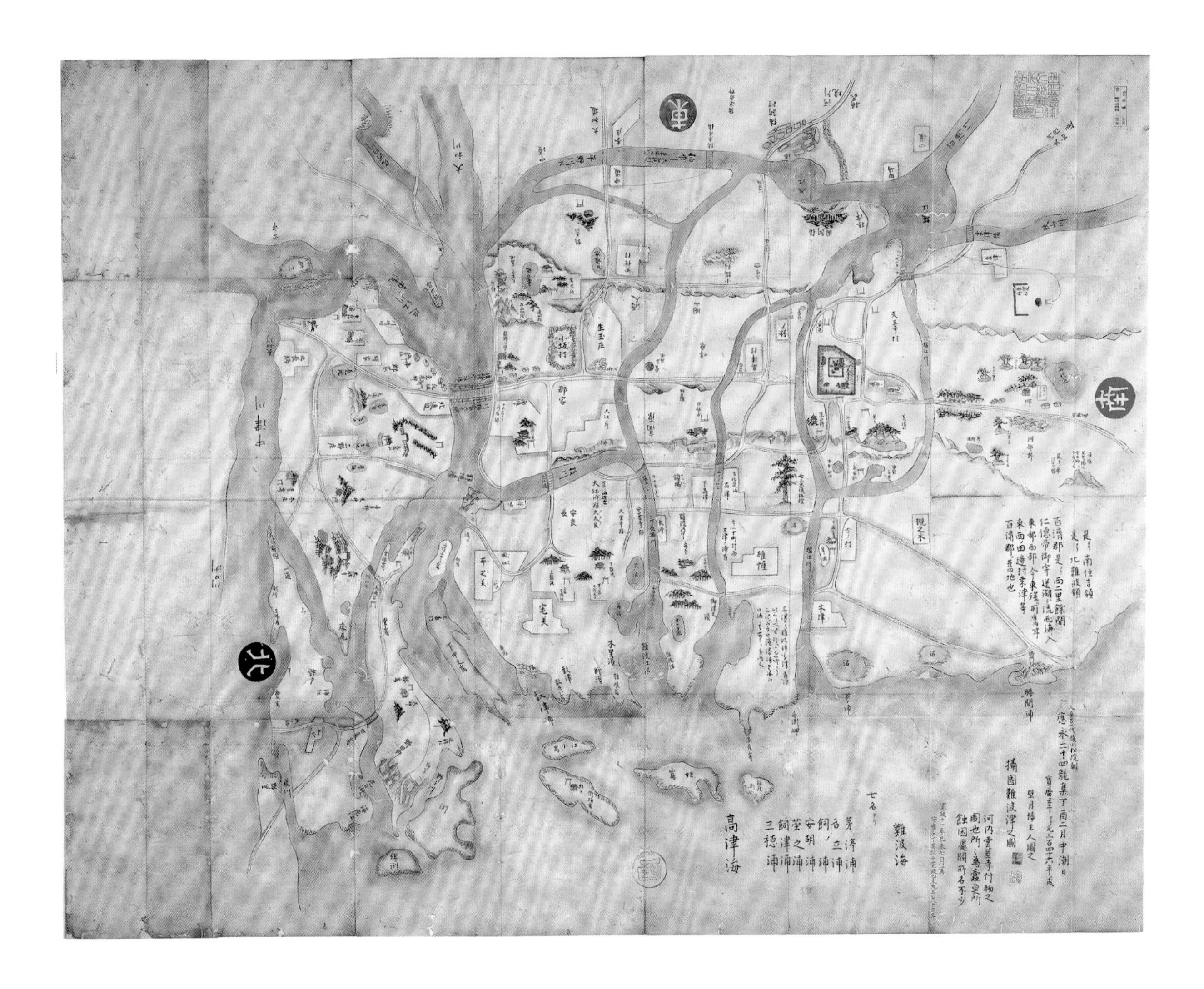

FIGURE 15.1 “Map of Naniwa Port, Map of Osaka, From the 24th Year of Ōei” [Ōei teiyū Osaka chizu Naniwatsu no zu 応永丁酉大坂地図難波津之図], 1799. Manuscript, 99 × 127 cm. Courtesy of University of Tsukuba Library. An example of an old map of Osaka [Naniwa kozu 難波古図]. The claim that it was originally made in 1417 and then traced in 1799 is written at the lower right.

A strong curiosity about local history was a hallmark of Japanese society in the eighteenth century. Intellectuals throughout the country had begun to investigate the history and geography of their homelands and other areas, producing an outpouring of topographies and guidebooks. Meanwhile, the Tokugawa shogunate and local domain governments conducted topographical research and undertook investigations to understand the geographic distribution of resources like medicinal herbs.[1] At the same time, many samurai became curious about locations connected to their masters’ ancestors.[2] Such cultural movements required taking note of change over time in each area, as well as an understanding of how past landscapes had been configured. So it is not surprising that many historical maps were made during this time.

In addition to intellectual interests driving this development, popular entertainments with history-inspired themes created further demand for historical maps. For example, popular kabuki and puppet plays during the period featured a mixture of history and current events, often deploying previous centuries as a setting for thinly veiled retellings of sensational stories or scandals. A similar blending of past and present influenced pictorial forms like ukiyo-e, whose artists often satirized present-day trends by likening them to past phenomena. The common people who enjoyed these popular entertainments had enough historical knowledge to appreciate the cleverness of historical metaphors in contemporary pictures and plays, and some historical maps were published in the latter half of the eighteenth century specifically to help the general public understand the historical context of popular works.

Historical maps of Osaka, of which we can find more than one hundred examples, constitute one of the best-developed illustrations of this genre from the eighteenth century. In fact, they are sufficiently prominent and distinctive to merit a name of their own: *Naniwa-kozu*. Deriving from Naniwa (the original name of Osaka) and *kozu* (meaning historical maps), this broad category has been divided into twelve different subtypes based on content.[3] Some of these were printed, while others were drawn by hand. Adding to these features, ten of the twelve map types were sometimes treated as a series in historical atlases. We do not know of any other historical maps having such distinguished quality and quantity as *Naniwa-kozu*.

Why Osaka? Part of the explanation must lie in the sheer dynamism of the city at the time. Nicknamed the "kitchen of the realm" for its central role in the nation's rice trade, Osaka was Japan's second largest city and in many ways its merchant capital. At the same time that it served as the nation's commercial hub, it was also a hotbed of popular culture, celebrated especially for its puppet plays and comic entertainments. In addition, Osaka had a long and illustrious history that made it worthy of historical excavation. Examining the contexts in which it featured in historical atlases shows one of the reasons why historical maps were so well developed in Osaka. This venerable city had been the site of key events in the Japanese past. Home to palaces in the ancient period, it emerged as a prominent temple town in the medieval era, was remade as the castle town of one of Japan's great sixteenth-century unifiers (Toyotomi Hideyoshi), and finally supplied the battlefield on which the Tokugawa won their final military victory in 1615. In a cultural context marked by the rise of interest in the past, the residents of Osaka understandably wanted to visualize their hometown landscape at each of these moments and were keen to see the way the city's morphology had evolved over time. Historical maps and atlases answered their demands.

Broadly speaking, the resulting historical maps fall into two types. Some have the character of *pseudoarchaic maps* in that they masquerade as products of an earlier era. For example, one group of *Naniwa-kozu* claims to be based on a 1417 original, as seen in figure 15.1. The colophon reads:

> 1417, the era of the retired Emperor Gokomatsu.
> It was 346 years [later], 1762, when this edition was
> traced by master Heki-getsu-rō
> And this copy was retraced in 1799.

In fact, the original of this map could not have been made in the medieval era, because several elements of the landscape it depicts date back only to the early modern era. Furthermore, no tracings that were clearly made before the eighteenth century have been located.

Nevertheless, the attempt of an unknown producer to create a new map with an apparently genuine landscape from the past and a specious date seems to have succeeded to some degree. We can find many copies of this type of *Naniwa-kozu* from the latter half of the eighteenth century through the nineteenth century. The wide circulation of this type of map tells us that it was popular among the people of that time. Indeed, some intellectuals mentioned it as a good historical resource for reconstructing the past landscape of Osaka. Even Motoori Norinaga (1730–1801), one of the most famous and influential intellectuals of eighteenth-century Japan, used a pseudoarchaic map of Naniwa as one of his primary sources for reconstructing the world of the eighth-century *Kojiki* (Record of ancient matters), Japan's first written chronicle.

Exploiting the wave of popular culture, some publishers tried to publish woodcut versions of similarly faux antique maps. For example, a historical atlas called

Naniwa jōkozu, showing the birth of Osaka in ancient times, was published in 1800. It is difficult to say whether this atlas sold well during this era. The fact that it can still be found in various libraries and archives suggests that the print run, at least, was not small.

While maps that claimed a bogus historical authenticity were gaining in popularity, some intellectuals looked upon such artifacts critically. Having researched their contents thoroughly, they finally concluded them to be forgeries, made not in the medieval era but in the early modern period. Among these intellectuals was Mori Yukiyasu (b. 1701), who moved to Osaka after retiring from a Kyoto incense shop in 1730. Producing more than four hundred maps over the next decade, Mori wanted to make a large topographical atlas of Japan that would bring together maps made at a variety of scales, ranging from city to world and celestial maps.[4] He also sought to describe Japan and its major cities, notably Kyoto and Osaka, from a historical perspective. In this effort, Mori became a maker of what we might call *retrospective maps*.

Of four images of early Osaka that Mori made in the 1750s, one of the best known is the "Historical Map of Naniwa, Settsu Province," reproduced here as figure

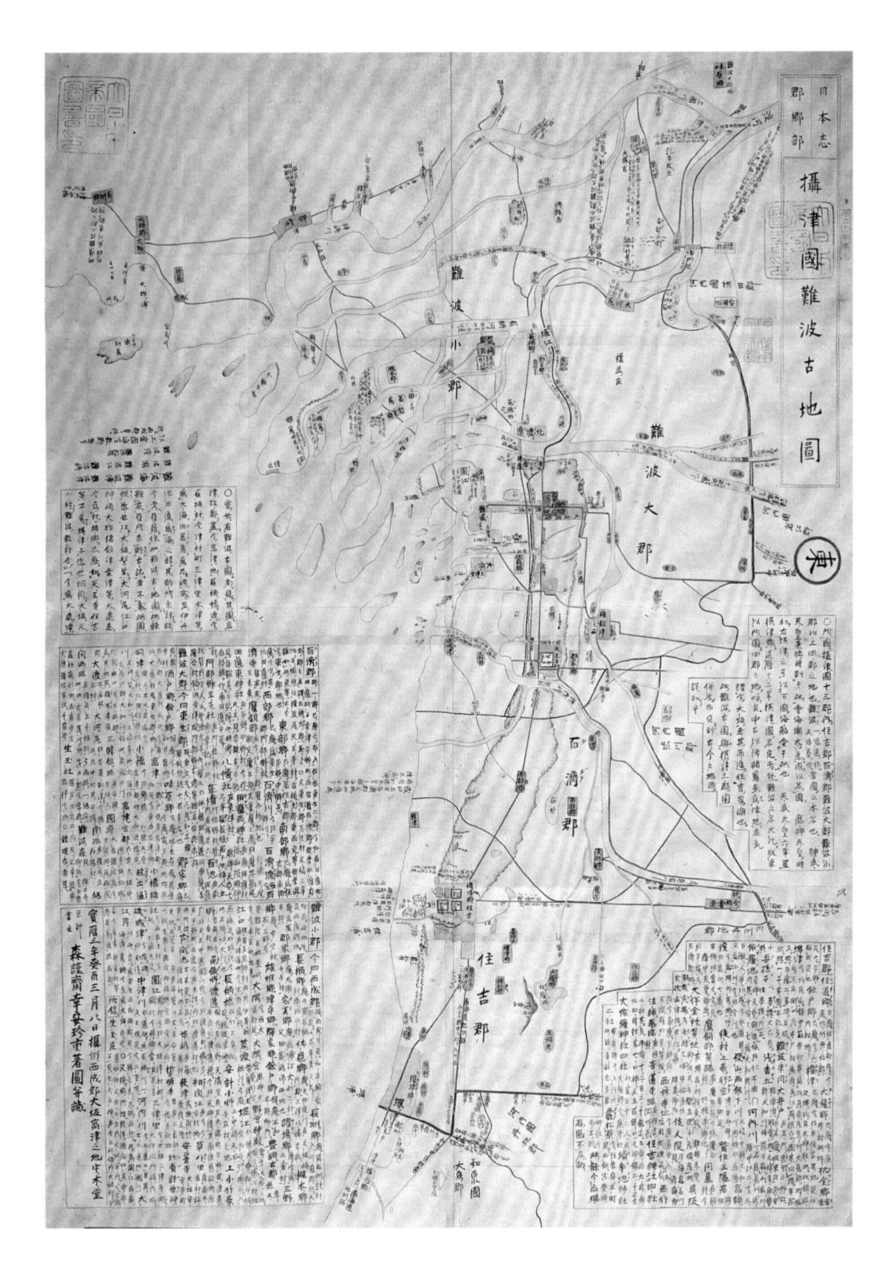

FIGURE 15.2 "Historical Map of Naniwa, Settsu Province" [Settsu koku Naniwa kochizu 摂津国難波古地図] by Mori Kōan 森幸安, 1753. Manuscript, 109 × 77 cm. Courtesy of National Archives of Japan.

15.2. In his explanation of this map, Mori stated clearly that because there were numerous problems with existing *Naniwa-kozu*, he had tried to reconstruct the past landscape of the city by consulting textual documents. Thus, every viewer of this map understood that it was the product of historical reconstruction. To be sure, some of Mori's reconstructions appear inaccurate from a present-day perspective.[5] However, one should remember his challenges as an amateur scholar of history and geography. A retired merchant in eighteenth-century Osaka did not have access to the advanced reference tools and archival systems for historical documents available today. Still, Mori's attitude toward history was essentially the same as that of present-day historical geographers. It might even be fair to say that the beginnings of modern historical-geographic thought took place around Mori's time, for eighteenth-century Japan witnessed an impressive blooming of interest in historical cartography. Ordinary people and intellectuals alike had begun to enjoy making, seeing, tracing, and using historical maps.

Notes

1. Uesugi Kazuhiro, *Edo chishiki-jin to chizu* (Kyoto: Kyoto University Press, 2010), 239–80.
2. Yamamura Aki, "Kaioku no naka no sengoku jōka machi," in *Daichi no shōzō*, ed. Fujii Jōji, Sugiyama Masaaki, and Kinda Akihiro (Kyoto: Kyoto University Press, 2007), 224–46.
3. Uesugi Kazuhiro, "Kinsei ni okeru Naniwa-kozu no sakusei to juyō," *Shirin* 85 (2) (2002): 33–74.
4. Uesugi, *Edo chishiki-jin to chizu*, 113–98.
5. Uesugi, *Edo chishiki-jin to chizu*, 179–91.

Suggested Reading

Platt, Brian W. "Elegance, Prosperity, Crisis: Three Generations of Tokugawa Village Elites." *Monumenta Nipponica* 55 (1) (2000): 45–81.

16 The Urban Landscape of Early Edo in an East Asian Context

TAMAI Tetsuo 玉井哲雄

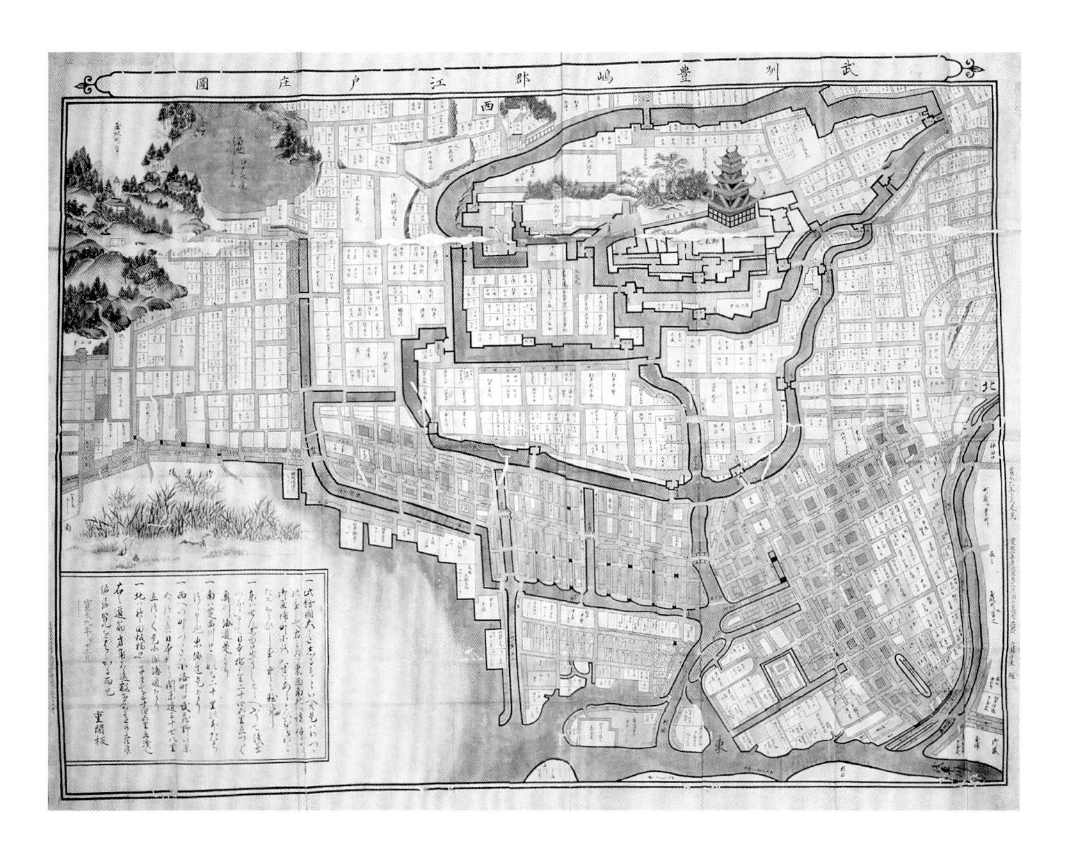

FIGURE 16.1 "Map of Edo in Toshima County, Musashi Province" [Bushū Toshima gun Edo no shō zu 武州豊島郡江戸庄図]. Hand-colored manuscript copy of a printed map from 1632, also known as the Kan'ei Edo Map; 100 × 130 cm. Courtesy of Tokyo Metropolitan Library.

The Kan'ei Edo map (fig. 16.1)–so called because it is dated "the twelfth month, ninth year of the Kan'ei era" (1632)—is our most accurate image of the shogun's capital in its early heyday. Made during the reign of the third shogun, Tokugawa Iemitsu, it represents the townscape after Edo Castle had been largely completed but before the great Meireki fire of 1657, when much of the central city was destroyed. Although the main castle tower was never rebuilt, the fire led to a period of great expansion and redevelopment based on the framework initially established in the Kan'ei era.

Of course, this map is not without its challenges. The version currently in existence is a copy that was produced during the later Edo period; none of the original prints, which are assumed to have been issued during the Kan'ei era, have been found to date. Nor is anything known about the copying and transmission process of the surviving map. Nonetheless, many details depicted here can be verified from other written and illustrated sources to clarify the characteristics of early seventeenth-century Edo.

FIGURE 16.2 Annotated version of figure 16.1, identifying Mt. Atago, Zōjō-ji, Hie Shrine, Tenshu tower, Edo Castle, Nihonbashi, and Honchō Street. English labels added by author.

Structurally the Kan'ei map combines pictorial elements with a top-down view of the city's core. Oriented toward the west (the direction of distant Kyoto), it takes Edo Castle as its focal point. A handful of key landmarks—the main tower of Edo Castle, Momijiyama within the castle grounds, the Hie Shrine, the city reservoir, Mount Atago, and Zōjō-ji Temple—stand out clearly in pictorial form. The bulk of the city, however, is depicted in plan view, including the castle grounds as well as the surrounding residential districts of samurai and townsmen. The cartographer accurately shows how the city was divided into urban blocks by roads, which are colored yellow. Moats and canals are equally clearly marked with bold blue lines and black edges. Finally, the bay shoreline (which would be altered significantly after the Meireki fire) also helps to clarify the shape of early Edo.

Within the map, one of the first things to catch the eye is the outsized depiction of Edo Castle's five-tiered Tenshu tower, which, at forty-five meters in height, was the tallest donjon in the country. Although stylized, the representation seems congruent with contemporary depictions found in other sources, from folding screens to carpentry manuals. This visual emphasis on the Tenshu tower is one of the most striking characteristics of the Kan'ei Edo map, underscoring how the tower served as the center of Edo at the time. Outside the castle walls, the majority of the urban area was taken up by samurai residences, left colorless but labeled with the names of their chief occupants. It should be noted that on the Kan'ei map the names of samurai are all written facing the same direction, whereas in later maps such labels were aligned to reflect the orientation of individual properties.

A number of temples and shrines can be seen at the margins of the map. The shrine labeled "Sannō" (*center top*) is the Hie Shrine; Zōjō-ji, the Tokugawa family temple, is visible on the far left. Pushing shrines and temples out to the urban periphery was a typical feature of castle towns during the Tokugawa era, and Edo was no exception. The remaining areas, drawn in shades of pink, are residential neighborhoods for townsmen. Because these areas were bases of commerce and industry, it was necessary that they be readily accessible by both land and water.

The map shows Edo's commercial districts as unfolding from the area in front of the main gate to Edo Castle, extending northeast along Honchō Street toward Asakusabashi (*lower right*). This whole area can be considered the core of Edo's commercial and industrial zone. Beyond Asakusabashi lies the Sensō-ji Temple. During medieval times, this area was home to the important Ōshū Road connecting Edo to Asakusa. Honchō Street can be considered the successor of the Ōshū Road. Two other important turnpikes, the Tōkaidō (leading south) and the Nakasendō (leading north), are depicted as originating at Nihonbashi, the "bridge of Japan." Both of these routes were referred to within Edo as the "Nihonbashi roads," and we can see from the map how each was lined with commercial and industrial blocks. It seems apparent that once the construction of Edo was complete, the connection through these Nihonbashi roads to distant parts of the country (particularly the Kyoto-Osaka area) had come to be more important than the connection to Sensō-ji Temple along Honchō Street. In this way, the Kan'ei Edo map captures a moment of transition from Honchō's medieval route network, anchored by temples, to the more commercially oriented Nihonbashi routes.

Edo Castle was still being constructed during this era, and the Kan'ei map underscores just how important waterways were for bringing construction materials into the city. Many harbors are depicted within the commercial and industrial neighborhoods, with waterways entering from the south in the direction of the Honchō Street area. (The Yoshiwara pleasure district, which would later move to Asakusa, was originally established on an island amid these waterways.) Some canals actually reach beyond Nihonbashi to connect with the moat of Edo Castle itself. With their many wharves, these canals were important pathways along which goods could be brought into Edo. But they would largely disappear in later maps as the city's expansion pushed wharf functions further out to peripheral areas.

Turning our attention once again to the commercial areas located along Honchō Street north of Nihonbashi, we can see that many city blocks appear to contain darker squares at their center. According to the urban plan for Edo, the homes of tradesmen and artisans were meant to face the roads, with undeveloped spaces in the interior of their residential blocks. It seems that the dark areas of the Kan'ei map are meant to represent these empty spaces. That many such areas were actually left undeveloped throughout Edo is confirmed by ordinances forbidding the dumping of trash within their confines. Careful inspection, however, reveals a few blocks along Honchō Street north of Nihonbashi that have been left white in their centers, with names inscribed therein. Many more examples can be found along Nihonbashi Street south of the bridge. Judging from the names, it seems that these plots were bestowed by the shogun on various purveyors of goods and services to the government, including physicians and actors. This unplanned development contributed to a shift in the nature of the commoner districts.

In most castle towns other than Edo, square blocks with empty centers were a rarity, the basic unit of the urban grid being a rectangle. This was certainly more practical; one might even argue that it was irrational to leave undeveloped spaces within square blocks. The reason such spaces existed in Edo was essentially symbolic, harking back to the nature of the urban landscape of Japan's original capital. In ancient Kyoto, the basic unit of city planning had been a square plot measuring forty *jō* (120 square meters). The fact that the area of Edo's empty lots is nearly identical to the Kyoto measurement is certainly no accident and shows the degree to which city planners kept the Kyoto example in mind.

It is noteworthy in this connection that Edo's main thoroughfare (leading north from Shinbashi through Nihonbashi Honchō to the main entrance of the shogun's castle) is the one part of town that was consistently lined with square (rather than rectangular) city blocks. There is a distinctive historical explanation for this anomaly, related to international diplomacy. During the Kan'ei era, Korean diplomatic missions and Ryukyu envoys both visited Edo and formed processions to enter the castle. It was only along this route that arcade-style eaves were permitted in Edo. Three-story buildings (later forbidden on corner plots) were also common here. Since the shogunate used diplomatic processions as a type of public spectacle to underscore its influence over foreign powers, it is impossible to understand the existence of square city blocks along Edo's "Main Street" as anything other than a stage for the shogunate's displays of its diplomatic influence overseas. At the time, most commoners in Edo thought of the Korean and Ryukyu diplomats as envoys from Qing China. The area of Edo displaying these square city blocks thus helps us appreciate Edo's place in a larger East Asian context.

(Translated by Bret Fisk)

Suggested Reading

Tamai Tetsuo. *Edo: Ushinawareta toshi kūkan o yomu*. Tokyo: Heibonsha, 1986.

17 Spatial Visions of Status

Ronald P. Toby

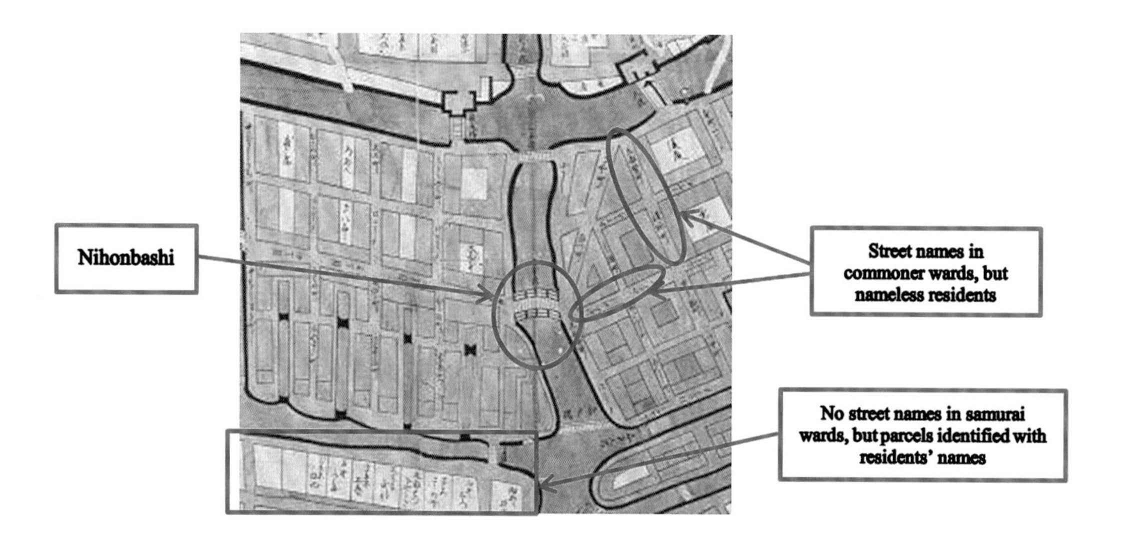

FIGURE 17.1 Detail from figure 16.1, "Map of Edo in Toshima County, Musashi Province," showing the Bridge of Japan (Nihonbashi). Copyists color-coded the merchant wards near Nihonbashi as well as writing in street names, but still did not identify residents or shops. In samurai wards, parcels are still labeled with the name of the resident daimyo; streets are unlabeled. English labels added by author.

The society of early modern Japan was organized into *mibun*, hereditary caste or status groups that were roughly analogous to the estates in France under the ancien régime. In Japan, the four major status categories were samurai, peasants, artisans, and merchants (the latter two often referred to jointly as "townsfolk"), but there were several other status positions beyond the bounds of the "four folk" as well. At the top of the social pyramid, the imperial and courtier houses, perhaps a thousand or so people in all, floated metaphorically "above the clouds." Buddhist and Shinto clerics and other religious professionals likewise stood apart. Finally, two distinct outcaste groups, known by the derogatory terms *eta* (or *kawata*) and *hinin*, lay beneath—and outside—the other *mibun* groups, performing essential but ritually defiling functions such as the disposal of dead animals, leather tanning, making footwear, handling executions, and serving as urban guards.

For the most part, cities and villages across Japan were laid out with an eye to segregating these groups from each other. Castle towns large and small had clearly demarcated samurai districts (about 60 percent of Edo's area, less in other towns) and districts for Buddhist temples (accounting for another 20 percent of Edo); merchant and artisan quarters were crowded into the remaining space, while entertainment districts for theaters and licensed prostitution, along with outcaste sections, were typically located on the outskirts. Even city administration was divided along status lines, with separate magistrates handling the commoner wards, shrine and temple districts, and outcaste communities. In areas of the country with large outcaste populations, peasant villages might have a contiguous *eta* hamlet subordinated to them (see chap. 33).

Such segregation was never complete. In Edo, for example, small shops lay scattered among the samurai districts, serving the large complements of samurai on temporary tours of duty in their lords' Edo residences.[1] Prior to the great Meireki fire of 1657, there were even merchant blocks alongside daimyo compounds in the outer precincts of Edo Castle. Nonetheless, the principle of segregation by status was deeply etched into the urban landscape. *Mibun* had to be visible in order to sustain the complex webs of duties and privileges accorded to each group. Distinct hairstyles and clothing made it possible to pick out samurai, commoners, religious professionals, and outcastes on the street. Architecture performed a similar function. Both in Edo and in castle towns around the country, *mibun* categories were mapped onto the landscape by means of distinct architectural features that included walled, gated compounds for the mansions of daimyo and bannermen, shop fronts open to the street in merchant and artisan districts, and distinctive gates and enclosures for Buddhist temples and Shinto shrines.

In a society based on such pervasive status distinctions, it is no surprise that cartographers too felt compelled to render *mibun* visible on the face of their maps. Consider the earliest printed image of Edo, the woodblock print "Kan'ei Edo zu" of 1632 discussed by Tamai Tetsuo in the preceding essay. The designer of this map—which was originally printed in black and white—used a number of graphic and pictorial devices to distinguish areas of samurai residence from merchant and artisan quarters, on the one hand, and Buddhist temple districts, on the other. Commoner wards, for instance, had their names inscribed directly on the street running through each ward; the wards themselves were labeled simply by the generic term "townsfolk houses" (*machiya*). Clustered overwhelmingly on low-lying land reclaimed from the wetlands of Edo Bay, these crowded neighborhoods were laid out in square or rectangular blocks, with shop fronts facing streets that were plotted along a rough grid pattern. The late Edward Seidensticker referred to this as the "low city": both physically low-lying and occupied by folk of lower status (fig. 17.1). As Mary Elizabeth Berry explores in her essay (chap. 19), commoner wards often bore names that reflected the artisanal trade or merchandise on offer there. Among those featured on this map are "lumber ward," "blacksmith ward," "helmet makers' ward," "stable ward," "clothiers' ward," "indigo dealers' ward," and "doll makers' ward."

Most of early Edo's warrior districts, by contrast, sat on higher ground, on ridges and hills that often defied layout in rectilinear blocks. The sprawling Edo mansions of hundreds of daimyo and hundreds more bannermen, as well as the more modest residential compounds of lower-ranking samurai, took up three-fifths of the city, even though samurai accounted for only half of Edo's population. The walled daimyo estates were complex constructions that ranged from an acre or two, for the lord of a small domain, to over eighty acres for the truly wealthy. The compounds of the grandest daimyo included not just residential palaces for the lord and his family but extensive gardens, arsenals, and barracks for his samurai retainers. In contrast to the named streets and wards of the commoner neighborhoods, the mapmaker inserted no toponyms on the streets in warrior districts, instead inscribing the family name or courtesy title of each lordly proprietor within the outline of his individual parcel. The high status of the occupants was effectively signaled precisely by offering no detail except the identity of the daimyo or bannerman who lived there. And for the highest ranking daimyo of all—the Tokugawa collateral families known as the Three Lords—not even their surnames, only their respective provinces, are identified. The mansions of these exalted Tokugawa kin are labeled simply "The Senior Councillor of Owari [or Kii];" "The Middle Councillor of Mito."

Wards reserved for Edo's dozens of Buddhist temples, great and small, were treated much like warrior neighborhoods. As in daimyo districts, there was no indication of toponyms at the street or neighborhood level. Only the label written inside the space occupied by "Such-and-Such Temple" or "So-and-So Cloister" told the reader of the map that this was a neighborhood of religious compounds. It is clear, however, from the many hand-colored copies of the map that readers and users recognized the distinct *mibun* nature of these Buddhist temple districts (fig. 17.2).

The mapmaker who executed the original printed map did not supply an explicit key to his graphic codes, but he clearly had distinctions of *mibun* in mind in the treatment of the areas reserved for samurai, commoners, and Buddhist clergy. (This early map does not deal with Edo's outcaste communities, which were sited outside

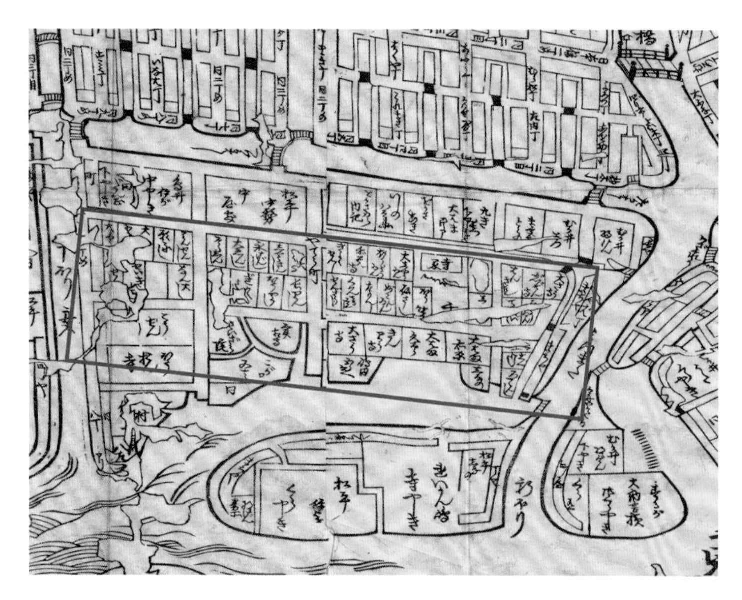

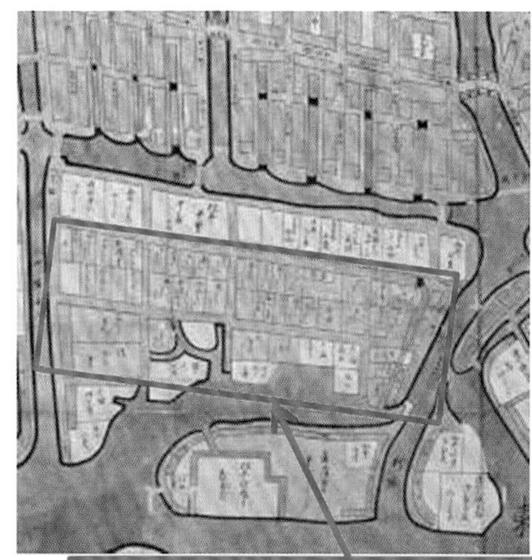

FIGURE 17.2 (*left*) Detail from "Map of Edo in Toshima County, Musashi Province" [Bushū Toshima gun Edo no shō zu 武州豊島郡江戸庄図]. 1682 woodblock reprint of the "Kan'ei Edo zu" of 1632 (fig. 16.1); 85 × 120 cm. Courtesy of Geography and Map Division, Library of Congress. Temple wards were hard to distinguish from samurai neighborhoods on the original and subsequent monochrome printed maps, since streets were unlabeled and each parcel was identified only by the name of the temple.

FIGURE 17.3 (*right*) Detail from figure 16.1, showing how later users and copyists enhanced the distinction by color-coding temple districts in pink.

the area covered, but many later maps clearly labeled these outcaste neighborhoods with the word *eta*.) The medium of black-and-white printing limited what the designer could do to show status distinctions. But users of his map, and the many people who made hand-drawn copies of it, enhanced the original to make status distinctions stand out more clearly, as is evident from the prevalence of color-coding in most surviving manuscript copies of the original 1632 map.

Maps of other cities, including Osaka, Kyoto, and Kanazawa, evince a similar concern to render *mibun* visible, whether by color-coding or other graphic devices.[2] Precisely because inherited status was the most salient reality of early modern Japanese society, mapmakers and map readers demanded its representation as a central feature of their maps.

Notes

1. James L. McClain, *Kanazawa: A Seventeenth-Century Japanese Castle Town* (New Haven: Yale University Press, 1982), chap. 2, describes this process in early Kanazawa.

2. Interested readers can readily find additional Edo-period maps online that use similar conventions to show distinctions of *mibun*. Examples include Kōno Dōsei, *Meireki shinpan Ōsaka no zu* (1657); Hayashi Yoshinaga, *Shinsen zōho Kyō ezu* (1696); and *Kaga-no-kuni Kanazawa no zu* (1800). These are available at the Japanese Historical Map Collection of the University of California, Berkeley, http://www.davidrumsey.com/japan/.

Suggested Readings

Coaldrake, William H. *Architecture and Authority in Japan*. London: Routledge, 1996.

Howell, David L. *Geographies of Identity in Nineteenth-Century Japan*. Berkeley: University of California Press, 2005.

Seidensticker, Edward. *Low City, High City: Tokyo from Edo to the Earthquake*. New York: Knopf, 1983.

18 The Social Landscape of Edo

Paul Waley

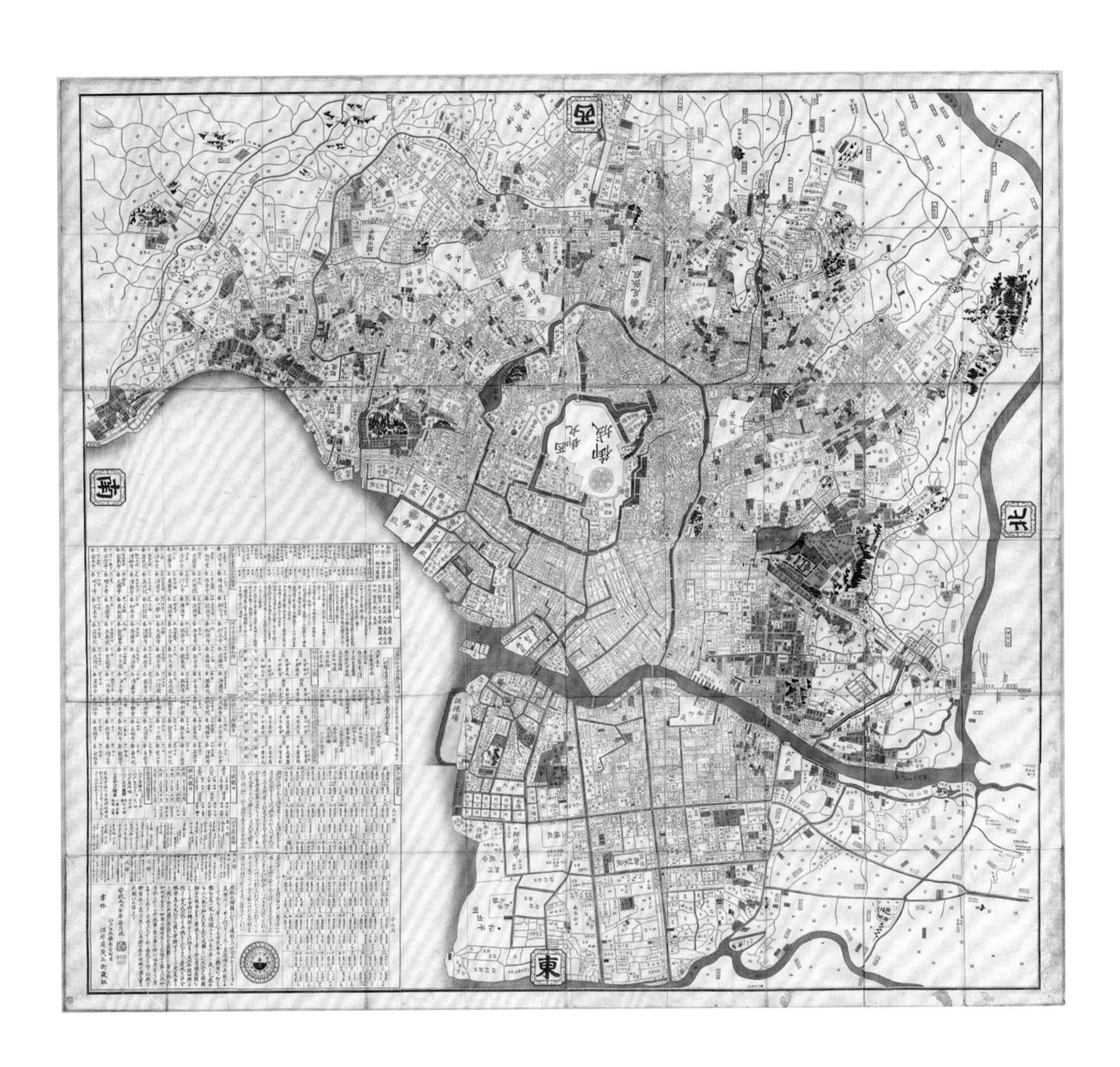

FIGURE 18.1 "Proportional Map of Edo" [Bunken Edo ōezu 分間江戸大絵図] by Mori Fūsai 森楓斎 and Suharaya Mohē 須原屋茂兵衛, 1858. Woodblock print, 179 × 198 cm. Courtesy of the C. V. Starr East Asian Library, University of California, Berkeley.

By 1858, Japan's political and military capital was in decline. That state of affairs is not apparent in figure 18.1, however, which shows the city as it appeared in its prime a few decades earlier, when it was probably the largest settlement in the world, with over one million people. Built from scratch by order of Tokugawa Ieyasu, Edo represented the culmination of an epochal period of city building in late sixteenth-century Japan. The shogun strategically sited his castle town next to the sheltered waters of a large bay, on a narrow coastal strip that opened out to the Kanto Plain, Japan's largest expanse of flat land. Laborers dug a port out of coastal marshes and just behind it constructed a grid of streets on low-lying alluvial land to house the merchants, artisans, and other members of the commoner (nonmilitary) population, whose presence was essential for provisioning and servicing lords and their retinues.

Edo's design combined pragmatic concerns for logistics and defense with a nod to cosmographic principles of city building. At the symbolic center of the city stood the castle, supreme expression of Tokugawa power, depicted on the map as an empty space surrounded by a spiral of moats and earth embankments. The castle was in turn surrounded by the compounds of other great lords, the daimyo, whose domain crests form a colorful component on many maps of Edo. This political checkerboard around the castle represents the shogun's attempt to freeze political power equations. The senior wives and eldest sons of the daimyo lived permanently as virtual hostages in their Edo compounds, while the lords themselves spent alternate periods in the city and in their fief capitals. While these and similar practical considerations drove the city's design, the four directions of traditional Chinese-inspired cosmography were observed as well, albeit with a sleight of hand. The emperor in distant Kyoto should properly have been able to look down over the shogun's headquarters from the north, but since Kyoto was in fact far to the west of Edo, most representations of the shogun's city, including this one, were oriented so as to put west—and the emperor—at the top of the map.

The Tokugawa's new city burned down in a terrible conflagration in 1657 but was soon rebuilt in an expanded form that endured over the following two centuries. Throughout this period, commoners remained largely locked into their crowded bayside quarters even as their numbers increased, although some were able to carve out new settlements along the bustling roads leading out of the city. Meanwhile, the compounds of the samurai and their lords, who numbered less than half the population, dominated well over half of Edo's urban space. Temple and shrine precincts, the third element in the tripartite administrative division of the city, were moved outward to form a suburban ring, shown here clearly in red. While this urban morphology remained stable for two hundred years, the lack of significant change in the city's structure belied the development of an increasingly fluid and mobile society. Fluidity manifested itself primarily in the growth of an urban consumer class composed of both commoners and samurai. Mobility was expressed in the large number of people moving around the city on daily errands as well as travelers from further afield, creating what the great historian of Edo, Nishiyama Matsunosuke, called "a culture of movement."

Before we discuss the resulting social landscape in more detail, we should pause to reflect on the map itself, an intriguing cross between conventional representation and trigonometric accuracy. This is but one of many splendid cartographic representations of the city from above, images that were fleshed out by the evocative portrayals of Edo's topographical features executed by Andō Hiroshige and other print artists, using the same woodblock technology.[1] The map makes manifest many of the most important aspects of the city's social geography. Yet is can also be misleading. The true center of Edo, for most residents, was not the castle, which played little part in city life and was entered only by a small coterie of daimyo, their leading retainers, and military bureaucrats. Rather, it was a small bridge called Nihonbashi, scarcely visible here, which lay about five hundred meters east of the castle's main gate. Literally translatable as "the bridge of Japan," Nihonbashi was the starting point from which distances were measured on all the country's main highways. More important, Nihonbashi, located next to one of the city's main markets, was the epicenter of mercantile, plebeian Edo and its distinctive urban culture.

Life in Edo was played out in a landscape of contrasts. The physical contrast between the low-lying central and eastern districts, in the middle and lower parts of the map, and the hillier areas to the west, on either side of the castle and above it, crudely corresponded to a fundamental social division. As Ronald Toby discusses in detail above (chap. 17), the hilly districts were given over mainly to the compounds of the daimyo, while the lowlands contained the houses and backstreet shacks of commoners, including merchants and artisans, servants and menials, entertainers and "outcastes" who fell outside the social hierarchy. This dynamic juxtaposition of a western "high city" with an eastern "low city" structured much of Edo life. As can be discerned on the map, however, this socio-geographic division was far from rigid. Daimyo storehouses were located east of the city's main river, the Sumida, as well as in more central and northern areas, while the roads that led through the valleys between the western hills were lined with the shops and workshops of merchants and artisans.

Other contrasts might be more apparent than real. Clearly the life of a military retainer in a compound in the green hills west of the castle differed in significant respects from the daily grind for a lowly artisan in a back alley in a crowded central block. Yet by the time this map was printed, life had become almost as precarious for the former as for the latter. The greater contrast was increasingly between wealth and the lack of it. The richest merchants, for example, although still considered commoners in this stratified society, came to possess rice warehouses along the serrated moorings visible on the banks of the Sumida. As their ever-increasing control over the price of rice brought them wealth—and economic power over the samurai lords and their retainers—they began to live in an opulent style that outshone the more austere elegance of daimyo life.

A further contrast involves the nature of the life spaces of the city's inhabitants. Most Edoites' days were spent in confined and crowded conditions. This applied almost as much to the samurai as it did to the commoner classes. The constraints were social but also physical, with cramped accommodation indoors and gates to compounds, city blocks, and streets that closed at dusk. The "culture of movement" that developed in such an environment of regulated mobility was channeled into roadways, marketplaces, riverbanks, and firebreaks. Such places, in the words of historian Henry D. Smith II, were "denser, more active, more sociable than others" and as such became spaces of convergence, transgression, and release.[2]

The biggest throngs in Edo could be found at Nihonbashi and other market areas. Crowds also converged on the open spaces created as firebreaks at the foot of major bridges, and on the spacious grounds of the city's main temples and shrines, notably the temple to Kannon at Asakusa (visible in the green and yellow area west of the Sumida in the bottom right of the map). As we read in contemporary topographical accounts, temple grounds provided the same sort of entertainment as did the firebreaks: freak shows, animal shows, conjurers, acrobats, and booths hosting the bawdy and lewd.[3] These places of transgression found their most extreme form in the licensed quarter of Yoshiwara, a small grid of exurban streets clearly visible to the right of the Asakusa temple. The teahouses and brothels of the licensed quarter were officially out-of-bounds to samurai, as were the kabuki theaters, which had been moved to Asakusa by the time this map was made. Yet everyone knew that transgressions occurred on a regular basis, as samurai sought respite alongside commoners from the stifling etiquette of daily business. Perhaps the most celebrated of such spaces were the Sumida River itself and its embankment outside the city (in the bottom right corner of the map). A walk along the embankment became a sort of pilgrimage marking the changing seasons, while the pleasure of boating on the river was a highlight of the summer months.

The convergence was also social, despite the contrasts discussed above that emerged between samurai and commoners and between rich and poor. Edo's cultural geography was animated by small but significant circles of literati, drawn from the ranks of samurai and commoners. These salons met to indulge in literary and artistic activities. They patronized certain eating establishments and cultivated each other's company in walks to outlying temples and shrines that can be seen ringing the north of the city. Through prints, paintings, verse, and topographies, such groups enriched the cultural geography of the city, of which this map is a fine expression.

This image is in some senses an idealized, even a nostalgic portrait of Edo. By the time of its publication, the city had been struck by a damaging earthquake (in 1854) and by social unrest, harbingers of the upheaval to come. In the same year, Commodore Matthew Perry and his Black Ships made their second appearance. The shogunate was thrown into confusion, unable to decide whether to reform or batten down. Its system of control, exercised through the enforcement of alternate periods of daimyo attendance in the city, disintegrated. While its power base weakened, a number of daimyo in the west of the country sensed the opportunity to remove the shogunate and set up a new system of government with the emperor at its head. Only ten years later, they had effected this move, and the Tokugawa shogun was overthrown. After a slight hiatus, Edo would be renamed Tokyo and made into the national capital of a modernizing Japan.

Notes

1. For an insightful discussion of Hiroshige's best-known series of prints of Edo, see Henry D. Smith II's introduction and com-

mentaries to *Hiroshige: One Hundred Famous Views of Edo* (New York: George Braziller, 1986).

2. Henry D. Smith II, "Sky and Water: The Deep Structures of Tokyo," in *Tokyo: Form and Spirit*, ed. Mildred Friedman (New York: Harry Abrams, 1986), 27.

3. For a detailed description, see Andrew Markus, "The Carnival of Edo: *Misemono* Spectacles from Contemporary Accounts," *Harvard Journal of Asiatic Studies* 45 (2) (1985): 499–541.

Suggested Readings

Fiévé, Nicolas, and Paul Waley, eds. *Japanese Capitals in Historical Perspective: Place, Power and Memory in Kyoto, Edo and Tokyo*. London: RoutledgeCurzon, 2003.

Guth, Christine. *Art of Edo Japan: The Artist and the City, 1615–1868*. New Haven: Yale University Press, 2010.

Hur, Nam-lin. *Prayer and Play in Late Tokugawa Japan: Asakusa Sensōji and Edo Society*. Cambridge, MA: Harvard University Asia Center, 2000.

Jinnai Hidenobu. *Tokyo: A Spatial Anthropology*. Translated by K. Nishimura. Berkeley: University of California Press, 1995.

McClain, James, John M. Merriman, and Ugawa Kaoru, eds. *Edo and Paris: Urban Life and the State in the Early Modern Era*. Ithaca: Cornell University Press, 1994.

Naitō Akira. *Edo, the City That Became Tokyo: An Illustrated History*. Translated by H. Mack Horton. Tokyo: Kodansha International, 2003.

Nishiyama Matsunosuke. *Edo Culture: Daily Life and Diversions in Urban Japan, 1600–1868*. Translated and edited by Gerald Groemer. Honolulu: University of Hawai'i Press, 1997.

19 What Is a Street?

Mary Elizabeth BERRY

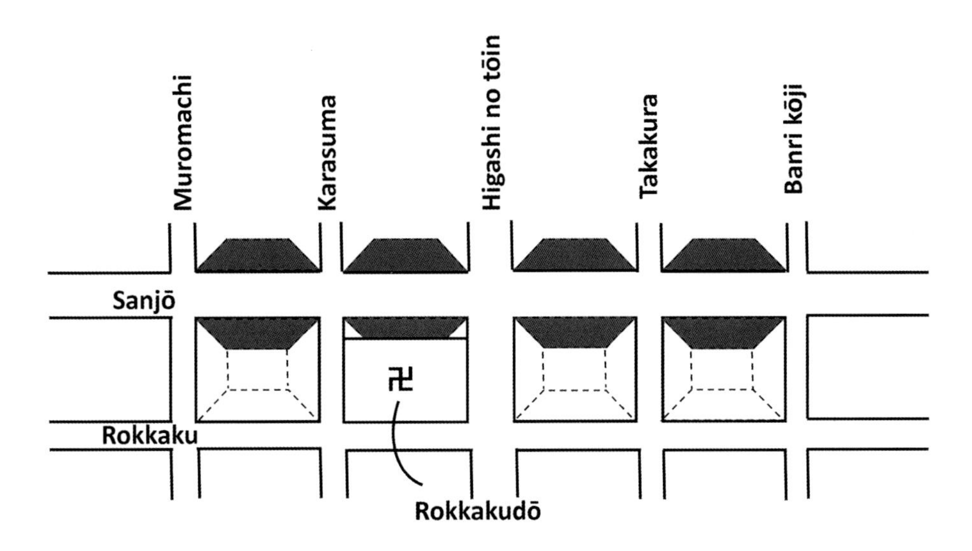

FIGURE 19.1 The configuration of medieval urban blocks. The shading indicates four two-sided blocks (*ryōgawa chō*) facing Sanjō Avenue in Kyoto. From Mary Elizabeth Berry, *The Culture of Civil War in Kyoto* (Berkeley: University of California Press, 1994), 211. Courtesy of the publisher.

Street names are scarce in most Japanese cities, a point of marvel for foreign visitors and web pundits alike. In a riff on the resulting difficulty of wayfinding in Tokyo, Roland Barthes grew famously cranky: "the largest city in the world is practically unclassified; the spaces which comprise it in detail are unnamed." Well aware of highly classified and detailed addresses, Barthes nonetheless dismissed them as bearing "only a postal value" in a system "apparently illogical, uselessly complicated, curiously disparate."[1]

Monsieur Barthes was too hasty. Streets play (at least) two roles in the city and inspire (at least) two naming practices that, equally "logical," disclose variable understandings of the urban body. On the one hand, streets provide the armature of the city's physical organization: naming them emphasizes this structural (and navigational) function. On the other hand, streets provide a matrix for the city's human activity: naming the communities that form among them emphasizes this social character. Often enough, of course, names attach both to the roadways themselves and the neighborhoods they engender. Yet in insisting on the singular cogency of street names, which proliferated across the globe only in the nineteenth century (when new forms of municipal administration and communication favored them), Barthes ignored the powerful spatial logic still conveyed by the *quartiers* of his own Paris and the similar districts of Delhi, New Orleans, Istanbul, Rio, and countless other cities.

True enough, though, Japan does remain something of an outlier, since neighborhood names (the core feature of those "uselessly complicated" addresses) continue to supplant street names almost everywhere. One factor behind the mystery is a history both long and uncommonly strong. It begins in medieval Kyoto, where mounting commercial traffic converted streets from boundaries that enclosed property (often with fortifying earthen walls) into links that connected the residents facing them. Trade-savvy commoners along the two sides of busy arteries came to form voluntary associations, principally to oversee their shared interests in safety, fire prevention, and property transactions. By the sixteenth century, the resulting *ryōgawa chō,* or two-sided neighborhoods, numbered in the hundreds, each with a distinctive name inspired by a local landmark or notable personage or dominant enterprise (fig. 19.1). Although Kyoto did have street names, an exceptional legacy of its Chinese model, they were soon surpassed in both official and vernacular use by the names of the neighborhoods where the lives of commoners centered.

A bottom-up form of medieval alliance, the two-sided neighborhood of Kyoto became a top-down form of early modern control throughout the cities that exploded in Japan around 1600. Leading the urban migration was a vast population of samurai. After more than a century of civil war, the conquest regimes of the Toyotomi and Tokugawa houses hoped to tame these volatile combatants through forced relocation from villages to castle towns. In their wake came waves of entrepreneurial commoners—merchants, artisans, and providers of myriad services—who settled the often-rudimentary castle towns in areas reserved for them by local authorities.

Take the case of Osaka. Development there proceeded, first under Toyotomi and later under Tokugawa

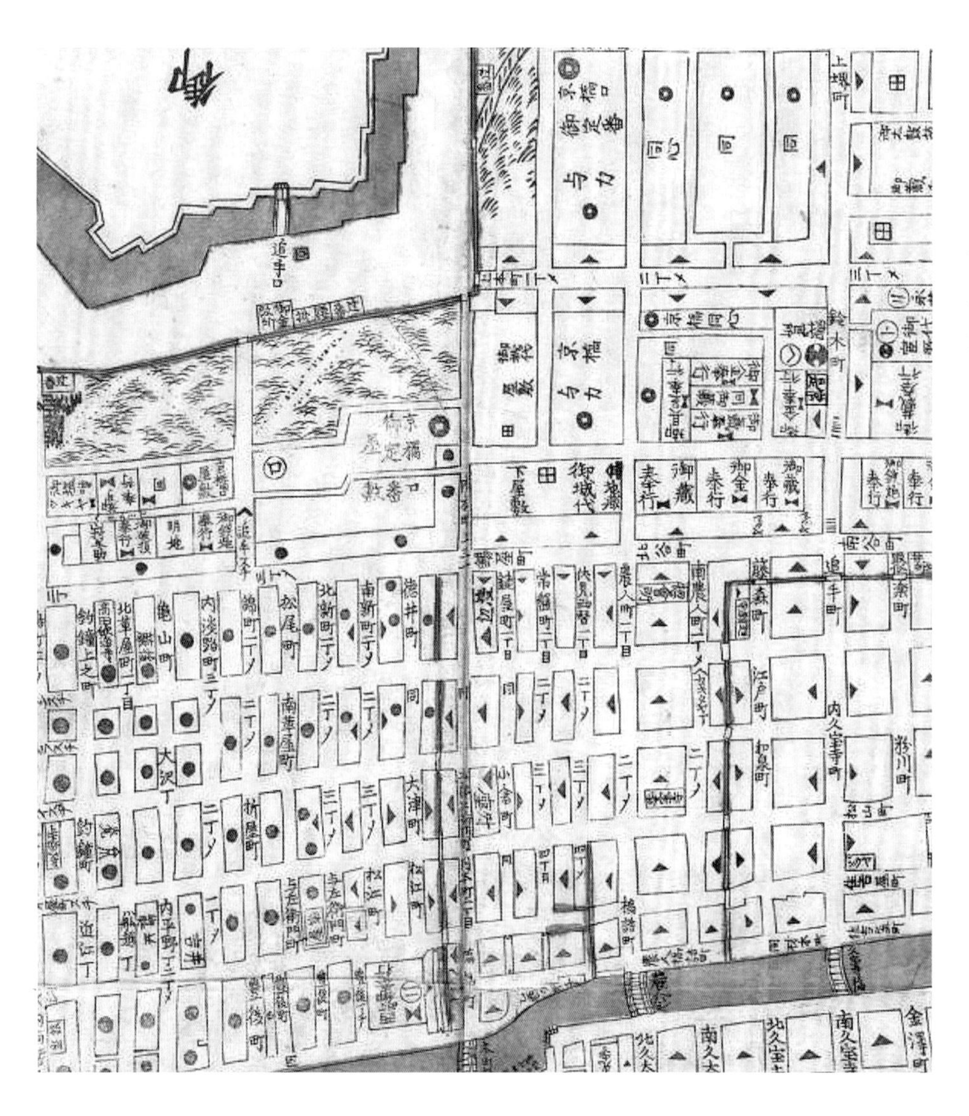

FIGURE 19.2 Detail from "Revised Map of Osaka, Settsu Province" [Zōshū kaisei Sesshū Osaka chizu 増脩改正摂州大阪地図], by Okada Gyokuzan 岡田玉山, 1806. Woodblock print, 153 × 142 cm. Courtesy of the C. V. Starr East Asian Library, University of California, Berkeley.

leadership, with the construction in the northwest of a mighty castle compound and nearby quarters for the martial elite. Commoners concentrated originally in the commercial zone laid out just west of the military precinct, where they were assembled into two-sided neighborhoods for purposes of official surveillance. But as Osaka became the national hub of commodity shipping, and a multiplying population approached four hundred thousand by 1700, commoners spread throughout the expansive "new lands" cleared and built up within the network of the city's wealth-giving canals.

And everywhere they went, neighborhood creation followed. Figure 19.2 is taken from an exceptionally detailed and accurate map (considered a masterpiece of legibility by cartographic scholars) that shows the mature city in 1806. Like seventeenth-century models, the map marks military offices and residences with both labels and icons. The Buddhist temples zoned in the southeast bear individual names, as do the tinted canals and pictorial bridges. While a very few streets along the (secondary) north-south axis bear names as well, those along the (primary) east-west axis do not. Instead, the labels cramming the map are the names of the two-sided neighborhoods—620 of them, according to a colophon—that were organized into three larger collectivities identified by black circles, black triangles, and white triangles.

Six hundred twenty neighborhood names? Why not substitute, or at least add, street names? After all, commercial interests prevailed in this "merchant capital of Japan," where tens of thousands of daily clients, many from afar, could do with navigational help.

Key, clearly, was so insistent an association between streets and social formation that rival naming practices gained little traction. This coincidence depended, in part, on the continued centrality of neighborhoods to Osaka's military administrators, who never ceased relying on neighborhood leaders to announce and enforce their many, many regulations. (The extant fraction of official directives from the eighteenth century alone numbers over ten thousand.)[2] More important was the voluntary extension of neighborhood governance into most aspects of communal life and a resulting convergence between neighborhood membership and social identity. Over time, internally selected officers (including councilors, officiating elders, deputies, monthly functionaries, and various servitors) developed portfolios as copious as their staffs:[3]

They undertook essential policing, especially at night, when they closed and guarded the gates erected at intersections.

They organized fire-prevention and fire-fighting protocols.

They maintained local streets and bridges; they cleared debris from the waterways.

They authorized property sales and rentals, routinely restricting access to enterprises that were hazardous because of fire (bathhouses and blacksmiths, for example) or raucous (bars and samisen schools) or just too dependent on heavy but low-profit traffic (dealers in firewood and charcoal).

They recognized marriages, adoptions, succession decisions, and wills.

They assigned and collected the levies due both the military magistrate and the neighborhood itself (which were several times heavier).

They stipulated the "ritual" fees payable upon property transactions, marriages, adoptions, comings-of-age ceremonies, and appointments of surrogate householders.

They made loans to needy residents, kept obligatory records for diverse purposes (including religious registration), and staged major seasonal celebrations (most splendidly at the New Year and Feast of All Souls).

This startling list reminds us, I think, that cities are places of constant migration and anonymity and, consequently, places of suspicion, loneliness, ambiguous morality, and latent lawlessness. Complex institutional services can help relieve the fear and abate disorder, but most early modern urbanites had to look to themselves, if anywhere, for the resources to fashion those services. Across the urban spectrum in Japan, commoners found the necessary organization, labor, and revenues in their neighborhoods—a response notable in both the intimacy of its focus and the ambition of its reach.

There were human costs. Neighborhood associations put residents under heavy surveillance and made fine hierarchical distinctions among them. Membership in the governing council and eligibility for office, for example, devolved solely on property owners. And holders of frontage properties, renters and owners alike, outranked renters in the longhouses that typically crowded the rear (and drove neighborhood populations as high as five hundred to seven hundred persons). Administrative burdens grew sufficiently onerous, moreover, to dispose wealthy commoners to rent rather than buy property and, hence, disqualify themselves for appointment as (unsalaried) neighborhood elders. Additional disadvantages seem likely but hard to prove—from braking commoner mobility and entrepreneurial innovation to foiling the centralization of urban services at higher levels.

Even so, neighborhood associations apparently provided enough advantages to keep them going for an extravagantly long time. They delivered indispensable services. They maintained acceptable levels of safety and order. They offered sociability, a ceremonial life, and frameworks of trust. They also accorded remarkable governing continuity, not least because some offices became hereditary. It was neighborhood administration, not incidentally, that helped preserve stability during the long decades of upheaval attending the collapse of the Tokugawa shogunate and the construction of the Meiji state.

My mission, however, is not to pass judgment on neighborhood formation but to elucidate its primacy in the spatial order of early modern Japan. Neighborhood names mattered because they defined urban attachment. As vital as the commercial traffic of strangers surely was, the public character of streets ceded to communal identity in the fundamental practice of naming.

So, too, today. But if scholars have helped us understand the importance of neighborhoods well into the twentieth century, the persisting scarcity of street names—in giant metropolises with skyscraper apartment blocks and swarms of outsiders—baffles not just Monsieur Barthes. I would predict an increase in street names and a slow vanishing of neighborhoods. Except for a revelatory development at the shelters that housed refugees from the Fukushima disaster. The survivors who gathered in those large school gyms and auditoriums organized themselves into *chō*, not so much "two-sided neighborhoods" as equivalent communities of succor, support, and, again, copious responsibility (from draining fuel from damaged cars to massaging the shoulders of elders in pain). The inertia of history was hardly the point. Ascendant once more was the spatial logic of social attachment.

Notes

1. Roland Barthes, *Empire of Signs,* trans. Richard Howard (New York: Hill and Wang, [1982] 1983), 33.

2. Kuroha Hyōjirō, ed., *Ōsaka machi bugyōsho ofuregaki sōmokuroku* (Osaka: Seibundō Shuppan, 1974).

3. These and the following remarks rely on Miyamoto Mataji, *Kinsei Ōsaka no keizai to chōsei* (Tokyo: Bunken Shuppan, 1985); Shinshū Ōsaka Shishi Hensan Iinkai, ed., *Shinshū Ōsaka shishi*, vol. 3 (Osaka: Osaka City, 1989); and Ōsaka Shishi Hensanjo, ed., *Ōsaka no machi shikimoku* (Osaka: Osaka City, 1991).

Suggested Readings

Berry, Mary Elizabeth. *The Culture of Civil War in Kyoto*. Berkeley: University of California Press, 1994.

Bestor, Theodore C. *Neighborhood Tokyo*. Stanford: Stanford University Press, 1989.

Hauser, William B. "Osaka Castle and Tokugawa Authority in Western Japan." In *The Bakufu in Japanese History,* edited by Jeffrey P. Mass and William B. Hauser, 153–72. Stanford: Stanford University Press, 1985.

20 Locating Japan in a Buddhist World

D. Max Moerman

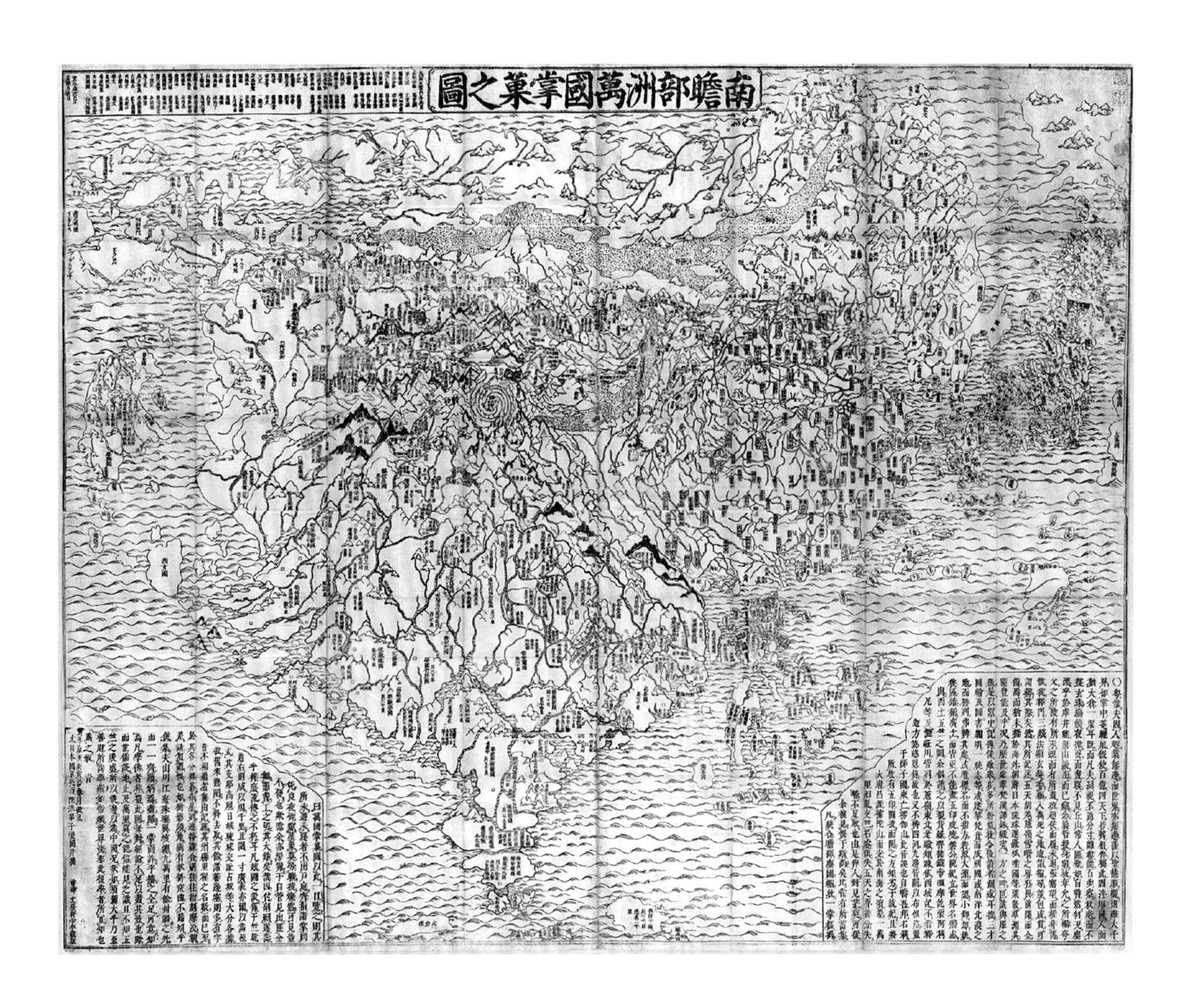

FIGURE 20.1 "Map of the Myriad Countries of Jambudvīpa" [Nansenbushū bankoku shōka no zu 南胆部州万国掌菓之図], by Hōtan 鳳潭, 1710. Woodblock print, 115 × 145 cm. Courtesy of the C. V. Starr East Asian Library, University of California, Berkeley.

The place of Japan within the Buddhist world was a central concern in both cartographic and Buddhist discourse as early as the eighth century. Yet the nature of Japan's position within this realm was neither singular nor final. From at least the twelfth century, Japan was seen as at once marginal and central, simultaneously a "peripheral country as small as a millet grain" and a "great kingdom within the Buddhist world." The unsettled status of Japan's position relative to India, to China, and, in later periods, to Europe and the Americas as well persisted as a source of creative tension. The definition and defense of this remote archipelago as an integral part of the Buddhist realm would remain a fundamental task for Japanese Buddhist discourse into the late nineteenth century (as discussed in chap. 24). Maps reveal the ways in which that discourse was articulated visually and verbally as well as the role of cartography in asserting and contesting religious worldviews.

The "Map of the Myriad Countries of Jambudvīpa" (fig. 20.1) was drawn in 1709 by the monk Hōtan, a leading Buddhist intellectual of the day, and published in Kyoto the following year. The Japanese title of the map speaks to the relations between Buddhist cosmology and world cartography that its author had to negotiate. *Nansenbushū*, the Japanese translation of the Sanskrit *Jambudvīpa*, refers to the classical Indian view of the world as consisting of four great continents surrounding Sumeru, the cosmic mountain at the center of a flat earth. *Bankoku*, by contrast, literally "ten thousand lands," was a familiar term taken from the title of European-style world maps, which were printed in large numbers in Japan throughout the late seventeenth century. Hōtan's map was indeed a work of cartographic bricolage. Combining elements from Indian, Chinese, and European traditions, it represented a Buddhist view of the world that was designed to combat the competing claims of Europeanist, Neo-Confucian, and Nativist cosmologies. And it was evidently a success. Appearing some sixty-five years after the first European-style world maps were published in Japan, the "Map of the Myriad Countries of Jambudvīpa" remained in print for over one hundred years and was also copied and issued in various simplified and reduced editions through the late nineteenth century. The persistent presence of this Buddhist world map forces us to recognize Japan's cartographic pluralism as well as the continuing role of the Buddhist geographic imaginary in the construction of early modern and modern Japanese worldviews.

Although produced in the early eighteenth century, Hōtan's map reaches far into the past and draws on multiple traditions. The upper margins of the sheet list the titles of over one hundred works consulted: Buddhist sutras, commentaries, monastic regulations, histories, hagiographies, encyclopedias and dictionaries, Chinese dynastic histories and gazetteers, geographic compendia, encyclopedias, and literary works. Yet the primary sources of Hōtan's cartography remain unacknowledged even within so exhaustive a bibliography. The image derives ultimately from a depiction of the Buddhist world based on the journey of Xuanzang (600–664), the famous Chinese pilgrim whose travels through India and Central Asia were recorded in his *Great Tang Record of the Western Regions*. A fourteenth-century painting of Xuanzang's text, with a red line marking his itinerary within an ovoid continent labeled *Nansenbushū*, is the earliest Japanese world map. It includes every placename mentioned in Xuanzang's account, as well as their size and the distance between them. For five hundred years, this early world map was faithfully reproduced by other monks as a devotional object and a ritual means of reenacting Xuanzang's journey. The toponyms and topographic data that Hōtan lists for India and Central Asia are identical to those of the fourteenth-century painting. Perhaps the most distinctive detail of the prototype to be copied by Hōtan is the square lake of Anavatapta at its center, where four great rivers emerge from the mouths of four legendary beasts, wind around their source, and flow to the four corners of the continent. The only element of this early map that is not drawn from the record of the Chinese pilgrim is the Japanese archipelago, which Hōtan has expanded, in both scale and detail, based on other seventeenth-century maps of Japan.

In the final decade of the seventeenth century, however, a new Buddhist world map that combined the sacred geography of Xuanzang's pilgrimage with more recent cartographic sources appeared in manuscript. To the classical shape of Jambudvīpa it added the continental and peninsular forms from a Ming encyclopedia, the shape of Europe from Japanese editions of Matteo Ricci's maps, and the names and distances of European and Southeast Asian countries from still another source, Nishikawa Joken's *Thoughts on Trade and Communication with the Civilized and the Barbarian*. Hōtan relied heavily on this late seventeenth-century manuscript, giving form to Nishikawa's notion of Europe as an archipelago consisting of France, England, Hungary, Holland, Iceland, Scotland, Norway, Denmark, Russia, Poland, Turkey, Albania, Greece, and Italy. With Europe thus depicted as a cluster of small islands at the western periphery, while South America was sketched as a minor archipelago in the southeastern seas and North America remained unnamed and barely visible in the northeastern margins, Hōtan's Indo-centrism provincializes Europe and the Americas while aggrandizing Japan.

However exhaustive his sources, Hōtan was less concerned with contributing to a cumulative body of cartographic data than with asserting a Buddhist view of the world. His title, which might be translated literally as "A Map of the Myriad Countries of Jambudvīpa Like a Fruit Held in the Hand," uses a religious vocabulary to evoke not only a Buddhist cosmology (*Nansenbushū*) but a Buddhist vision. The phrase "fruit held in the hand" is a Buddhist term for something that is easy to see and understand. Hōtan expands on this image in the map's preface, where he writes, "The wisdom eye of the sage is far more powerful that the human eye and sees the boundless ten-thousand-fold world just like a fruit held in one's hand." He then goes on to describe the qualitative difference between Buddhist insight and ordinary human vision:

> There are innumerable realms, as countless as the leaves of mustard grass, beneath the four heavens. Our realm of Jambudvīpa is like a single grain within a great storehouse of millet. The ordinary person can see no more of the world than someone inside a cave peering through a tiny hole. Not even Lizhu, who could see something as minute as a single hair from the distance of a hundred paces, had such vision. Human vision is as limited as that of the horned owl who can catch a flea at night but cannot see a hill at midday. The vision of an ordinary person is as far from the vision of the "wisdom eye" as that of a blind person is from the sighted. He can say nothing of the worlds as numerous as atoms. He is like a frog in a well discussing the vast oceans.

Hōtan's map is thus as much about epistemology as it is about cartography. His argument, that the view of the world from the perspective of Buddhist wisdom is qualitatively different from that of human vision, relies on the classical Buddhist understanding of vision as a root metaphor for knowledge. The "wisdom eye" refers to the power of vision that correctly understands the world and is a technical term for one of the five levels of vision within the traditional Buddhist classification of discriminative knowledge. To this religious vocabulary Hōtan adds allusions to classical Chinese thought, invoking a mythic exemplar celebrated for his power of vision and citing allegories from the *Zhuangzi* to illustrate the comparative limitations of conventional perception. Hōtan concludes his preface with a vindication of Buddhist cartography. For Hōtan, cartography is essential not only for scholars but also for the very survival of Buddhism within an increasingly competitive marketplace of ideas.

> If Buddhist scholars do not examine this map when they consult the sutras, their investigations will be incomplete. Confucian scholars have debated geography and discussed distances for generations. [If Buddhists do not also pursue cartographic studies] our knowledge will be as insufficient as that of a frog in a well. We must seek as much understanding of distant lands as we do of our own, and even more so of Mount Sumeru at the center of the universe and the vast trichiliocosm itself.

The production, popularization, and persistence of Hōtan's "Map of the Myriad Countries of Jambudvīpa" reveals the history of Japanese cartography to be neither empirical nor linear. It presents a challenge to a positivist model of intellectual history in which earlier forms of knowledge and modes of representation are cast aside with the advance of scholarship and scientific accuracy. With its origins in a fourteenth-century image of a seventh-century pilgrimage to India, Hōtan's Buddhist view of the world was reproduced in popular Japanese encyclopedias throughout the eighteenth and nineteenth centuries. And as Sayoko Sakakibara's essay (chap. 24) shows, it continued to inform Japanese cartographic, print, and visual culture long after its initial publication. The fact that thousands of Buddhist world maps, based on ancient Indian and Chinese scripture rather than recent European science, were printed and purchased well into the Meiji period calls into question the standard narratives of Japanese modernity. Hōtan's map reveals not only the intellectual and cartographic pluralism of Tokugawa Japan but also the instrumental role of cartography in asserting and contesting ideological claims.

Suggested Readings

Moerman, D. Max "Demonology and Eroticism: Islands of Women in the Japanese Buddhist Imagination." *Japanese Journal of Religious Studies* 36 (2) (2009): 351–80.

Muroga Nobuo and Unno Kazutaka. "Nihon ni okonawareta Bukkyō kei sekaizu ni tsuite." *Chirigakushi kenkyū* 1 (1957): 64–141.

———. "Edo jidai kōki ni okeru Bukkyō kei sekaizu." *Chirigakushi*

kenkyū 2 (1962): 135–229.
———. "The Buddhist World Map and Its Contact with European Maps." *Imago Mundi* 16 (1962): 49–69.

21 Picturing Maps: The "Rare and Wondrous" Bird's-Eye Views of Kuwagata Keisai

Henry D. Smith II

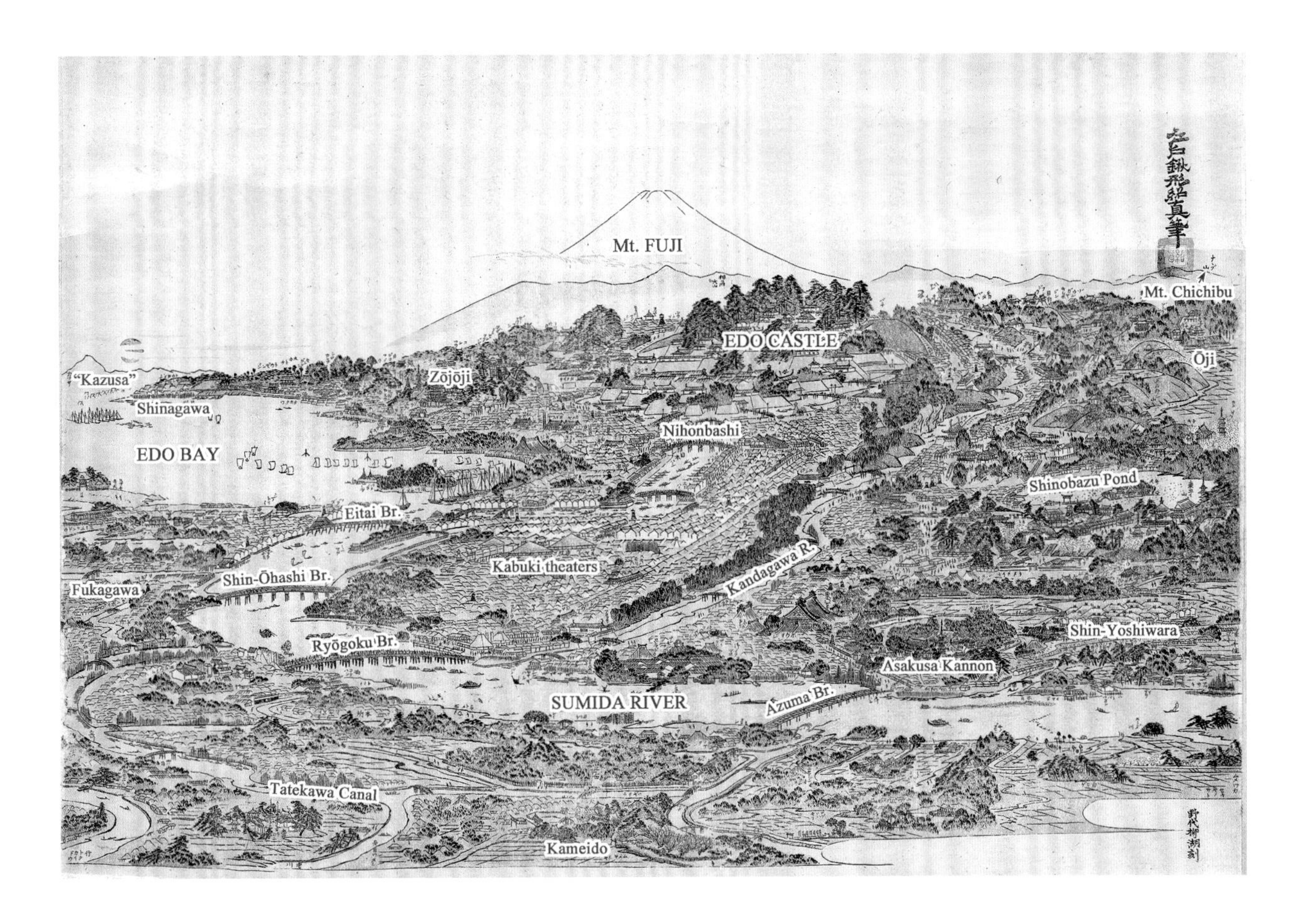

FIGURE 21.1 "A Picture of the Famous Places of Edo" [Edo meisho no e 江戸名所の絵], by Kuwagata Keisai 鍬形蕙斎, 1803. Woodblock print, 42.1 × 58 cm. Signed "Kuwagata Shōshin of Edo (江戸鍬形紹真)"; seal "Shōshin"; carved by Noshiro Ryūko 野代柳湖. Courtesy of the C. V. Starr East Asian Library, University of California, Berkeley. English labels added by author.

In the autumn of 1803, Tokugawa officials in charge of licensing maps gave permission to three mapmakers to produce and sell a color woodblock print by Kuwagata Keisai (1764–1820) entitled "A Picture of the Famous Places of Edo" (fig. 21.1). The promotional blurb on the wrapper in which the folded map was sold was more expressive than the bland title:

> Master Keisai has contrived anew to produce this unusual view of the bustling scenic and historic sites of the capital of Edo, spread out from a single viewpoint. It truly offers the delights of wandering from one place to another, and is suitable for framing or mounting as a scroll, or for presentation as a timely gift or souvenir. We offer for your display this most rare and wondrous view.

This "rare and wondrous view" raises a pivotal issue in the history of Japanese mapmaking: the distinction between a "map"—as the print was officially handled—and a "picture"—as the image presents itself to our eyes. Maps throughout world history have commonly incorporated pictorial elements with plans that rely on outlines, words, and symbols. Indeed, the most common word for maps in premodern Japan was a compound of terms for "picture" and "chart" (*e-zu*). The Keisai view, however, together with a sequel view of all Japan (fig. 21.2), reveals a relationship between maps and pictures that brings the very distinction into question. From where did this conception emerge, and what was new about it?

The print features a view of Edo looking west-southwest, toward Mount Fuji (which rises majestically above the city), with the Tanzawa range and the peak of Mount Ōyama just below. The spacious grounds of Edo Castle lie right of the central axis, and directly below this are the two bridges, Nihonbashi and Edobashi, that lay at the center of the city. Along the lower third of the print, the great Sumida River flows right to left, southward into Edo Bay, spanned by four bridges. The low-lying area east of the Sumida is laced by a system of canals.

The geography is in most ways plausible if not particularly true to the kind of scale used on the detailed single-sheet printed maps of the city available in this era. Keisai certainly had such a map close at hand when he constructed his view, consulting it often to plan the layout. Still, it is immediately apparent that although the major sites are properly connected, the artist indulged in many willful manipulations of both scale and direction. This was purely practical, since, as advertised, it is a "picture" of the "scenic and historic sites of the capital" that "offers the delights of wandering from one place to another."

The emphasis on each separate "famous place" yields numerous charming vignettes, often in considerable detail. These reveal iconic features of each place that had long been established in countless paintings, books, and single-sheet color prints. Further visual interest and vitality is provided by over five hundred tiny figures walking to and fro, only two to three millimeters high but recognizable as clear human forms; sometimes, you can even make out the swords worn by samurai. At the same time, map-like textual information is provided in the 270 inscriptions of place-names, shrines and temples, and other geographic features, a valuable source for identification of lesser places.[1]

Drawing back again to the overall view, we can see still more manipulation, particularly for symbolic and expressive intent. Most obvious is Mount Fuji, which was famously visible on the far horizon from many points in the city on a clear day. While its distance of some one hundred kilometers from the capital made its apparent height quite small, painters had long compensated both by exaggerating the mountain's size and by simplifying its form, as we clearly see here.

Other spatial manipulations in which Keisai indulged were less conventional. Most of the canal network east of the Sumida, for example, lay on a rectilinear grid oriented to the cardinal directions, but the artist radically transformed it into a curvaceous pattern. The effect is to naturalize the landscape by making it appear less man-made, even bucolic. Most provocative, however, is the inclusion on the far left of the morning sun rising over mountains labeled "Kazusa" (the province in the center of the Bōsō Peninsula, across Tokyo Bay to the south-east). This leftward spread of over ninety degrees beyond the central axis toward Mount Fuji is a big geographic stretch, but it enables a powerful symbolic statement of Edo as the capital of Japan, land of the rising sun, and of the auspicious prosperity that it had come to suggest—particularly when seen at the New Year.

The roots of Keisai's vision are complex, and only a brief outline is possible here. Two logical precedents may be ruled out: first, the great screen paintings of Kyoto known as "Views In and Out of the Capital" that appeared in the sixteenth century, and second, the European city views copied onto screens in the same era (for which fig. 1.1 above provides a spectacular example). More such views were available in books imported by the Dutch in Keisai's time, but no direct influence is apparent.

Western perspective did, however, strongly influence Keisai's way of seeing. The "perspective pictures" (*uki-e*) that appeared in Edo about 1740—using single-point Western perspective to create a sense of dramatic recession—were a Western genre in which Keisai, who began his career as an ukiyo-e designer in the 1780s, became adept. He underwent a dramatic career change in 1794, when he was appointed an official painter by the lord of the Tsuyama domain in western Japan. This rise in status probably led him to a second lineage of Western perspective derived from the innovative painter Maruyama Ōkyo (1733–95). Ōkyo's panoramic painting of the city of Kyoto in 1791—featuring a rising sun to the east and translating rectilinear streets into elegant curves—may well have influenced Keisai's Edo view.

Whatever the specific inspiration, Keisai's view of Edo became widely known and replicated. Perhaps its most important legacy was a second effort by Keisai at pictorial mapping, this time escalating the scale by a full order of magnitude to the entire Japanese archipelago (fig. 21.2). No documentary evidence survives to date the print, but it must have come closely on the heels of the Edo view, which was so popular that a second edition was published in spring 1804, just months after the first.

Just as East Asian written texts read from right to left, so Keisai's distinctive and ingenious projection of the archipelago of Japan begins from his bold signature at top right and moves clockwise. The eye is led from the southern tip of Hokkaido (Ezo), down through the precipitous mountains of Tohoku, across central Japan (with the pristine form of Fuji centered near the bottom and Edo Bay to its lower right), then curving upward through Kyushu, where a tiny Dutch ship flying long pennants marks the approach to Nagasaki. Above, the islands northwest of Kyushu recede in the distance toward the most striking feature of the composition: a long horizon into which a waning moon sinks through bands of clouds, as the silhouette of Korea to the right identifies Japan's immediate neighbor on the Asian continent.

Even more than with his bird's-eye view of Edo, Keisai must have constantly consulted a detailed map, for he carefully recorded an immense amount of information: each of the sixty-eight provinces is identified, and scattered throughout the map are over seven hundred minutely inscribed place-names. About 150 castle towns are indicated by an icon of a tiny castle tower, which co-

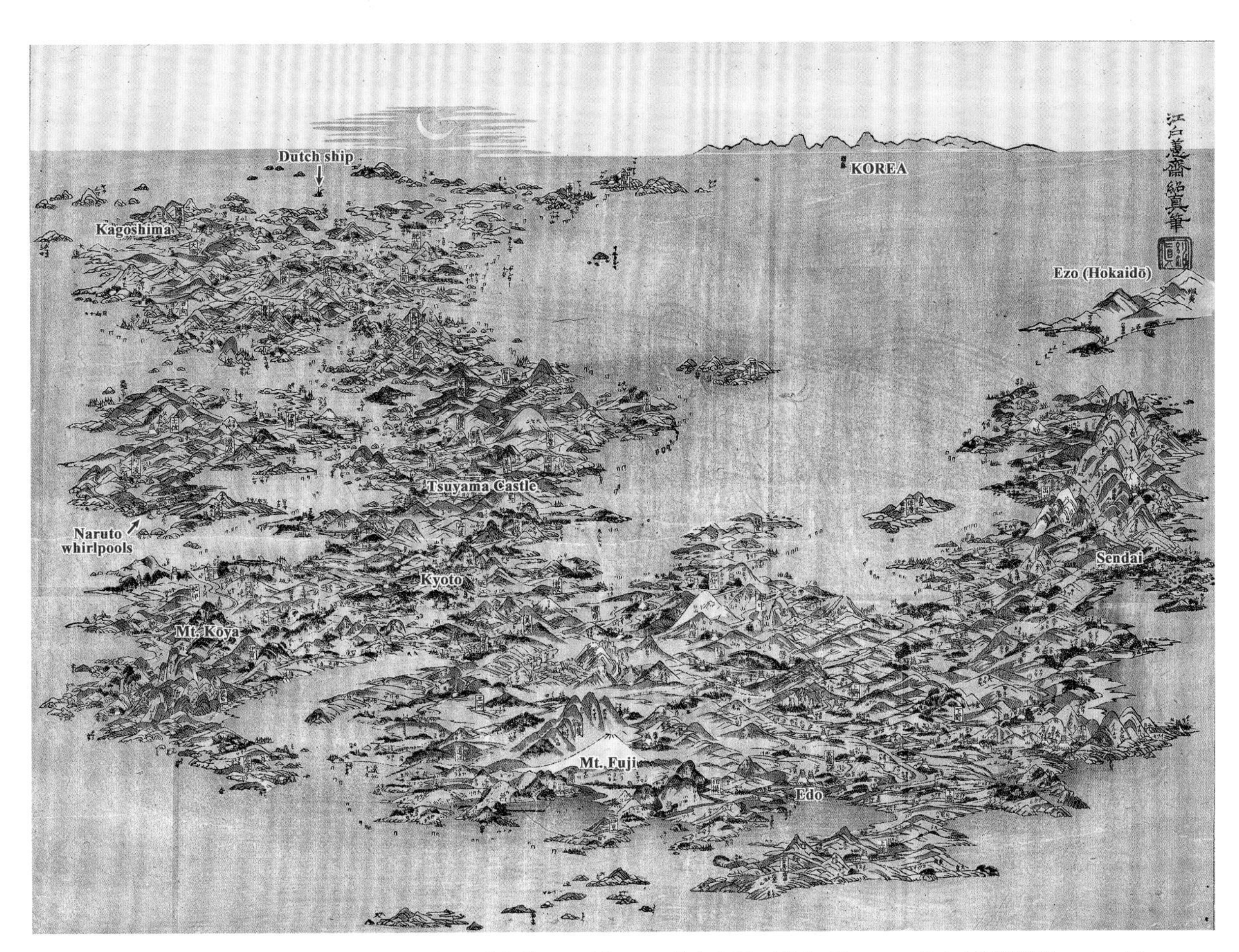

FIGURE 21.2 "A Picture of the Famous Places of Japan" [Nihon meisho no e 日本名所の絵], by Kuwagata Keisai 鍬形蕙斎, undated (ca. 1804). Woodblock print, 41.6 × 56.1 cm. Signed "Kuwagata Shōshin of Edo (江戸鍬形紹真)"; seal "Shōshin." Courtesy of the C. V. Starr East Asian Library, University of California, Berkeley. English labels added by author.

FIGURE 21.3 "Clear View of Japan," April 30, 1999. Satellite image. Cropped and rotated 78 degrees clockwise for a level horizon. Provided by the SeaWiFS Project, NASA/Goddard Space Flight Center, and ORBIMAGE.

incides roughly with the number of daimyo known as "castle-owners" (*jōshu*, who had proper castles, in contrast to the 100-odd lesser lords who lived in ordinary mansions), and conveys an effective pictorial sense of a land under pervasive military rule. His greatest challenge was to give the entirety of Japan a sense of both spatial and pictorial continuity.

Keisai's attempt at a unified and homogenous picture of Japan from on high was paralleled by attention to the discrete "famous places" of the title. A few of these are charmingly depicted, with unique features such as the whirlpools of Naruto, the temple complex of Mount Kōya, and the famous red *torii* of Miyajima. But there are only a dozen such examples of any real visual interest, and even those lack the tiny human figures that so animate the view of Edo.

Whatever it lacks in delicacy of pictorial detail, however, Keisai's view of Japan more than compensates with the very daring of the overall conception. I have searched widely but can find no example elsewhere in the early modern world of any attempt by any artist to show an entire nation as if actually seen from such a distance. A comparison of Keisai's work with a satellite photo of Japan almost two centuries later (fig. 21.3) suggests how long it took for science and technology to catch up with his imagination.

Where did Keisai get the idea? I believe the answer is simple: inspired by his inventive inclusion of the rising sun from the east in his view of Edo to the west, he depicted a view of Japan *as seen by the rising sun*. In an age when no man could aspire to distant views from any point higher than a mountain peak, only a heavenly body could provide the inspiration. And in the land of the rising sun, it could only be a "sun's-eye view," here working in poetic concert with a setting moon.

Note

1. For high-resolution images that provide access to the fine details of both prints described in this article, see the website "Japanese Historical Maps" from the East Asian Library, University of California, Berkeley, from David Rumsey's Cartography Associates: http://www.davidrumsey.com/japan/. Search for the artist under the name Kitao Masayoshi.

Suggested Readings

Screech, Timon. *The Lens within the Heart: The Western Scientific Gaze and Popular Imagery in Later Edo Japan.* Honolulu: University of Hawai'i Press, 2002.

Smith, Henry D., II. *Hiroshige, One Hundred Famous Views of Edo.* New York: George Braziller, 1986.

———. "World without Walls: Kuwagata Keisai's Panoramic Vision of Japan." In *Japan and the World: Essays on Japanese History and Politics in Honour of Ishida Takeshi,* edited by Gail Lee Bernstein and Fukui Haruhiro, 3–19. London: Macmillan Press, 1988.

22 An Artist's Rendering of the Divine Mount Fuji

Miyazaki Fumiko 宮崎ふみ子

FIGURE 22.1 "Portrayal of Mount Fuji" [Fujisan no zu 富士山之図], by Utagawa Sadahide 歌川貞秀, preface by Sawaguchi Seiō 澤口正應, 1848. Woodblock print, 91.4 × 96.5 cm. Courtesy of the University of British Columbia Library, George H. Beans Collection. English labels added by author.

Mount Fuji, with its enormous size and conic shape, is a notable landmark in Japan. Over the years it has inspired poets, novelists, and painters, and fascinated many with its mysterious beauty. Simultaneously a rich source of water and a dangerous volcano, the mountain also has long been an object of worship. Nearby communities long ago enshrined the female deity of the volcano, Sengen, on its flanks. The mountain also attracted devotees and pilgrims from afar, who revered it as either the incarnation or the residence of a divinity. Reflecting these different perspectives and interests, many kinds of images of the mountain have been produced since the late Heian period. Among them all, the panoramic portrayal of the mountain introduced here is certainly one of the most striking.

This nineteenth-century panorama (fig. 22.1) is a large multicolored woodblock print, about one meter square, based on a drawing by the painter Utagawa Sadahide (1807–78?). The print, which features no information regarding its publisher, was apparently published in 1848 or shortly thereafter.[1] Given that it also does not have a fixed title, in this essay I will call it "Portrayal of Mount Fuji" [*Fujisan no zu*], following the tentative Japanese title given by the University of British Columbia Library.[2] The bulk of the portrayal is similar to what we now call a map, depicting a large area covered by Mount Fuji viewed from above. "North," "south," "east," and "west" are written inside a circle drawn in the center of the sheet, which represents the crater. The names of sites and landscape features are given in rectangular cartouches.

Since Utagawa Sadahide had previously collaborated with others on drawing provincial maps, he certainly possessed knowledge about map representations.[3] Yet in some ways the image clearly differs from maps familiar to us. Buildings, rocks, waterfalls, and trees are shown pictorially rather than by abstract symbols. Furthermore, the roofs of buildings and tops of trees stretch toward the summit, located in the center of the sheet. In this sense the print is a sort of a bird's-eye view, depicted from multiple viewpoints facing toward the peak. These features put the "Portrayal of Mount Fuji" in an intriguing border zone between "pictures" and "maps."[4]

What is still more unusual is the artist's way of presenting the body of the mountain. Traditionally painters and cartographers focused on only one side of Mount Fuji at a time, depicting it in the shape of an isosceles triangle (for an example, see fig. 26.1 below). A newer development in the mid-nineteenth century was to depict Mount Fuji as a complete circle viewed directly from above.[5] The depiction of the body of Mount Fuji in this portrayal differs from either of these approaches. The mountain is presented here in the shape of an open fan, with a span of three hundred degrees. This novel view of the mountain's conic body is evidently driven by a desire to depict a three-dimensional object on a flat sheet of paper. The painter's interests in three-dimensional representation can be seen in a smaller detail, too. An irregularly shaped small piece of paper, on which rocky peaks are illustrated, is folded and attached to the circle in the center of the sheet (1). When one pulls it open, the peaks of the outer rim rise up to surround the part of the map that depicts the crater.

Another unusual feature of "Portrayal of Mount Fuji" is that it depicts not only what is visible on the outer surface but also what cannot be seen but is supposed to exist inside the mountain. In the triangular space outside the fan-shaped picture-map of the mountain are images of two figures sitting in caves: Kakugyō Tōbutsu (?–1646) on the right and Jikigyō Miroku (1671–1733) on the left (2). Both men were founding figures revered by the Mount Fuji cult. Kakugyō was the progenitor of a lineage of lay ascetics dedicated to Sengen, the deity of the mountain; Jikigyō was a lay ascetic belonging to Kakugyō's lineage, who developed teachings about the ideal world based on an oracle received from Sengen. Praying for the realization of the ideal world and the salvation of all people, Jikigyō conducted a fast to the death in the mountain. After his death, his followers formed groups of Mount Fuji devotees, giving rise to one of the largest popular religious organizations of the Tokugawa period, known as Fujikō.

Another unusual feature of this map is a monochrome picture covered by a flap, in the form of a small rectangular piece of paper, on which trees and clouds are drawn (3). When one raises the flap, a glimpse of a cave network appears (4). This is one of many caves close to Mount Fuji's surface that were formed by streams of lava. The particular one shown in the picture was located close to Yoshida, the gateway to the northern climbing route, and was a major pilgrimage site at the time. The faithful believed that the cave represented a human womb, and that one could experience rebirth by passing through it. The painter recorded his name inside this womb-cave with the pious declaration, "I humbly reproduce the true features of the mountain."

"Portrayal of Mount Fuji" includes other images separate from the mountain—a stone lodge for pilgrims (5), pilgrims dressed for climbing (6), and a huge torch featured in the famous fire festival of Sengen Shrine at Yoshida (7). There are also Japanese and Chinese po-

FIGURE 22.2 "Complete Portrayal of the True Features of Mount Fuji" [Fujisan shinkei zenzu 富士山真景全図], by Utagawa Sadahide 歌川貞秀. Folded replica, published by the Kanagawa Prefectural Museum of Cultural History, 41.7 × 44.2 cm. Photograph courtesy of Kohga Communication Products Inc., Yokohama, Japan.

ems, as well as passages copied out from topographic books. The Japanese poems written in the central part are believed to have been composed by Jikigyō and were chanted at Fujikō religious rites (8). Quotations from literary works by well-known contemporary and ancient authors are recorded in the margins (9). In its overall design, the portrayal emphasizes places and features connected to Fujikō and the Yoshida entrance, while also depicting various other matters related to Mount Fuji.

How were prints of this map made and distributed? A clue is found in the introduction to a book called *Maps of Japan Divided into Regions*, which the same painter, Sadahide, published in 1855. Here he wrote that he had summited Mount Fuji several times and sketched what he had seen there, and that the blocks for a large panoramic picture-map based on his sketches had been carved some years earlier and were currently held by the *oshi* at the Yoshida entrance. From this statement we can guess that one or several religious figures called the Yoshida-entrance-based *oshi* were engaged in printing and distributing the "Portrayal of Mount Fuji."

The *oshi* of Yoshida were originally ritual performers who spread the cult of Sengen as they journeyed through the eastern provinces. Their travels took them through many villages and towns, where they performed rituals for lay believers, distributed amulets, and collected offerings. In the summer climbing season they received affiliated pilgrims in Yoshida, offered them lodgings, held purification rites for them, helped them in various ways to safely climb the mountain, and collected offerings and fees. It is highly possible that the *oshi* made use of this image as a means to attract Fujikō believers and encourage them to make the pilgrimage to the mountain. The detailed depiction of the Yoshida entrance and Yoshida Sengen Shrine, as well as the *oshi* offices there (10), would have allowed the *oshi* to use a print like this to explain the sacred sites affiliated with the cult, particularly the ones accessible from the Yoshida entrance.

If so, what was the most effective way to display the "Portrayal of Mount Fuji"? My assumption is that the *oshi* developed it into a portable three-dimensional model of the sacred mountain, transforming the flat image into a three-dimensional cone. Most likely, they folded the paper along the two edges of the fan-shaped picture-map—which the painter-cartographer indicated as representing both sides of Ōsawa Valley (11, 12)—since there is in fact a long, deep erosional valley running from the summit down the western flank of the real Mount Fuji. Folded in such a way, the flat sheet becomes a conic body with the crater at the top, and what initially appeared to be two caves, each with one ascetic inside it, are fused into a single cave holding two holy men (fig. 22.2).[6]

Assuming that "Portrayal of Mount Fuji" was intended to be folded to produce a three-dimensional body, we can understand why the painter drew the images of the two great ascetics where he did. In the three-dimensional model, the images of Kakugyō and Jikigyō are no longer exposed on the surface but hidden inside the crease of the paper. Seeing these revered holy men inside the innermost part of the mountain would leave a strong impression on Fujikō believers. Likewise, folding the image into a cone helps us understand why the roofs of buildings and tops of trees around the crater stretch toward the center of the sheet. Once the mountain is raised up in this way, the trees and houses appear to be standing on the ground.

The resulting three-dimensional model of Mount Fuji must have been more appealing than a flat woodblock-printed picture-map of the sort that many shrines and temples of the Tokugawa period published for the purpose of propagation. It may have appealed particularly to people who could not go to see the mountain. For most of the Tokugawa period, women were excluded from the middle and higher zones of Fuji because of traditional taboos. Many male devotees of the cult could not climb the mountain either, whether for physical or economic reasons. For such people, the picture-map could function as a medium for imagining more fully the sacred places that they could never see in reality. Although "Portrayal of Mount Fuji" drawn on a flat sheet looks strange at first glance, it turns out to be a brilliant device for portraying a sacred mountain, conveying vital messages to believers from the men who designed and produced the image—both the painter and the *oshi*.

Notes

1. The preface, by Sawaguchi Seiō in the form of a Chinese poem, is written in the upper left corner.

2. Apart from *Fujisan no zu*, or "Portrayal of Mount Fuji"—the title it has been given at the University of British Columbia Library in Canada—the same map is referred to as *Fujisan shinkei zenzu* (Complete portrayal of the true features of Mount Fuji) at Kanagawa Prefectural Museum of Cultural History and Fujisawa City Board of Education, and as *Fujisan-zu* (Portrayal of Mount Fuji) at the University of Tokyo Library in Japan. The author thanks Ms. Katherine Kalsbeek of the University of British Columbia Library and Ms. Donna Kuwayama of Kanagawa Prefectural Museum of Cultural History for giving her a chance to examine their copies.

3. From the late 1840s, Sadahide worked together with a geographer, Tsurumine Shigenobu, to produce maps of the six eastern provinces. Sugimoto Fumiko, "Bōsō no kūkan o egaku," in *Chibaken no rekishi, tsūshihen kinsei* 2, ed. Chiba-ken shiryō kenkyū zaidan (Chiba: Chiba Prefecture, 2008), 682–84.

4. Henry D. Smith II, "World without Walls: Kuwagata Keisai's Panoramic View of Japan," in *Japan and the World: Essays on Japanese History and Politics in Honour of Ishida Takeshi*, ed. Gail Lee Bernstein and Fukui Haruhiko (London: McMillan Press, 1988), 4.

5. The new development is represented by the famous *Fujimi jūsanshū yochi no zenzu* (Map of the thirteen provinces where Mount Fuji can be seen), ed. Funakoshi Seizō, with illustration by Akiyama Nagatoshi, 1843.

6. It is possible that the image of Kakugyō in a cave reminded some viewers of the fact that he practiced asceticism in a cave called *hitoana*, which is depicted near the lower left corner of the map. It is unlikely, however, that the *oshi* of Yoshida suggested that revered ascetics were staying in the cave when they showed this map to the believers. Pilgrims to the cave would probably not come to the Yoshida entrance, which is far away from the cave, but rather go to entrances on the southwestern flank of Mount Fuji, which were under the control of their rivals.

Suggested Readings

Earhart, H. Byron. *Mount Fuji: Icon of Japan*. Columbia: University of South Carolina Press, 2011.

Miyazaki Fumiko. "Female Pilgrims and Mount Fuji: Changing Perspectives on the Exclusion of Women." *Monumenta Nipponica* 60 (3) (2005): 339–91.

23 Rock of Ages: Traces of the Gods in Akita

Anne Walthall

FIGURE 23.1 "Rocks from the Age of the Gods" [Jindai ishi hakkenchi ryakuzu 神代石発見地略図], by Chita Kaji 千田嘉治, 1894. Manuscript in the collection of letters addressed to Masaki Yūsuke [真崎勇助宛書状：落737], 24 × 35 cm. Courtesy of the Akita Prefectural Archive, Akita City. English labels added by author.

In 1894 a man by the name of Chita Kaji sent a map to one Masaki Yūsuke, an antiquarian with a specialty in rocks. In the letter that accompanied the map, Chita explained that after hearing Masaki lecture on his rock collection, Chita had gone home to search for significant stones in his neighborhood. Sure enough, he found some, which he enclosed with his letter, and he drew the map to verify their original location. The rocks he deemed to be from the age of the gods (fig. 23.1).

By 1894, the Meiji government had carried out reforms to centralize administration, build industry, require elementary education for all children, and conscript men for its modern military, but the map shows little sign that these changes had had an impact on Chita's life. Even though the metric system had become the new way to measure distances, for instance, the legend at the top of the map delineates distance in *chō*, a standard of measurement that harked back to the Tokugawa period.

Village names also resonate with earlier times, ignoring the recent administrative mergers. Chita probably knew that these changes had taken place, but he remained more comfortable with the older names.

If an outsider happened upon this map without knowing its origin, she would have no way of ascertaining where it was from. Nothing about the map situates it inside Akita Prefecture, located high on the Japan Sea coast north of Tokyo. There is not even an indication of how far these villages were from the former castle town, now the prefectural capital of Akita city. Since the map was sent from one local resident to another, this lack of a larger context might not seem surprising, but it also suggests that Chita's vision did not extend much beyond his own locality.

Why did Chita bother to insert a map with his letter? It is designed to mark the three places where he had found the rocks: one in a field next to a shrine, one on a hummock, and one on a hill. The drawing melds a sketch of the region, showing topographical features such as trees and mountains, with a map showing the location of the remains of an old fortress, fields, a shrine, an ancient watercourse, and a road, and naming the villages where the rocks were found. By depicting a precise location, Chita aimed at scientific accuracy in locating rocks that defied dating. The letter reflects this mixture of precision and ambiguity, as when Chita describes the shrine by writing, "The place itself is not large, with many zelkova trees forming a dense forest in the midst of which is a small shrine. According to what is carved on a stone stele, it is the shrine for the Hakusan god. Since there are no old records or anything of the sort, I could not discover how many years had elapsed since it had been erected, but given the size of the trees, in my humble opinion it is by nature old."[1]

What made these particular rocks redolent of the age of the gods? Chita's map shows where the rocks were discovered, but they were surely not the only rocks around. Masaki, the map's recipient, sketched the rocks on the right-hand side of the first page of Chita's letter, noting that one was shaped like a dagger with a knob at the top. While the shape was natural, in his judgment the stone was of poor quality. Two were nodular. The third was larger with three sets of indentations. White, banded lightly with pale blue, it was translucent with a fine luster—in other words, a crystal.

Although crystal balls for the telling of fortunes are seldom found in Japan, a crystal's visual qualities make it a fitting object to stand in for the gods. Neither opaque nor transparent, it marks the border between the visible world of people and the invisible supernatural world. The use of bronze mirrors to represent a god's body at Shinto shrines is well known; crystals, too, performed the same function. When people in the Ina Valley of central Japan decided to raise a shrine to the four great teachers of native Japanese history and culture, the adopted son of the last teacher, Hirata Kanetane, sent a blue crystal surrounded by a ring of white ones to represent his famous father, Hirata Atsutane.

Hirata Atsutane had a long-standing connection with Akita. He had been born there in 1776, and after spending most of his life making his name in Edo as a student of the age of the gods, he had returned to Akita as an exile in 1841 (the year that Masaki was born). Before Atsutane died, he acquired rank and status as a domain retainer, a position inherited by Kanetane and his son Nobutane. Nobutane maintained close connections with Akita, corresponding with his grandfather's disciples as well as domain officials and even traveling there in the lord's retinue in 1868. In 1872 the Hirata School had around three hundred disciples in Akita, most of them shrine priests and samurai.[2]

Although neither Chita nor Masaki ever joined the Hirata School, Masaki had ample opportunity to become acquainted with its teachings. Born into a high-ranking samurai family, he spent his childhood in the domain's castle town. At the end of 1867, he visited Nobutane in Edo, just after Nobutane had received a report that Tokugawa Yoshinobu had returned his powers as shogun to the emperor.[3] Following the Meiji Restoration, Masaki became an attendant in his former daimyo's house, where he took charge of preserving old documents. He also made a survey of Akita's geology in search of iron, a valuable commodity in modern Japan. As he aged, he became more attached to recollecting the past by participating in his local historical society. The same collection of letters he received that includes this map also contains correspondence with members of the Hirata School and the Hirata family. Thus the notion that rocks had significance because they came from the age of the gods would have made sense to him given his connections to the Hirata School.

Japanese people who believe that the creator gods produced all things also believe that what distinguishes each bit of matter made manifest (rocks, trees, mountains) is the spirit of the god that animates it. According to imperial mythology, the age of the gods lasted until the sun goddess sent down her grandson to rule over Japan. To clear the way for him, the gods who had previously inhabited the land went into hiding, either of their own volition or by force (different shrines give different accounts, depending on which god is enshrined). In 1940, Japan celebrated the 2,600th anniversary of this descent. This would indicate that Chita's rocks were over 2,600 years old: not old in geological time, but old in terms of Japan's history.

At the beginning of the letter accompanying the map, Chita wrote a poem that further confirms how older understandings of Japan's history lingered into the late nineteenth century. He begins with a phrase—"ikubaku mo" (no matter how many)—that comes straight out of Japan's oldest poetry collection, *The Man'yōshū*, from the eighth century. His second line—"ame ga shita ni wa" (everywhere in the realm)—also derives from the same collection, although a different poem. By citing these ancient poems, Chita showed off his erudition and associated himself with a particular poetic tradition, one that disdained the elegant courtly style found in imperially commissioned poetry anthologies in favor of an earlier text that the Hirata disciples celebrated as closer to the immediate expression of the gods.

Trying to approach the gods through the artifacts they left behind had motivated much scholarship on early Japan in the first half of the nineteenth century. The coming of the West in the century's middle decades brought a turn to modern science and political philosophy based on the compelling need to learn the secrets of the West's technological superiority. Chita created this map after a conservative resurgence had set in, one that reminded Japanese people of their own worthy heritage. In so doing, he exposed how beliefs propagated before the impact of modernity continued to inform the way at least some people saw their world. While Japan changed, he clung to what he knew: his locality and the eternal evidence for the age of the gods.

Acknowledgments

My thanks to Miyachi Masato for help in deciphering Chita Kaji's letter.

Notes

1. "Masaki Yūsuke ate shojō," Akita-ken Kōbunshokan, *Ochibo bunko, raku* 737.
2. Yoshida Asako, "Akita no Hirata monjin to shomotsu, shuppan," *Nihon shisō shigaku*, no. 39 (September 2007): 130.
3. Miyachi Masato, ed., "Hirata kokugaku no saikentō (II)," *Kokuritsu rekishi minzoku hakubutsukan kenkyū hōkoku* 128 (2006): 370.

Suggested Readings

Harootunian, H. D. *Things Seen and Unseen: Discourse and Ideology in Tokugawa Nativism*. Chicago: University of Chicago Press, 1988.

McNally, Mark. *Proving the Way: Conflict and Practice in the History of Japanese Nativism*. Cambridge, MA: Harvard University Asia Center, 2005.

Pyle, Kenneth B. *The New Generation in Meiji Japan: Problems of Cultural Identity, 1885–1895*. Stanford: Stanford University Press, 1969.

Ruoff, Ken. *Imperial Japan at Its Zenith: The Wartime Celebration of the Empire's 2600th Anniversary*. Ithaca: Cornell University Press, 2010.

24 Cosmology and Science in Japan's Last Buddhist World Map

Sayoko Sakakibara

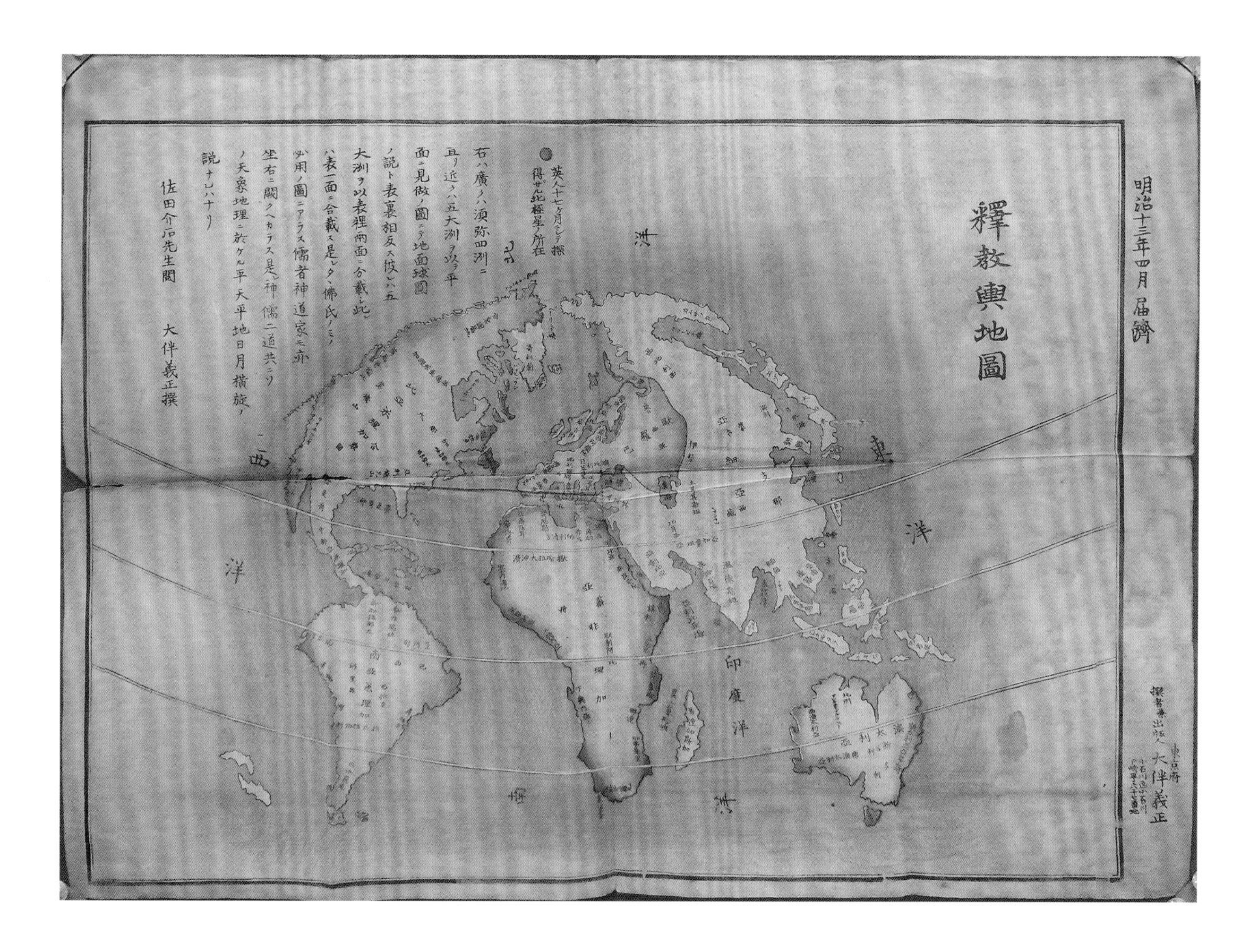

FIGURE 24.1 "Buddhist World Map" [Shakkyō yochizu 釈教與地図], by Ōtomo Yoshimasa 大伴義正, 1880. Woodblock print, 36.4 × 50.2 cm. Courtesy of Kobe City Museum, Hyogo, Japan.

In 1876, the Meiji government's Ministry of Religious Affairs forbade the preaching of "Sumeru theory," a Buddhist cosmology that posited the world as a constellation of four continents centered on the sacred Mount Sumeru (see chap. 20). This ban—part of a broad campaign to stamp out Buddhism in Japan—compelled Buddhist monks and scholars to reframe their traditional cosmology as religious *myth*, rather than doctrinal *truth*. A Buddhist astronomer named Sata Kaiseki was one of many scholars directly affected by the new law.

Although Kaiseki was better known for his writings on political economy,[1] he was also an avid proponent of Sumeru theory, having published several research articles on astronomy and geography based on the Buddhist understanding of the world.[2] His publisher Ōtomo Yoshimasa was likewise a Buddhist scholar and had studied Chinese learning and astronomy under Kaiseki. In 1880, teacher and student collaborated to create a new "Buddhist World Map" (*Shakkyō yochizu*, fig. 24.1). In a prominent note at the upper left, they explained their intent:

> [This map] can be understood abstractly as a map of Sumeru and the four continents surrounding the mountain, or concretely as a map of the terrestrial globe. Although a flat map may appear to contradict the spherical theory of the earth, this one attempts to show the curvature of the earth. [For that reason], it should serve as a useful reference not only for Buddhists but also for Confucian and Shinto scholars. For when it comes to astronomy and geography, both Confucians and Shintoists [wrongly] take the position that the sun and the moon move between a flat heaven and a flat earth.

This passage shows an intriguing attempt to forge a compromise between Sumeru theory and spherical-earth theory. It goes so far as to claim that such a compromise would be instructive to Confucians and Shintoists as well.

At first glance, Kaiseki's map appears to be drawn from a Western point of view. Its depiction of India shows no Buddhist influence; north of India, we only see the label "Himalaya Mountains." This is striking because the Buddhist term "Tenjiku" and the mythical Lake Anavatapta had appeared on virtually all Japanese premodern world maps, even non-Buddhist ones. Scholars as diverse as Nishikawa Joken, the Western-studies astronomer, and Koyano Yoshiharu, a Confucian doctor, had included such Buddhist terms and symbols on their world maps, alongside new information about the world gleaned from Western geography books.[3] Likewise, Kaiseki seems to side with European cartographers in the way he frames the world. Whereas Japanese mapmakers had traditionally located the three sacred lands of the Buddhist realm—India, China, and Japan—at the center of the cartographic frame, Kaiseki's "Buddhist World Map" pushes China and Japan off to the rightward (eastern) edge.

Other place-names are equally telling. Many of the toponyms used throughout this map are relatively new terms, transliterated from Western gazetteers and geography texts. The choice of "Shina" for China is a particularly interesting case. World maps made in pre-Meiji Japan had always used either traditional dynastic names (Tang, Ming) or a Buddhist term (Shintan) to refer to China. For Kaiseki to identify Japan's continental neighbor as "Shina" again showed the direct influence of Western geography.

Yet a closer look at Kaiseki's map reveals other ways in which its designer stubbornly adhered to a Buddhist worldview. Consider the large labels appearing around the edges, identifying East, West, South, and North "Oceans" (*Tōyō, Saiyō, Nan'yō, Hokuyō*) framing a central "Indian Ocean" (*Indoyō*). This labeling might appear to indicate five global seas—the basic maritime taxonomy taught in elementary geography readers of the Meiji era. But a closer look shows that the "Western Ocean" and "Eastern Ocean" labels are located not in distinct bodies of water, but on opposing shores of the Pacific. Moreover, whereas textbook terminology always identified ocean basins with the term "Great" (*Taiheiyō, Taiseiyō, Dainan'yō, Daihokuyō*), Kaiseki does not do that here. Instead, he chooses the terms that were commonly used during the Meiji period to designate "Orient" and "Occident." In other words, what appear to be ocean labels on Kaiseki's map should be understood instead as locating civilizations.

Reinterpreted in this light, figure 24.1 begins to reveal its essence as a Buddhist map. This becomes even clearer as we consider its structure more closely. In traditional Buddhist representations of the world, India was always located at the center, with the Buddhist lands flanking it to the right (east), and the later-discovered continents (Europe, America, Australia, and Africa) on its left (west). The Kaiseki map roughly follows this format; the "Indian Ocean" lies at the center, with the Occident to its west and the Orient to its east. Buddhist geographic ideas are further underscored by Kaiseki's fourfold color scheme. The whole Buddhist Orient, from India to Japan, is tinted yellow; Europe and Australia are shown in brown, Africa in pink, and the Americas in green. The effect is to show the world as consisting of four major continents centered on India—as Sumeru theory claims.

But perhaps the most conspicuous Buddhist element on this image is the otherwise puzzling appearance of the North Star, shown as a prominent red circle at the top of the map. In Japanese Buddhism, the North Star was widely believed to be the avatar of an ancient Babylonian bodhisattva called Myōken. Simply by including this sacred star, Kaiseki may have signaled to the faithful that his was a Buddhist world map. To underscore the point, he adds a caption boasting that "even after seventeen months [of searching], the British were unable to nail down the position of the North Star." As it happens, Kaiseki himself wrote an astronomical essay in the same year that this world map was published, arguing that the North Star, the earth, and Mount Sumeru were directly connected by one straight axis. This brief, pointed note would have reminded Kaiseki's readers that astronomical questions could not be resolved by Western knowledge alone.

It is worth stressing that the creator of this map did not necessarily mean to *merge* Buddhism with Western science. Rather, he saw the two as coexisting systems of knowledge, separate but compatible. This is what made it possible for him to accept Western scientific knowledge about global geography while retaining core elements of the Buddhist worldview. For Kaiseki, Buddhism retained a vital role in protecting the Japanese nation, and Sumeru theory remained viable within a Buddhist framework. The concept of three sacred countries gave Japan special status on the globe, making it the ultimate goal of Buddhism's eastward diffusion from India. By depicting Japan at the far eastern periphery of the Orient, Kaiseki was able to show his homeland's special position as the ultimate Buddhist realm. Yet he was happy to include elements from Western geographic science as well, for Buddhism alone could neither account for the structure of the planet nor explain the cosmos.

In order to appreciate Kaiseki's simultaneous appropriation of Buddhism and modern geographic science, it may be useful to compare him to another scientist who grappled with clashing worldviews: Isaac Newton. We now know that Newton, while embracing the scientific method, was an enthusiastic student of the occult with a lifelong interest in alchemy and biblical interpretation. For Newton, the Christian God was a dynamic force in the universe, keeping the planets in orbit and sustaining the world. The parallels with Kaiseki's combinatory mode of thinking are striking. Buddhist scholars like Kaiseki and Yoshimasa, just like early Christian scientists in the West, were compelled to offer a theory that could bridge between their traditional religious worldviews and modern universal science.

From our vantage point today, Kaiseki's attempt to reconcile these clashing cosmologies may not be very convincing. For years, in fact, his scholarship was criticized for its "anti-Western" elements. But recent studies of Kaiseki—like new scholarship on Newton—have begun to shed a more sympathetic light on his struggle to reconcile religion with science. Especially in his later works, a compromising attitude with Western astronomy is clearly on view. This 1880 map can thus be best appreciated as a sign of the epochal struggles of Buddhist scholars at the end of a tradition—a turning point in the country's geographic imagination.

Notes

1. Calling for a boycott on Westernization, Kaiseki became a person in the news by publishing his writings on political economy. These included "Ranpu bōkoku ron," in *Tokyo nichinichi shinbun,* July 16, 1880, and "Baka no banzuke" (originally written in 1868, republished in *Shakai keizairon,* annotated by Honjō Eijirō [Tokyo: Nihon Hyōronsha, 1941]).

2. As a Buddhist astronomer, Kaiseki wrote many essays including *Tsui chikyū setsuryaku*, 1862, in which he explained the cosmic structure by both spherical earth theory and Sumeru theory.

3. Arano Yasunori, "Higasi Ajia no hakken: 'Sekaishi no seiritsu' to Nihonjin no taiō," *Shien* 61 (1) (November 2000): 100; and Ishizaki Takahiko, "'Gotenjiku' ni okeru Indo ninshiki," *Indogaku bukkyōgaku kenkyū* 60 (1) (December 2011): 558.

Suggested Readings

Umebayashi Seiji. "Sata Kaiseki bukkyō tenmon chirisetsu no kattō." *Kumamoto kenritsu daigaku bungakubu kiyō* 13 (2007): 31–56.

Yamashita Kazumasa. "Kochizu wan bai wan: Shakkyō yochizu." *Chizu chūshin* 445 (October 2009): 26–27.

Yonemoto, Marcia. "Maps and Metaphors of the 'Small Eastern Sea' in Tokugawa Japan (1603–1868)." *Geographical Review* 89 (2) (1999): 169–87.

25 Fun with Moral Mapping in the Mid-Nineteenth Century

Robert Goree

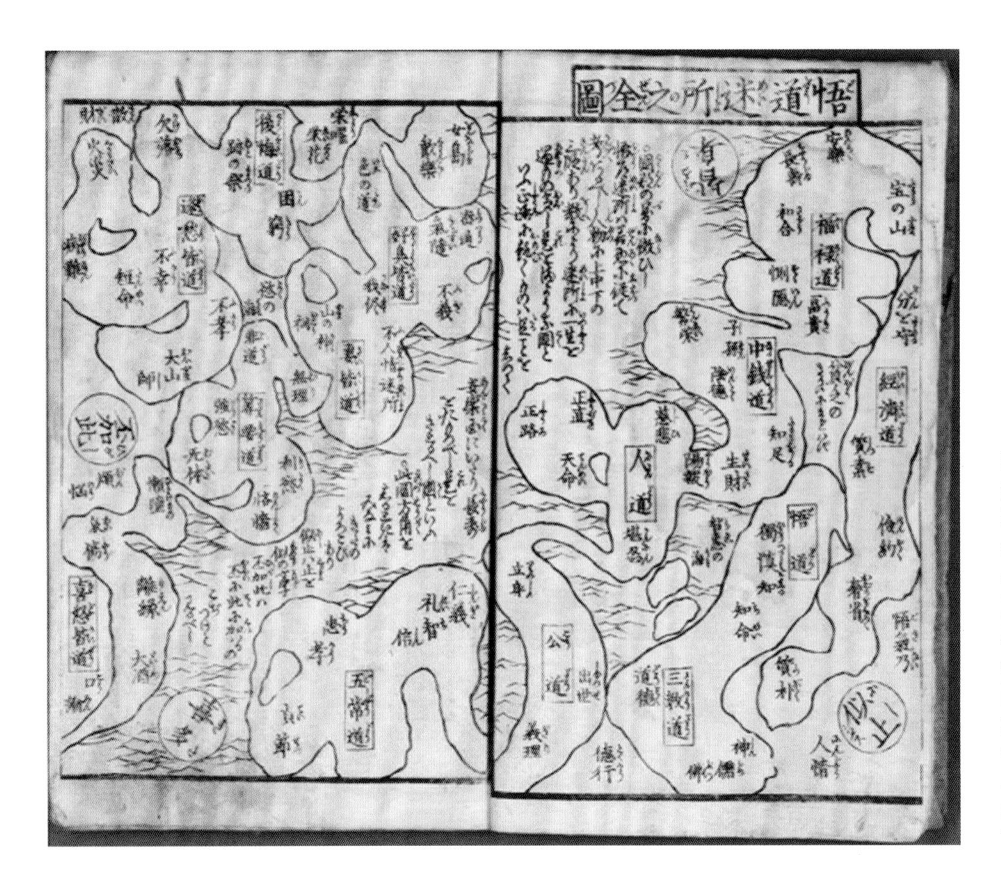

FIGURE 25.1 "The Complete Map of Delusional Landmarks along the Road to Enlightenment" [Godō meisho no zenzu 悟道迷所の全図], from *Illustrated Landmarks of Delusion in the Realm of Good and Evil* [Zen'aku meisho zue 善悪迷所図会], by Keisai Eisen 渓斎英泉, 1846, p. 4 (leaf 4). Woodblock print, 12 × 18 cm. Courtesy of Waseda University Library.

Maps in Tokugawa Japan could be very serious. From boundary disputes to land-yield estimates, maps made at every level within the political settlement helped to document expectations about how society was to relate to geography. Maps could be serious business, too. Commercial publishers sold portable maps to travelers wishing to orient themselves as they took to the road for pilgrimage, pleasure, work, and duty. As the tastes of geographically inclined consumers grew more adventuresome during the period, maps, which appeared alongside books and artistic prints in shops throughout Japan, could be serious, commercially viable, *and* entertaining at the same time. Some maps could even be useful in their very preposterousness.

"The Complete Map of Delusional Landmarks along the Road to Enlightenment" (fig. 25.1) is a case in point. Produced by the prolific woodblock artist Keisai Eisen (1790–1848) in the mid-nineteenth century, it depicts two oddly shaped landmasses separated by a channel of water.[1] Place-names like "Long Life," "Wastefulness," and "Disaster" dot this imaginary land of practical morality, while rectangular labels indicate various paths of human experience, such as "The Road of Happiness" and "The Road of Regret." Wordplay abounds in these toponyms. Viewers at the time would have had no trouble reading "The Broad Road of Fond Fragrance" (Kōshū kaidō), for instance, as a pun on the Kōshū Post Road (also Kōshū kaidō), one of the five turnpikes linking the city of Edo to other parts of the archipelago.

Like many maps of the time, this one depends on extensive annotation. The large block of text in the channel of water explains why the landforms look the way they do. Descending on the left side, an emphatic command written in kana—"Do not get lost" (Mayouna)—outlines a morally precarious land riddled with pitfalls. Travelers on life's journey would do well to proceed with extreme caution when passing through such dangerous sites here as "Adultery" and "The Sea of Desire." In clear contrast, the equally emphatic command on the right side—"Be enlightened" (Satorubeshi)—spells out the borders of a paradise flush with places well worth a visit, such as "Loyalty" and "The Sea of Wisdom." As if to suggest the challenges of finding this virtuous course, Eisen makes it hard to decipher the encouraging command by positioning it upside down at a diagonal.

"The Complete Map of Delusional Landmarks" was not a stand-alone document. It appeared a few pages into an otherwise map-free book of humor titled *Illustrated*

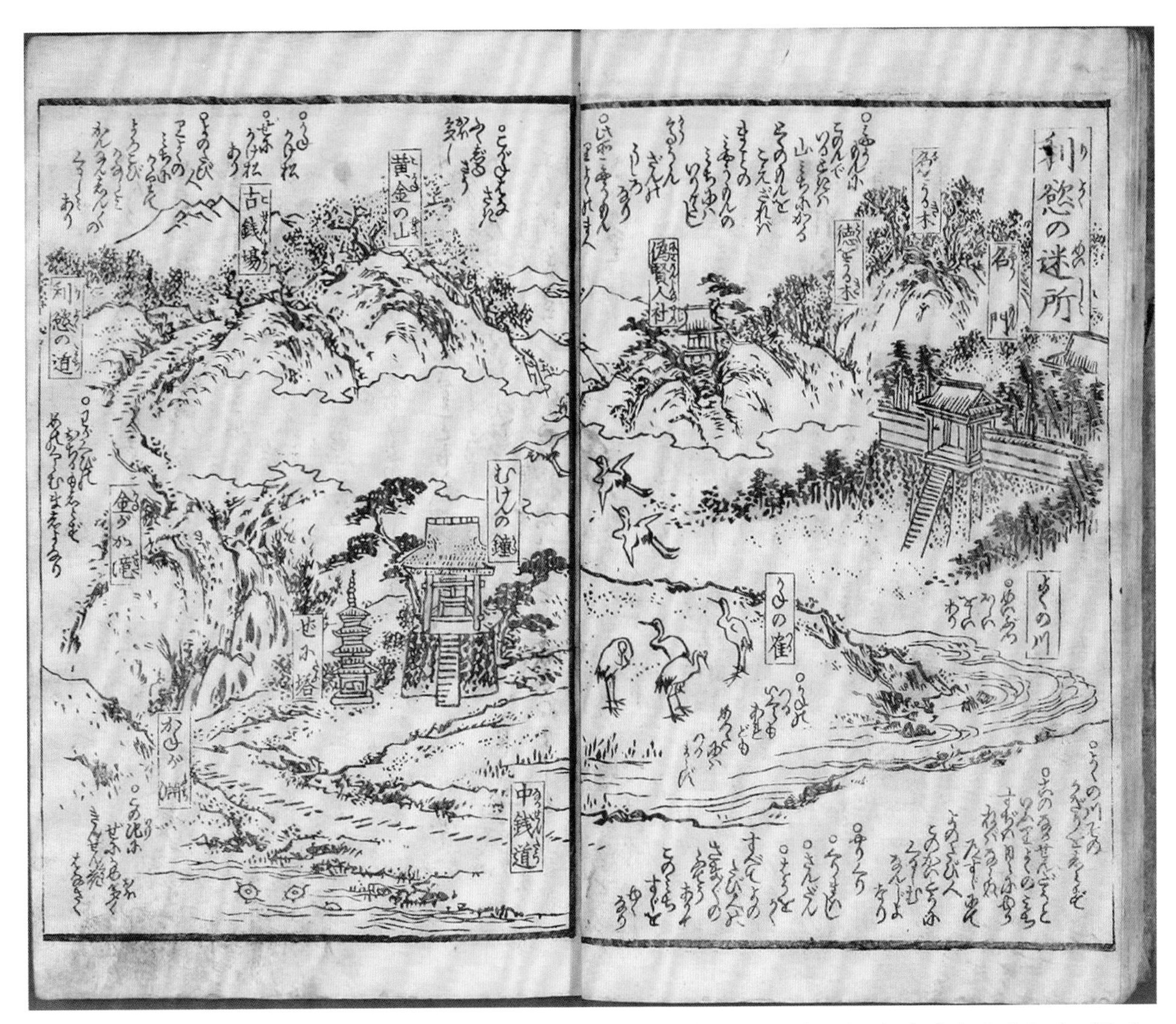

FIGURE 25.2 "Delusional Landmarks of Greed" [Riyoku no meisho 利慾の迷所], from *Illustrated Landmarks of Delusion in the Realm of Good and Evil* [Zen'aku meisho zue 善悪迷所図会], by Keisai Eisen 渓斎英泉, 1846, p. 18 (leaf 18). Woodblock print, 12 × 18 cm. Courtesy of Waseda University Library.

Landmarks of Delusion in the Realm of Good and Evil (1846). Part of a still-longer work called *The Travel Record of Good and Evil: A Personal Guide to Life* (published serially from 1844 to 1862 as a joint project by several publishers, editors, and artists), *Illustrated Landmarks of Delusion* is a send-up of the illustrated gazetteers that were popular at the time. For over fifty years prior to the map's publication, these compendious guidebooks had fed the imaginations and reference needs of readers with encyclopedic information about the cultural geography of cities like Kyoto, Osaka, and Edo, as well as of roads, domains, and regions around the Japanese archipelago and beyond. The key category within the guidebooks—"famous places" (*meisho*)—is recast here by Eisen as "delusional landmarks." Confident that his readers were familiar with the popular illustrated gazetteers and therefore would get the joke, Eisen twists the notion of *famous* places into a meditation about those *infamous* places along the path of life that threaten happiness at every turn.

Illuminating the map's cryptic overview of this moral universe, the remaining pages of *Illustrated Landmarks of Delusion* single out specific human hazards and havens for detailed treatment in word and image combinations that hew closely to the representational strategies found in illustrated gazetteers. At first glance, the moralistic toponyms seem to be the spitting image of conventional famous places. The "Delusional Landmark of Greed" (fig. 25.2), for example, looks much like a temple complex in the mountains. On closer inspection, however, it is no ordinary place at all. Appearing on the right page, and descending in a gradual zigzag starting from the upper right corner, are "Reputation Gate," "Name-for-sale Tree," "Virtue-for-sale Tree," "Phony Wiseman Shrine," and "The River of Desire." Next comes a small flock of "Money Cranes," punning symbols for all those sources of financial revenue that people wish to secure for themselves. Like the three cranes that have taken flight from the banks of the "River of Desire," Eisen suggests, cash is likewise fleeting and not easily grasped.

At the bottom of the left page, and marking the beginning of an upward sweeping zigzag, is "The Road of Extra Fees," which refers to the often exorbitant costs charged to visitors of Japan's licensed entertainment districts. This is a pun on a major inland turnpike connecting Kyoto and Edo called the Middle Mountain Road (Nakasendō). The dangerous costs of pleasure seeking are then given tangible expression in "The Deep Pool of Money," in which float three turtles with shells in the shape of coins. Note that one of the turtles is sinking (or perhaps rising), which is Eisen's pictorial way of emphasizing the same point made with the cranes. Since money is in perpetual flux, chasing after it inevitably dooms one to a life of frustration. From here, the traveler passes by "Cash Stupa," then "Hell Bell." This bell, when struck, guarantees the striker immense wealth in this life but at the price of spending the next life in the Buddhist Hell of Constant Suffering. Then comes "The Waterfall of Gold," which has a ghoulish facial quality seen in Eisen's renderings of water and stones elsewhere in the book. Moving on, the traveler enters the "Road of Greed" which leads to "Ancient Coins" (or, in a looser translation, "The War Zone of Money," since this toponym is a pun on the word for historic battlegrounds that frequently appears in illustrated gazetteers). Lastly, "The Golden Mountains" rise above this seemingly tranquil but entirely fraught landscape of greed.

That Eisen was the one who composed such moral landscapes is ironic—and indeed part of the book's appeal—since his other works celebrate the very pastimes and priorities he warns against here, most notably those associated with money, sex, and drinking. Eisen is best known today for his woodblock prints of beautiful women, illustrated erotica, and illustrations for works of fiction. Whether Eisen meant his map and landscapes in *Illustrated Landmarks of Delusion* to strike the serious tone found in works by orthodox moralists of his day is highly unlikely, given his relentless tongue-in-cheek choreography of high-minded themes. But in an era when the boundaries between didacticism and entertainment were not always clearly drawn, does indulging in some fun with a map necessarily mean he was not in earnest? Might there be a subtle brand of pedagogy in the very act of playing around with a ridiculous map?

Eisen was an artist and writer who engaged in an astonishing range of themes that extended beyond the hedonistic, effervescent, and self-indulgent. A man with broad interests in historical topics and innovative depictions of landscape and the human body, he frequently collaborated with the luminaries of his day, including Utagawa Hiroshige and Shikitei Sanba. And like many of his contemporaries, he was remarkably prolific, trying his hand at a breathtaking array of artistic formats and literary genres. As editor, writer, or illustrator, his name is associated with nearly four hundred extant books. Adding to this prodigious output, he also designed hundreds of single-sheet prints for commercial publishers and private collectors alike. It does not seem far fetched, then, that such a capable and intellectually promiscuous artist would design a map with just the right amount of ambiguity to make it hermeneutically elastic and yet still easy enough to comprehend.

One thing can be said for certain about "The Complete Map of Delusional Landmarks": Eisen used its composition as the occasion to adapt a cartographic idiom to moral subject matter. By 1846, when the map went on sale, an enthusiasm for geography had already penetrated deeply into popular consciousness. Readers were ready for maps of lands nearby and far away. They also seem to have had an appetite for strange new forms of cartographic signification. Beyond inviting us to consider the meaning of Eisen's impish treatment of life as a spatial game, "The Complete Map of Delusional Landmarks" also asks us to stretch our imaginations when thinking about what counted for cartography in Japan on the eve of modernization. For Eisen at least, maps could do whatever he wanted them to do.

Note

1. Like many artists and writers of his day, he produced works under variant names, the most well known of which are Ishida Eisen and Ippitsuan.

Suggested Readings

Berry, Mary Elizabeth. *Japan in Print: Information and Nation in the Early Modern Period.* Berkeley: University of California Press, 2006.

Fahr-Becker, Gabriele. *Japanese Prints.* Cologne: Taschen, 2007.

Hayashi Yoshikazu. *Keisai Eisen, Katsushika Ōi.* Tokyo: Kawade Shobō Shinsha, 2012.

Izzard, Sebastian. *Hiroshige/Eisen: The Sixty-Nine Stations of the Kisokaidō.* New York: George Braziller, 2008.

Kornicki, Peter. *The Book in Japan: A Cultural History from the Beginnings to the Nineteenth Century.* Leiden: Brill, 1998.

Nenzi, Laura. *Excursions in Identity: Travel and the Intersection of Place, Gender, and Status in Edo Japan.* Honolulu: University of Hawai'i Press, 2008.

Traganou, Jilly. *The Tōkaidō Road: Traveling and Representation in Edo and Meiji Japan.* London: RoutledgeCurzon, 2004.

Yonemoto, Marcia. *Mapping Early Modern Japan: Space, Place, and Culture in the Tokugawa Period (1603–1868).* Berkeley: University of California Press, 2003.

26 A Travel Map Adjusted to Urgent Circumstances

Kären Wigen and Sayoko Sakakibara

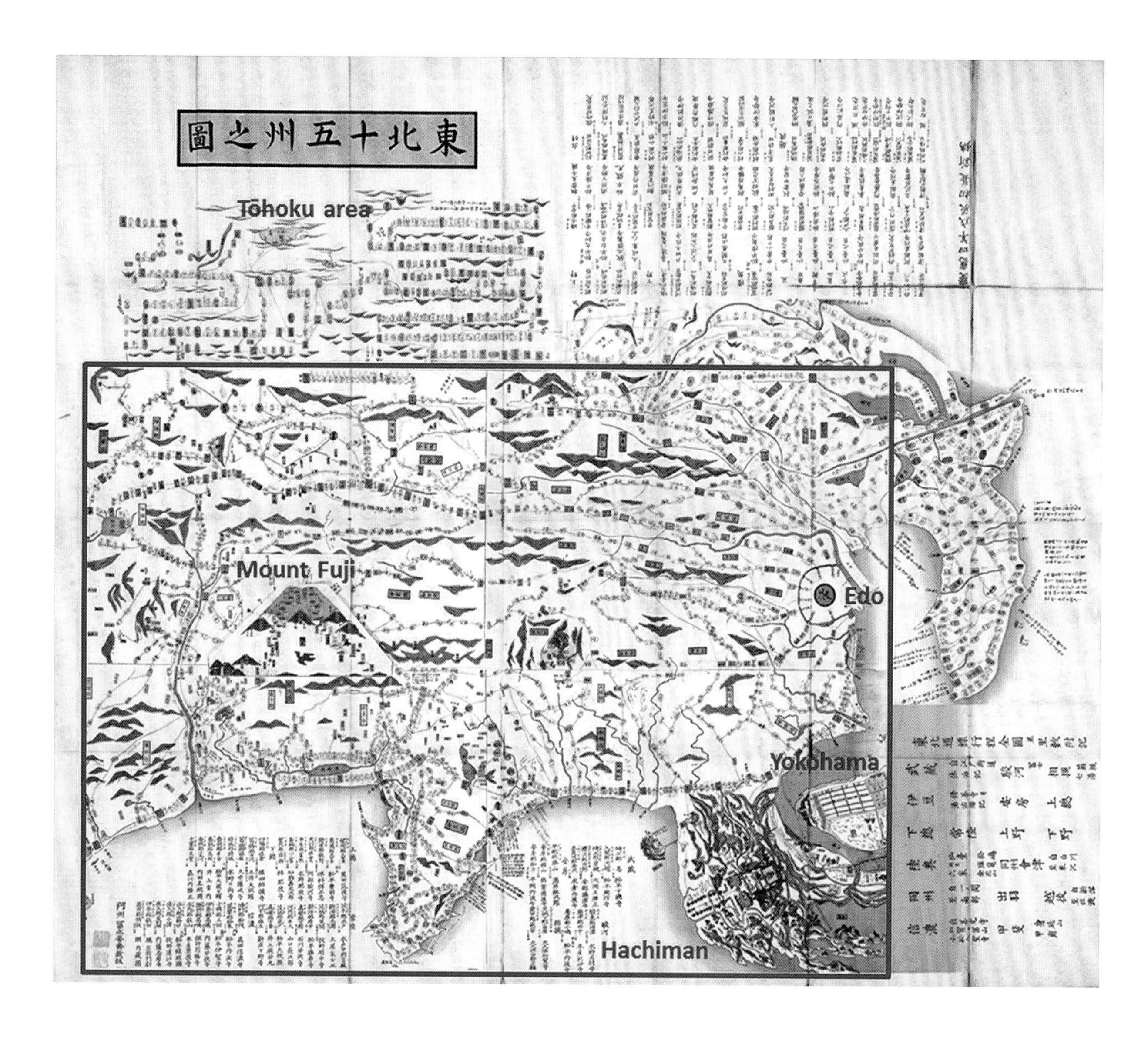

FIGURE 26.1 "Itinerary Map of Northeastern Japan" [Dai Nihon Tōhoku kōtei zenzu 大日本東北行程全図], by Tominaga Shinsai 富永晋斎, 1868. Woodblock print, 102.5 × 119.5 cm. Courtesy of the University of Tsukuba Library. English labels added by authors.

In early 1868, Japan was caught up in political turmoil. Impatient with the shogun—and furious with the foreigners who had begun doing business in Japan's treaty ports a decade earlier—a coalition of southwestern daimyo marched their combined armies east toward Edo, determined to seize power in the name of the emperor. Within weeks, the shogun surrendered and his capital was occupied. But thousands of Tokugawa allies held out in the northeastern backcountry known today as Tohoku, resisting the rebels for a year and a half until they were finally conquered in late 1869.

"Itinerary Map of Northeastern Great Japan" (fig. 26.1) was rushed to press in the midst of this civil war. Its author, Tominaga Shinsai, dated his composition "early summer of Keiō 4" (1868): shortly after the shogun had left Edo, but before the Meiji emperor had arrived. A close look at his composition suggests that it was a rush job: a mash-up of existing sources to suit the urgency of the times—and to answer the demands of a literate public eager to keep up to date.

The heart of Tominaga's map is a stripped-down version of an existing route map that had been published just eight years earlier (fig. 26.2). While decorative borders and mountains have been removed, the bulk of Tominaga's 1868 print faithfully follows the contours of its 1860 predecessor, entitled "Eight Provinces from Which to See Mount Fuji." Both the 1860 version and the 1868 revision highlight the main turnpikes through east-central Honshu, placing the Japanese Alps on the left and the Kanto Plain on the right. Neither indicates the name of a publishing house, but it is likely that the same publisher printed both, since Tominaga identifies his 1868 image as "newly carved."

Drawn in the familiar style of a late-Tokugawa travel map, this part of Tominaga's map catered to travelers and pilgrims along the bustling coastal road. Post stations (white or yellow ovals) crowd its surface; famous scenic spots (red circles), castle towns (red rectangles), and other landmarks (yellow rectangles) are marked. Distances (given in red) are calculated from Edo, which features prominently as the center of the road network. The large red dot on the far right treats the shogun's capital as a starting point for travel, real or imagined.

In keeping with its original title, this part of the map also gives special prominence to Mount Fuji. Befitting its status as a national symbol and magnet for pilgrims, the sacred volcano is not only enlarged but also shown in three dimensions; an outsized image of the iconic mountain on the lower left is cleverly glued on as a reversible flap, allowing the viewer to see climbing trails from both the northern and southern approaches. A separate attachment even offers an aerial view of the crater. A second foldout was glued along the left-hand edge of the original 1860 map to show another alpine pilgrimage route, the path to Mount Ontake. But the overwhelming emphasis here is on Mount Fuji. This undoubtedly reflects the influence of the Fuji cult, whose popularity swelled among the commoners of Edo during the last decades of the Tokugawa era (see chap. 22).

Still, Fuji is not the only feature to receive special treatment. Published in the spring of 1860—less than a year after Yokohama was designated for foreign trade—the original "Eight Provinces" map also highlighted the

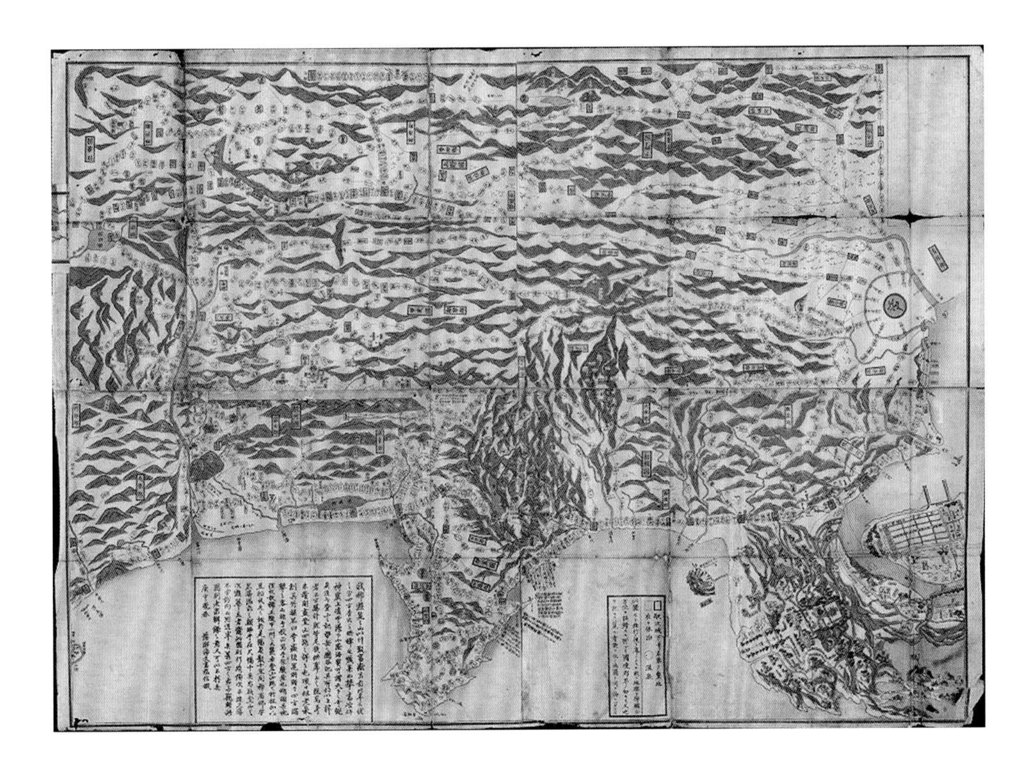

FIGURE 26.2 "Eight Provinces from Which to See Mt. Fuji" [Fujimi hasshūzenzu 富士見八州全図], by Takahashi Shin 高橋信, ca. 1860. Woodblock print, 93.0 × 66.3 cm. Courtesy of Shinshū University Library.

Kamakura-Yokohama area. The foreign settlement at the newly opened port of Yokohama—depicted as a virtual island in the lower right-hand corner—is blown far out of proportion. So too is the medieval warrior capital of Kamakura nearby. Here the designer adopts a birds'-eye view to draw a number of religious sites in a highly pictorial style. A viewer familiar with Kamakura might make out a number of familiar landscape features, from the Great Buddha to the shrine of Hachiman, the Shinto God of War. Since Hachiman was credited with the defense of Japan, the latter detail makes sense at a time when "repel the barbarians" was a rallying cry for many.

In all the features described so far, the "new carving" of 1868 stayed true to its 1860 prototype. Yet Tominaga did not stop at retracing the "Eight Provinces" outline. Working during a period of civil war, he found space for two sets of additions that would surely have appealed to consumers in the capital. One supplement was textual: a pair of lists (in the upper right and lower right) indicating the names of local daimyo, the size of their holdings, and their distance from Edo. This was not particularly privileged information, since it could be found in many publications. Nonetheless, its inclusion here would have enhanced this map as a handy guide for those following the news of the northeastern campaigns. The other additions were spatial: extending the original map to include regions east and north of its original boundaries. At a time of great public interest in the contested hinterlands of the northeast, Tominaga found a way to graft seven additional provinces onto the original eight-province view. Tellingly, however, these add-ons did not get equal treatment. While the Bōsō Peninsula (bracketing the eastern edge of Edo Bay) was depicted more or less at the same scale as the original "Eight Provinces," the deep northeast was treated in a starkly abbreviated style. Here Tominaga hastily sketched out the main roads, folded them in on themselves, and attached them as a boxy appendix along the upper left margin.

Lacking entirely in pictorial detail, this stripped-down box contains the entirety of the region that Japanese today know as Tohoku. In this one spare corner are condensed all the main highways of Mutsu, Dewa, and Echigo—homeland of the last Tokugawa loyalists, and site of the last Restoration battles. If drawn to scale, these three sprawling provinces would have doubled the size of Tominaga's image; instead, he cuts them to a fraction of their relative size and renders them in an abstract idiom almost like a modern subway map. Reduced to a string of place-names, the whole northeastern extension of Honshu is folded accordion style into a compact rectangle measuring just a few inches on each side. Marginal notes indicate the distances from the edge of the Kanto to major towns in the north, including several that would loom large in the battles of 1869.

The result of this cutting and pasting is an odd, hybrid image. The contrast between the exaggerated foreground (zooming in on pilgrimage destinations near the Pacific coast) and the attenuated background (zooming out on the battlegrounds of the deep north) may be disconcerting to modern viewers. Likewise, the prominence of the word "Tohoku" in the map's title might be off-putting, given Tominaga's cursory treatment of the region known by that name today. Yet what strikes us as inscrutable or ungainly clearly had market appeal in its day. In the spring of 1868, Japanese readers were eager to follow the course of a civil war that was raging north of Edo. By creating a quick pastiche of existing works, Tominaga managed to satisfy their curiosity in a timely way, creating a rough guide to a region that was much in the news. And while it may have short-shrifted the deep north in some ways, his map managed to highlight the crucial feature of its geography: the transport routes that linked it to the greater Edo region (where most map readers lived). However ungainly it may seem to modern readers, Tominaga's hodgepodge was a success in its time, judging from the large number of copies that survive to this day. Finally, it is important to note that Tominaga's title itself was ad hoc. "Tohoku" was an uncommon usage that had no fixed definition at the time; the routine use of this directional term (literally "east-north") to denote the upper reaches of Honshu is essentially a modern invention.

Viewed in this light, Tominaga's "Itinerary Map of Northeastern Great Japan" can be seen as a cartographic memento of a special moment in Japanese history. By featuring daimyo data alongside a thriving treaty port, and by grafting an abridged northeast onto an enlarged Tōkaidō, Tominaga turned a tourist map into a practical guide for those interested in following the news. The cartographer's compound design was good enough to serve a country on the threshold between regimes; his odd composition and hasty execution mirror the urgency of the times. A map like this was not exactly novel in its outlines, nor did it convey breaking news. But in its pragmatic assembling of existing information into a marketable collage, this repurposed itinerary map reminds us of the fast-moving moment in which it was made.

Suggested Readings

Steel, William. "Edo in 1868: The View from Below." *Monumenta Nipponica* 45 (2) (1990): 127–57.

———. *Bakumatsu kara Meiji no fūshiga/Poking Fun at the Restoration: Satirical Prints in Late 19th Century Japan.* Tokyo: ICU Yuasa Memorial Museum, 2013.

Sugimoto Fumiko. "Jiji o tsutaeru ezu." In Sugimoto Fumiko et al., *Ezugaku nyūmon* (Tokyo: Tokyo Daigaku Shuppankai, 2011), 84–89.

27 Legendary Landscape at the Kitayama Palace

Nicolas Fiévé

FIGURE 27.1 "Bird's-Eye View of the Golden Pavilion of the Northern Hills of Kyoto" [Kyōto Kitayama Kinkakuji zenzu 京都北山金閣寺全図], by Tōtōsha Shimotori Haryō 東涛舎霜鳥巴凌, 1895. Copperplate engraving, 32 × 46 cm. Courtesy of the C. V. Starr East Asian Library, University of California, Berkeley.

Japan's railway revolution, which began in the 1870s, led to unprecedented changes in the ways landscape was perceived and represented. The development of an extensive railway network inaugurated new forms of tourism, replacing the traditional practices of travel and pilgrimage that had been popular during the Edo period. The late nineteenth century also saw the construction of grand hotels at the most renowned historic sites across the country, including some, like the 1888 Kyoto Hotel, which went on to become tourist attractions in their own right. Most important for our purposes, the railway tourism of the Meiji era coincided with an upsurge of interest in Japan's urban architectural heritage as a whole, including not just ancient edifices with prestigious pasts but modern structures like industrial and port facilities as well.

Coming in the wake of an 1868 decree, separating Shinto shrines from Buddhist temples, that had placed Buddhist institutions in a disastrous economic situation, the development of architectural tourism provided a new source of funding and appreciation that Buddhist institutions were quick to use to their advantage. It was in this context that the "Bird's-Eye View of the Golden Pavilion of the Northern Hills of Kyoto" (fig. 27.1) saw the light. Designed as a souvenir to bring back from the temple, this print depicts Rokuon-ji Temple—a vestige of the retirement villa of Shogun Ashikaga Yoshimitsu (1358–1408), commonly known as the Golden Pavilion—and its surrounding grounds.

The "Bird's-Eye View of the Golden Pavilion" was printed in 1895 from a copperplate engraving, a process imported from the West and used mainly for city maps.[1] This new technique yielded finer results in terms of precision and detail than woodblock printing. The effect was further enhanced by the fact that the engraving was printed on larger sheets than most woodcut engravings produced during the Edo period. The resulting image depicts the complex, as it stood during the early Meiji period, in unprecedented detail.

The map centers on the Golden Pavilion and its surrounding gardens, stretching from the Sekika-tei Teahouse—built in the 1600s for an imperial visit—to the small Fudō-dō Temple. Mounts Kinugasa and Hidari Daimonji form the backdrop.[2] The lofty perspective highlights both the historic vestiges of Yoshimitsu's palace and later extensions, as well as the surrounding rural and mountainous landscape that offered opportunities for excursions beyond the temple grounds. The inclusion of an English title on the print is a clear indication of the development of international tourism in Japan at the time. An ornamental frieze along the bottom edge, composed of reclining deer and maple leaves, evokes the temple's Buddhist name, "Temple of the Deer Park," a reference to the site in India where the Buddha was said to have given his first sermon. The boxed text in the bottom left-hand corner briefly summarizes the temple's history since Yoshimitsu's time. The engraving's date appears to the left of the box, together with the name of the artist (who also created numerous engravings representing other famous sites).[3] Together these elements suggest that the engraving was designed to introduce the general public to the temple's architecture and landscape.

In order to fully appreciate the engraving, it is important to know that the Temple of the Deer Park was initially not a place of worship but a grandiose palace, built by a retired shogun. Together with its expansive ponds and gardens, this compound symbolized the apogee of the Muramachi shogunate's (1336–1573) military power, incorporating an austere aristocratic style with innovations in recreational architecture. Yoshimitsu was only thirty-seven in 1394 when he handed power over to his son, took the tonsure, and formally withdrew from political life. Three years later, he bought an estate in Kyoto's northern outskirts, site of an earlier temple and the remains of a palace built in 1224. Yoshimitsu kept some of the original buildings and built his own residence on the grounds, where he moved the following year. The complex—which became known as the Northern Palace—included buildings destined to house warriors and nobles of the court as well as Buddhist dignitaries.

In Yoshimitsu's time, the Northern Palace consisted of a series of impressive structures encircling a pond and connected by covered galleries. Yoshimitsu presided from the *shinden*, a replica of the imperial palace's main edifice. In 1408, he erected thirteen additional buildings, eight of which were destined to accommodate the emperor and his entourage. The buildings were decorated with artificial flower bouquets made of gold and silver thread, while cherry trees were planted on either side of the *shinden*, which stood at the center of a courtyard graveled in five colors to represent a fish-scale motif. The complex also included a number of other prestigious buildings, most of which have since disappeared: a seven-tiered pagoda, various shrines, and recreational pavilions. Chief among the latter was the Sacred Relic Pavilion (*shari-den*) or Golden Pavilion.

The Golden Pavilion was modeled after Saihō-ji monastery's Lapis-Lazuli Pavilion (*ruri-den*), whose name evoked the Immortals' Palace. The presence of a gilt phoenix representing the mythical bird *hōō* (Chinese Fenghuang) at the pinnacle is not fortuitous: this solar creature dwells on Mount Hōrai, the Taoist Island of Eternal Life. In Chinese tradition, the appearance of the phoenix heralds the reign of wisdom, the establishment of order and great peace; it is an imperial symbol that the retired shogun took up for his own purposes. With its elegant blend of distinct architectural styles, reflecting the fusion of the court and the military as well as of old and new schools of Buddhism, the pavilion may be seen as the epitome of Kitayama culture.

The Meiji-era engraving depicts this historical site with far greater precision than that found in earlier ren-

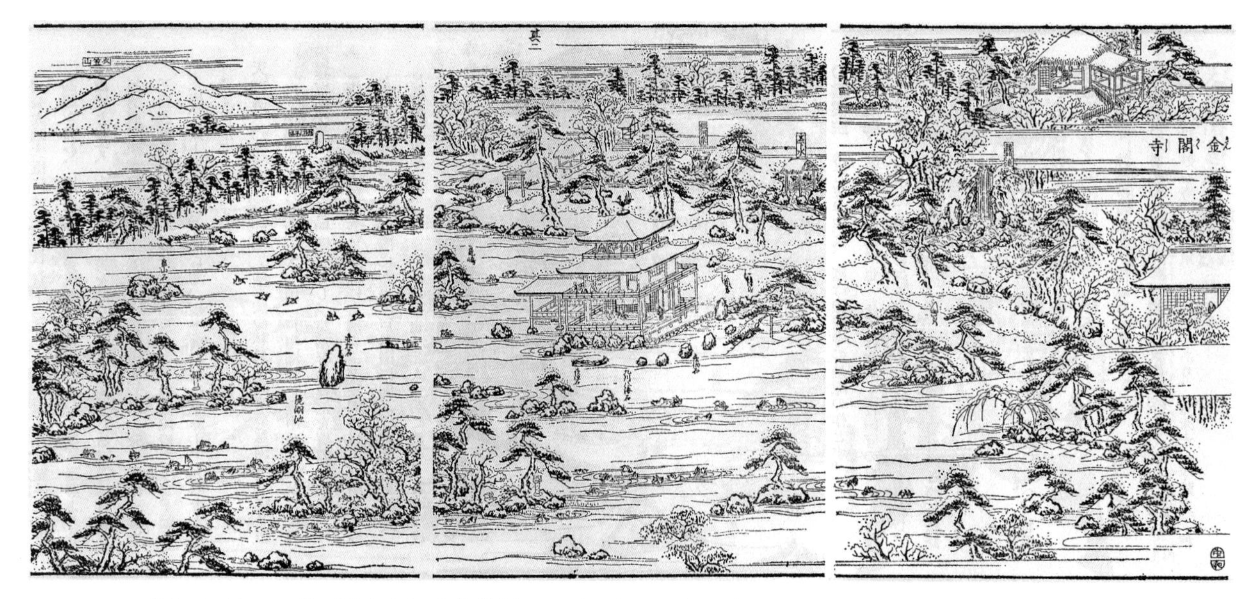

FIGURE 27.2 "The Golden Pavilion" [Kinkaku-ji 金閣寺], woodblock print from the *Illustrated Account of the Capital's Picturesque Gardens* [Miyako rinsen meishō zue 都林泉名勝図会], by Akisato Ritō 秋里籬島, published in Kyoto by Yoshinoya Tamehachi 吉野屋為八, 1799.

derings (compare fig. 27.2). Many of the individual islets and rocks are labeled. The largest island facing the pavilion—named Ashihara-jima, an abbreviation for Toyoashihara nakatsukuni, "the Middle Land of Reed Moors"—represents Honshu as described in the origin myths related in the *Kojiki* and the *Nikon shoki*. Clusters of rocks on the island's edges symbolize the Amitâbha Triad, figures of peace watching over the land of Ashihara. The inclined rock to one side of the island is named *Hosokawa ishi*, after the powerful ally clan of the Ashikaga who, at that time, occupied the high post of administrator of Kyoto. Also depicted are the *Hatakeyama* and *Akamatsu* rocks, representing two other clans allied with the shogunate. The rock to the island's left represents Awaji, the large island connecting Honshu to Shikoku. Thus, the landscape's layout was designed to represent the territory over which Yoshimitsu ruled supreme.

Between the central island and the Pavilion lie the "crane" and "turtle" islets, with twisted and midget pines, evoking the craggy peaks where the immortals dwell. (According to ancient Chinese tradition, the immortals traveled the East Seas on the backs of turtles—which, together with the crane, rocks, and twisted pines, symbolize longevity.) Since the Heian era, the Immortals' Isle of Eternal Life, represented by rocks and pine trees, had constituted a recurring theme in landscape design. Two other "turtle" islets lie to the pavilion's left: one "entering" and the other "exiting." Opposite the latter to the left of Crane Island lies the Kusenhakkai rock, representing the Nine Mountains and Eight Seas in Buddhist cosmology, where Mount Sumeru stands as sacred center of the universe. To the right of the pavilion lie the *yodomari ishi*, "rocks to spend the night," representing the Immortals' sailboats made of coral, gold, silver, and precious stones bound for the Isle of Eternal Life. The rocks served as a mooring site for the boat that was used to navigate the pond.

The landscape's particular design, visible from all the buildings surrounding the pond, was meant to symbolize the longevity of the political order imposed by Yoshimitsu. The Golden Pavilion symbolized the alliance among the imperial court, the military, and the Buddhist establishment. From the pavilion, Yoshimitsu, associated with the mythical phoenix *hōō*, could contemplate his territory: Ashihara and the Island of Awaji, consolidated through the alliance of the Hosokawa, Hatakeyama, and Akamatsu clans, and protected by Amitâbha's compassion. The garden allegorized an idyllic and paradisiacal world.

Thus, the 1895 "Bird's-Eye View of the Golden Pavilion" may be viewed in different ways: as a modern document providing information on the features of the temple and its grounds; as a pocket guide enabling the tourist or pilgrim to identify the main landmarks of this celebrated site; and, finally, as the resurgence of Shogun Ashikaga Yoshimitsu's eternal realm through a finely detailed rendering of an idyllic and long-gone landscape.

Notes

1. In 1868, the renowned publishing house Kyoto Takehara Yoshibē printed the first pocket map of Kyoto, a 34 × 41 cm copperplate engraving that was sold folded in an envelope: "A detailed map of Kyoto, a pocket treasure, new engraving on copperplate" (Dōban shinsen kaihō Kyōto meisai no zu).

2. The 1895 engraving provides a far larger view of the site than those published previously, which include only the grounds immediately surrounding the temple and Mount Kinugasa.

3. Engravings bearing Tōtōsha Shimotori Haryō's or Tōtōya shujin Shimotori Haryō's signature are found in many collections. The artist created a great number of copperplate engravings of renowned Japanese sites, namely the Bird's-Eye Views of Japan (*Nihon hakuran-zu*) printed throughout the Meiji era. Following Haga Akiko, Shimotori Haryō's real name should be Shimotori Haru. See Haga Akiko, "Meijiki keikan dōban-ga o megutte: Saitama o egaita 'Hakuran-zu,'" *Saitama kenritsu bunshokan kiyō* 26 (2013): 45.

28 New Routes through Old Japan

Roderick Wilson

FIGURE 28.1 "Kiso River" [Kisogawa 木曽川], from *Eight Renowned Landscapes of Japan* [Nihon hakkei meisho zue 日本八景名所図会], by Yoshida Hatsusaburō 吉田初三郎, 1930. Lithograph, 9.1 × 36.4 cm. Courtesy of Shufu no tomo sha, Tokyo.

Paintings and poems on the theme of "eight renowned landscapes" have been a motif in the arts of East Asia since as early as the eleventh century. In April 1927, not long after the Showa emperor ascended the throne, the editors at two of Japan's largest newspapers decided a new series was in order and set about coming up with eight renowned landscapes suitable for the new era. Embracing the democratic and popular spirit of the times, the editors asked the general public to vote by mail for the best landscape in eight different topographical categories ranging from mountains and gorges to rivers and hot springs. The campaign must have struck a chord, given that a month later, more than 93 million votes had been received (at a time when the population of Japan was only about 64 million). In the end, a blue-ribbon committee of forty-eight distinguished men (including top government and military leaders as well as respected scholars and artists) finalized a list of what turned out to include eight renowned landscapes, twenty-five famous sites, and one hundred vistas.[1]

With the official support of the Ministry of Railways, the newspaper companies promoted these landscapes by commissioning well-known painters and writers to depict each of the selected areas. While the newspapers dispatched eight pairs of artists to render the individual landscapes in word and image, Yoshida Hatsusaburō alone was commissioned to provide a bird's-eye view for all of the selected landscapes. Known as the Hiroshige of the Taisho period (1912–26), Hatsusaburō produced nearly sixteen hundred such panoramic images during his long career. What made his renderings of the eight landscapes particularly new was their commissioning through the newspapers and Ministry of Railways, their target audience of middle-class women (the whole set was published together as a special insert in the leading women's magazine *Housewife's Companion* [*Shufu no tomo*]), and their montage representation of modern Japan. As with most things modern, however, this commercial artist's maps also drew on existing artistic notions and techniques, in this case Edo-period itinerary maps and bird's-eye views.

Hatsusaburō's "Eight Renowned Landscapes of Japan" were printed together on a single rectangular sheet of paper, measuring 21.5 × 179.4 centimeters, which was designed to be folded up and distributed as a magazine insert.[2] The eight individual bird's-eye views formed two rows of four, which when read from top to bottom and right to left transport the reader through the whole archipelago from the northern expanses of Hokkaido to the southern island of Kyushu. Red labels to the right of each map identify their subjects as follows: Karikachi Pass (Hokkaido), Towada Lake (Aomori and

Akita), Nikkō Kegon Waterfall (Tochigi), Kamikōchi Valley (Nagano), Kiso River Valley (Gifu and Aichi), Muroto Cape (Kōchi), Mount Unzen (Nagasaki), and the Beppu hot springs (Ōita). Like many mapmakers in Japan, Hatsusaburō oriented each of his eight maps in a way that allowed him to best capture that particular landscape. Hence, Towada Lake is seen from the north, the Kiso River from the west, and Beppu and Kamikōchi from an imaginary vantage point in the east. The four remaining maps are oriented in a more conventional fashion, with north at the top.

Pictured here, the bird's-eye view of the Kiso River valley (fig. 28.1) demonstrates a number of the visual techniques Hatsusaburō used in this series and throughout his long career. Being a panoramic image, his map is horizontal in format. The Kiso River forms a flat U shape with its midstream spreading wide across the foreground of the painting, its upstream portion emerging on the left from the mountains of Nagano, and its downstream portion flowing on the right almost imperceptibly into Ise Bay. Sharing the center of the painting with the Kiso River are the dark, hulking crags of the "Momotarō Alps," which together with the river form a landscape that the artist labels the "Japanese Rhine." Like his nineteenth-century predecessors in panoramic mapmaking (notably Kuwagata Keisai and Utagawa Sadahide; see chaps. 21 and 22), Hatsusaburō willfully eschewed a linear perspective and uniform scale, preferring to emphasize specific parts of the landscape. Hence, whereas Hatsusaburō used a low, wide angle for much of his Kiso River map, encompassing more than one hundred kilometers in one vast sweep (from the inland town of Kisofukushima on the left to the coastal city of Nagoya on the right), he exaggerated the size of the Japanese Rhine and the newly created tourist sites of the Momotarō Shrine and Inuyama Amusement Park, allowing those three locations to dominate the foreground. He compensated for the distortion this causes by shrinking less noteworthy places in the landscape and by wrapping the red rail lines and stylized meandering of the river around and away from the exaggerated foreground. The result is characteristic of Hatsusaburō's bird's-eye views: a subtle but increasing recession of the landscape as one moves from the center to the sides of the map.

Along with this purposeful distortion of space, Hatsusaburō's panoramas are notable for their stylized features and bright, clear color scheme. In his "Eight Renowned Landscapes" and similar images, Hatsusaburō and his studio of apprentice artists tended to use a fixed set of stylized elements, such as zigzagging rivers and almost cartoonish trains and automobiles, to depict the various places they had been commissioned to represent. These features effectively created an easily identifiable image that was closely associated with both the artist and the publishing company he founded. No doubt, the same stylized elements also allowed him to leave much of the detail work to his apprentices.

There seem to have been very few cloudy days in Hatsusaburō's repertoire. The only disturbance to the ubiquitous blue sky in his "Eight Renowned Landscapes" is the rare white plume of steam or smoke from one of the country's many volcanoes. Despite providing a consistently sunny appearance, Hatsusaburō nonetheless employed atmospheric coloration to add a greater sense of depth to his distorted landscapes. In his rendering of the Kiso valley, the horizon is a faint yellow, with the mountains leading to it becoming progressively hazy and indistinct, leaving the famous peaks of Ontake and Komagatake identifiable only by their white labels. In contrast, areas in the foreground are crisply colored, with the river itself bright blue, areas on both sides of the river a brilliant yellow, and the surrounding hills in shades of green. The use of similar hues in the other seven panels contributes to an overall sense of pictorial and spatial unity.

Given the painterly character of Hatsusaburō's bird's-eye views, one may be tempted to ask if they are maps at all. Certainly, the artist did not intend for the "Eight Renowned Landscapes" to be used strictly as a navigational tool. Nonetheless, with its portable format, labeling of places, accompanying explanation of fares and sights to visit, and official promotion by the Ministry of Railways, Hatsusaburō's magazine insert was clearly intended to provide prospective tourists with wayfinding cues for active travel. Tellingly, Hatsusaburō's map of the Kiso River for this insert was based in part on an earlier map he had made for a 1924 railway guidebook. From the accompanying text that Hatsusaburō wrote for the mostly middle-class female readers of *Housewife's Companion*, it is clear that he was inviting his audience to imagine riding the train from Nagoya to Imawatari, where they could step aboard a riverboat to float downstream through the Japanese Rhine and on to the nearby shrine and amusement park.

As with film viewing or a railway journey at the time, the reproduction and distribution of Hatsusaburō's "Eight Renowned Landscapes" in upwards of six hundred thousand copies of *Housewife's Companion* meant that these often remote landscapes contributed to what the historian Wolfgang Schivelbusch has described as the modern "perception of *montage*, the juxtaposition of the most disparate images into one unit—the new reality of annihilated in-between space . . . [that] brings things closer to the viewer as well as closer together."[3] In just this way, the "Eight Renowned Landscapes of Japan" offered readers a montage of interlocking maps that juxtaposed the mostly urban content of *Housewife's Companion* with the often rural and remote location of these scenic places. For just as the bending lines of recession in Hatsusaburō's bird's-eye views distorted space to create pictorial unity, so were the newly improved highways, railroads, and ferry lines that feature so prominently in his panoramas binding the nation together to provide modern routes through old Japan.

Notes

1. "'Nihon shin-hakkei' no sentei," *Tōkyō nichinichi shinbun* April 9, 1927; "'Nihon shin-hakkei' suisen tōhyō kekka," *Tōkyō nichinichi shinbun* June 5, 1927; "Gyōgi jitsu ni 13 jikan-han 'shin-hakkei' kettei su," *Tōkyō nichinichi shinbun* July 4, 1927.

2. Yoshida Hatsusaburō, "Nihon hakkei meisho zue." *Shufu no tomo,* August 1930.

3. Wolfgang Schivelbusch, *Railway Journey: The Industrialization and Perception of Time and Space in the Nineteenth Century* (Berkeley: University of California Press, 1986), 42.

Suggested Readings

Hanes, Jeffrey E. "Contesting Centralization? Space, Time, and Hegemony in Meiji Japan." In *New Directions in the Study of Meiji Japan,* edited by Helen Hardacre, 485–95. Leiden: Brill, 1997.

Hotta Yoshihiro. *Yoshida Hatsusaburō no chōkanzu o yomu: egakareta kindai Nihon no fūkei.* Tokyo: Kawada Shobō, 2009.

Sato, Barbara. *The New Japanese Woman: Modernity, Media, and Women in Interwar Japan.* Durham: Duke University Press, 2003.

Schivelbusch, Wolfgang. *The Railway Journey: The Industrialization of Time and Space in the 19th Century.* Berkeley: University of California Press, 1986.

Shirane, Haruo. *Japan and the Culture of the Four Seasons: Nature, Literature, and the Arts.* New York: Columbia University Press, 2012.

Smith, Henry D., II. "World without Walls: Kuwagata Keisai's Panoramic Vision of Japan." In *Japan and the World: Essays on Japanese History and Politics in Honour of Ishida Takeshi,* edited by Gail Lee Bernstein and Fukui Haruhiro, 3–19. London: Macmillan Press, 1988.

Wigen, Kären. "Discovering the Japanese Alps: Meiji Mountaineering and the Quest for Geographical Enlightenment." *Journal of Japanese Studies* 31 (1) (Winter 2005): 1–26.

PART III

Modern Maps for Imperial Japan

Introduction to Part III

Cary Karacas

The maps in this section appeared over a tumultuous century that ranged from the final decades of the Tokugawa shogunate to the emergence of the Japanese empire—an empire that by the early 1940s controlled much of Asia. This was a period of dizzying change brought on by hurried responses to the increasing demands of Western powers. The maps surveyed here reveal the convulsive industrialization, urbanization, and militarization of these years, as the Japanese leadership sought to secure great-power status for the country. Unlike the Edo-era maps discussed in previous sections of *Cartographic Japan*, the maps from this era resist ready classification. A potentially useful way to look at them, however, may be to distinguish among exploratory, evidentiary, and imperial cartography.

First, we witness the emergence of an exploratory cartography around the dawn of the nineteenth century. This movement was spurred by the desire to nail down the precise contours of the Japanese archipelago, part of the shogunate's response to foreign encroachment on its own claimed territory. Outstanding leaders of this exploratory mapping project were Inō Tadataka and Mamiya Rinzō (discussed in the chapters by Suzuki Junko and Brett Walker, respectively). Inō, a towering figure in Japanese cartography, led a seventeen-year survey of much of the coastline of the archipelago. While efforts to map the outline of Japan's main islands date back to the Kamakura period, and while their shape had come into much sharper focus in the early Edo period, Inō's technological achievement in precisely mapping the coastlines—together with establishing exact latitudinal and longitudinal positions and accurate distance and direction—constituted a significant advance. This and the sheer artistic beauty of his maps account for their high value to this day.

Given that Inō's efforts belonged to a new phase in the expansion of cartographic imaging and imagining of the Japanese archipelago, we can see why he chose to begin his survey in Ezo (modern Hokkaido), which still constituted terra incognita for most Japanese. Pushing beyond the familiar confines of Honshu, Shikoku, and Kyushu was one response to an emerging global dynamic in which powerful outsiders, with the assistance of indigenous informants, conducted geographic surveys in the northern Pacific and elsewhere in Asia (chapter by Tessa Morris-Suzuki). This concern to map the furthermost extent of Japan is further evidenced in the Tokugawa government's clandestine commissioning of another cartographer, Mamiya Rinzō, to explore Sakhalin. Here, as Brett Walker explains, we see the use of cartographic techniques to reimagine Japan's sovereignty in relation to the Russian and Chinese empires. In a distinct break with earlier practice, the resulting maps were jealously guarded by the Tokugawa shogunate.

The 1853 appearance of US warships near Edo led the shogunate to try a variety of measures to ward off the incursion of foreign powers. In addition to confining foreign visitors to a few trading ports, the shogunate undertook a kind of preemptive cartography as well. Government surveyors were ordered to map the sacred waters of Ise Bay in an effort to prevent the British from carrying out a hydrographic survey in this symbolically sensitive area (chapter 30, by Suzuki Junko). Effectively, the determination of British and other naval powers to develop an intimate understanding of Japan's coastal features forced the Tokugawa shogunate to carry out its own crash course in nautical cartography. Despite these and other proactive measures, however, the unequal treaties forced upon the Tokugawa shogunate proved deeply unsettling, initiating a period of tumult that led to the overthrow of the regime.

The forward-thinking samurai who led the Meiji Restoration set out to create a "rich country and strong army" (*fukoku kyōhei*), resulting in a spurt of what we might call evidentiary cartography. This broad genre took many forms. For starters, the Japanese countryside was blanketed with cadastral surveys, as the Meiji government struggled to put agrarian taxes on a standardized basis for the first time. Rural surveyors not only clarified the precise location and productivity of farm fields; as Daniel Botsman explains, they also helped lay the groundwork for siting modern facilities like workhouses, hospitals, and jails. Evidentiary cartography was also critical for the industrialization that was central to the "rich country" program, fueled by Japan's significant coal deposits. A pair of maps discussed by Brett Walker documents Japan's worst-ever coal mine explosion, graphic evidence of the human cost of the Meiji industrialization drive. Finally, urbanization marched in lockstep with industry, as laborers poured into the cities to take up jobs in the new capitalist enterprises. With the decision of the Meiji rulers to remake Edo as the "Eastern Capital," the city also became the "ecological nucleus of the nation," home both to the organs of a newly centralized state and to a new residence for the emperor, who was transferred from Kyoto to Tokyo to assume his position as head of the *kokutai*, or national polity.

Given Tokyo's preeminent political status, in addition to its continued position as Japan's most populous city, it is no surprise to see increasing attention directed toward the problem of fires, those so-called "flowers of Edo" that had repeatedly destroyed significant portions of the castle town, and which still threatened the highly flammable city. One way the pursuit of "civilization and enlightenment" (*bunmei kaika*) helped bring about new forms of cartography can be seen in Steven Wills's essay on Meiji-era attempts to apply geographic science to analyzing the pattern of historical fires. The devastating Tokyo earthquake of 1923, which destroyed much of the capital and neighboring Yokohama, led to another wave of evidentiary mapping, explored here by J. Charles Schencking. As in the case of the colliery disaster, here too an early manifestation of industrialization—namely, large populations of working-class families living in densely built wards replete with factories and with a dearth of open space—magnified the potential for disaster.

The destruction of Japan's capital provided an opportunity to rebuild a more modern city, with wider thoroughfares, iron bridges, ferroconcrete buildings, and urban parks (chapter by André Sorensen). It wasn't just buildings and avenues that rose from the ashes of the ruined capital. Modernity did as well. The word *modan* soon became a popular catchphrase connoting "technological progress, mass communication, rapid urbanization, consumerism, cosmopolitanism, self-indulgence, iconoclasm, diversity, and dissent."[1] While these features of modernity manifested themselves throughout urban Japan during the interwar period, Tokyo's large population made the capital the thriving center of the *modan*.[2] Tokyo intellectual Kon Wajirō initiated the influential new field of "modernology" studies, inspiring a young generation of scholars to map the moment. One such evidentiary map surveyed Shinjuku, a key rail hub that accommodated the transit, shopping, and entertainment needs of the ever-growing commuter suburbs (chapter by Henry D. Smith II).

Even as this fascinating micro-map of Shinjuku went to press in the fall of 1931, the scene was being set for the Japanese invasion of Manchuria, where the Japanese state was moving aggressively to create a new type of imperial space in the puppet state of Manchukuo. This gambit provided the content for another chapter in Japan's story as a modern Asian empire, one that had opened with the Sino-Japanese War of 1894–95.

Imperial cartography assumed various forms. In their first formal colony, Taiwan, the Japanese sought to build an agricultural powerhouse. This involved embarking on some of the most extensive water control projects yet to be carried out in Asia, a map-intensive undertaking (as discussed by Philip Brown). The next significant acquisition—the Korean Peninsula—followed after Japan's victory in the Russo-Japanese War (1904–5). Making colonial Seoul the site from which Japan administered Korea following its annexation in 1910, the Japanese engaged not only in monumental building projects but also in an ambitious attempt to restructure the ancient city's street system (chapter by Todd Henry). While urban planners had to limit themselves to piecemeal attempts at land readjustment in Seoul, they could let their imaginations run rampant in Manchuria and elsewhere on the Asian continent, incorporating Western planning principles into their dream of creating cities from scratch (chapter by Carola Hein).

In the 1930s Japan's economy became ever more connected to the mainland, as shown in a map highlighting the country's port capacity in the middle of what has been called Japan's "brittle decade" (chapter by Catherine Phipps). The symbolic denouement of Japan's empire-making project came in 1940 with the grand pronouncement of a Greater East Asia Co-prosperity Sphere. Even children's maps at the time sought to convey East Asian solidarity in the face of Western colonialism. As David Fedman argues, we can consider these maps a localized, propagandistic expression of imperial cartography.

Notes

1. John Dower, "Modernity and Militarism," in *The Brittle Decade: Visualizing Japan in the 1930s*, ed. John Dower, Anne Nishimura Morse, Jaqueline Atkins, and Frederic Sharf (Boston: Museum of Fine Arts 2012), 9–49; quote from p. 11.

2. For an important examination of so-called second cities in Japan during the interwar period, see Louise Young, *Beyond the Metropolis: Second Cities and Modern Life in Interwar Japan* (Berkeley: University of California Press, 2013).

29 Seeking Accuracy: The First Modern Survey of Japan's Coast

SUZUKI Junko 鈴木純子

FIGURE 29.1 Exhibition of the replicas of *Inō-zu* at Koganei General Gymnasium in Koganei City, Tokyo. Photograph by author.

At the beginning of the nineteenth century, a lasting impression was made on the history of Japanese mapmaking by Inō Tadataka (1745–1818) and his team of surveyors, who succeeded in creating accurate maps of the coastline of the Japanese archipelago from Hokkaido to Kyushu. Their work resulted in two hundred and fourteen large-scale maps (1:36,000), eight medium-scale maps (1:216,000), and three small-scale maps (1:432,000), which are all commonly referred to as *Inō-zu*.

In recent years, panels of *Inō-zu* replicas have sometimes been laid out and displayed on the floors of gymnasiums, and the public has been able to walk over the exhibits (fig. 29.1). These events have typically drawn crowds of several thousand spectators daily, demonstrating the fascination inherent in the maps as well as their continuing social impact even after two hundred years. One reason for the popularity of the *Inō-zu* is the interest and respect people continue to feel for their creator. For this reason, it is important to gain some insight into the life of Inō Tadataka himself if we are to truly understand the *Inō-zu* maps.

At least three reasons can be given to explain the widespread interest in Inō as a person. First, we must consider the life he led. Inō spent most of his life in a village on the outskirts of Edo. His early adulthood was spent actively participating in his family's successful sake-brewing business; only in retirement did Inō turn his attention to mapmaking. Inō retired at forty-nine years of age and moved to Edo, where he became a student of the prominent astronomer Takahashi Yoshitoki at the Astronomical Office of the Tokugawa shogunate. It was there that Inō learned the principles of astronomy and the study of the calendar while perfecting his skills in astronomical observation. By the time Inō had finished his studies and begun surveying, he was fifty-five years old.

Inō's surveys took a total of seventeen years to complete and were presented to the shogun in 1821. (Inō himself had died in 1818, but the maps were completed by his team members.) During their surveys, Inō and his colleagues traveled a total of nearly forty thousand kilometers. That Inō led such a rich life as a senior citizen and that his team walked such great distances have captured the imagination of the Japanese public.

A second reason for the popularity of these maps has to do with the wealth of historical records associated with them. Inō's descendants have preserved not only his personal diary, in which he detailed the entire journey, but also numerous letters and other records. In addition, records of encounters with the survey team are preserved in local areas throughout Japan. In some instances we even know the menu of meals consumed by the surveyors as they traveled. Fans of Inō are captivated by a rich historical record that includes everything from details of local Japan to life-sized images of Inō himself.

Finally, the *Inō-zu* remain popular today for their high degree of perfection. Aesthetically pleasing in and

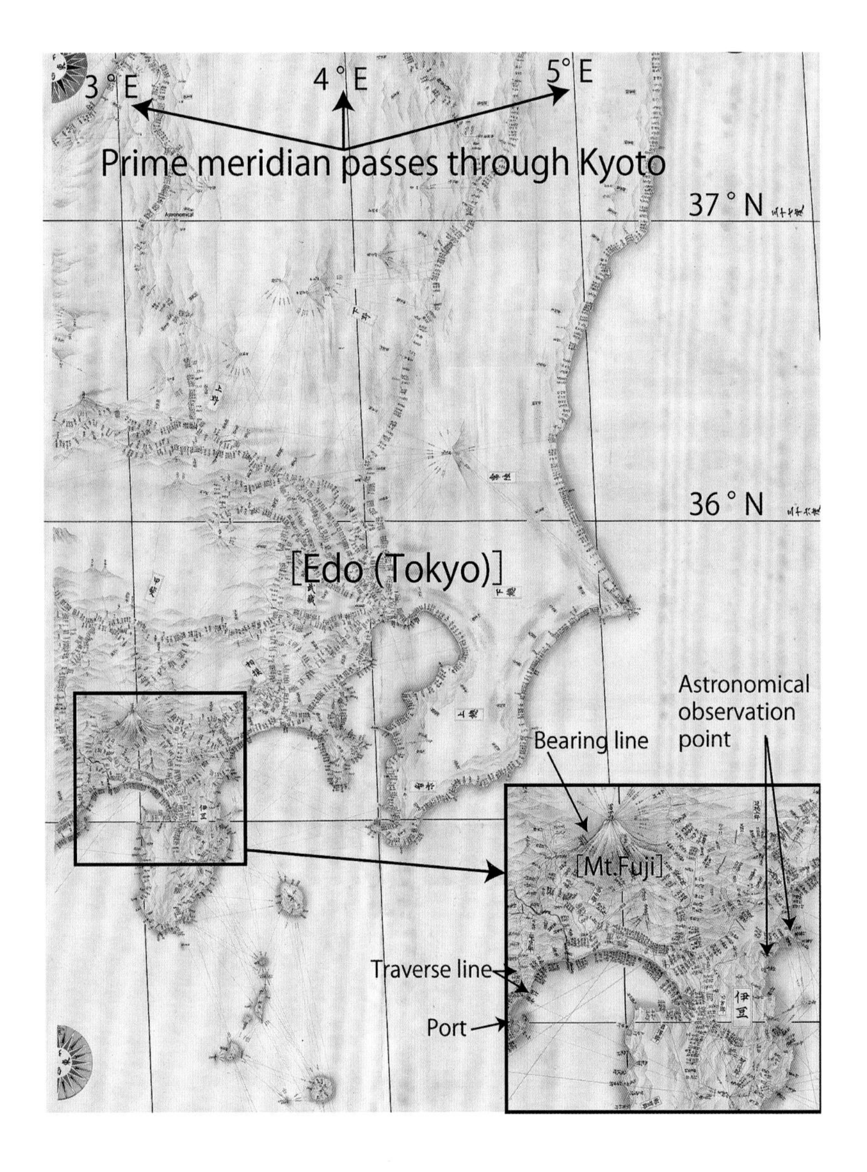

FIGURE 29.2 Detail from "Coastal Map of Great Japan" [Dainihon enkai yochi zenzu 大日本沿海輿地全図], by Inō Tadataka 伊能忠敬, ca. 1821. Manuscript, 280 × 159 cm. Courtesy of Nissha Printing Co. Ltd., Kyoto, Japan. English labels added by author.

of themselves, the maps lead viewers to a greater awareness of history, thanks to their rich information regarding past place-names and landscapes.

This leads us to the maps themselves. Figure 29.2 shows one of Inō's medium-scale maps, representing northeastern Japan as far south as the Izu Peninsula. This magnificent image can be said to represent a new development in modern Japanese cartography from two perspectives: the technical skill of its production, and the particular way in which it interprets the national territory.

The greatest technical characteristic of these maps is the way that each point corresponds to an absolute location on the earth's surface. At each stop during their travels, Inō's team would take astronomical readings on clear nights to establish latitude. The *Inō-zu* were the first maps produced in Japan to correctly show latitude through actual measurement in this way. Stars are included on the map to indicate places where astronomical observations were taken. Longitude, by contrast, was calculated by measuring distances traveled. Lines are included on small- and medium-scale series of the *Inō-zu* to indicate longitude and latitude. Latitude is expressed in parallel horizontal lines, to which the central meridian running through Kyoto forms a right angle. Although the maps' graticule resembles a Sanson-Flamsteed projection, it is in reality an original system of Inō's creation.

The distances and directions of coastlines and roads were measured in short linear segments according to the principles of traverse surveying. As the survey progressed, directional errors were corrected by comparison to locational measurements taken of prominent objects along the way. While such methods were conventional, Inō's use of trigonometry and specially ordered precision instruments conveys a sense of his commitment to accuracy, in line with modern standards of cartography.

Most of the coastlines and roads within the *Inō-zu* are represented by jagged red lines. These represent the traverse lines and form the framework of the maps. Correctly detailing the coastlines was the main goal of the *Inō-zu*, which can essentially be considered a coastal survey. Along the traverse lines are labeled the names of villages encountered on the way. Interestingly, at each corner of the traverse lines on the original sheets, a small needle hole can be found. These holes were formed during the process of copying the maps. The original was placed on top and a needle was used to mark the traverse corner points onto a sheet of paper below.

In the areas between the traverse lines, the maps are beautifully illustrated with colorful figures of mountains and cartographic symbols. Mountains are depicted in light green, water in blue; beaches and riverbanks are painted yellow. Red is used for cartographic symbols like the stars mentioned above. Small circles represent post towns, and boats represent ports. Other figures are used for provincial and county borders, castle towns, temples, and shrines. Beautiful compass roses also adorn the *Inō-zu*. In addition to indicating magnetic north, the compass roses serve as junction points for adjoining sheets.

From various places on the map, bearing lines are drawn to connect with features like Mount Fuji that can be seen from great distances; each of these lines is labeled with its azimuth angle. These lines were used when initially establishing the location of each place and are unnecessary for the final map. However, the red bearing lines do add pleasant color to the map and impart to the viewer a greater sense of the effort that went into making them as accurate as possible. Like the jagged coastlines (which would normally be redrawn into smooth curves on a traverse map) and the stars marking where astronomical observations were taken, the bearing lines emphasize the empirical nature of Inō's cartographic process.

It is important to note that the nature of the *Inō-zu* led to a new way of thinking about the homeland itself. Previously, all official maps produced by the shogunate were created piecemeal, using the province as the basic unit. In contrast, Inō sought from the beginning to illustrate the country as a whole. Although the project began as a personal endeavor that received only minimal government funding, the shogunate gradually took interest in it and, from 1805, the remainder—especially the western part of Japan—was surveyed under direct control of the shogunate. The sections of *Inō-zu*, especially the large-scale series produced by dividing the original maps at lines of latitude and longitude, strongly resemble modern topographic base maps and again give the impression of an effort undertaken with an eye to grasping the nation in its entirety.

Why were these maps created at that moment in history? It is certain that the personal resolve of Inō Tadataka to embark on an unprecedented journey across the entire country was instrumental in bringing about the

Inō-zu. However, it can also be said that the time was ripe for such an endeavor. Toward the end of the eighteenth century, Western ships were stepping up activities in East Asia and the shogunate was struggling to find a way to cope with the continual arrival of foreign vessels. At the same time, "Dutch studies" was coming into its own, and Inō Tadataka had friendly relations with leading Japanese scholars of European science and art. This sort of interaction probably helped Inō see Japan as one nation among many in the world. Also, Inō did not come from a samurai background but belonged to the educated commoner class, who were beginning to lead society in new directions both economically and culturally.

The impact of this heroic cartographic effort was mixed. On the one hand, we can see from his diaries that as Inō's team traveled the country they associated with local technical experts, advancing the development of surveying techniques throughout the country. Furthermore, the Astronomical Office of the Tokugawa government continued to investigate the measurement of longitude and other topics after Inō's team completed its project. Ironically, however, the Tokugawa eventually shelved the *Inō-zu* and halted the development of modern cartography. It was not until the closing days of the shogunate, a half century later, that the *Inō-zu* began to be widely seen and appreciated.

The new Meiji government thought highly of the *Inō-zu* and used them as a basis to develop even more accurate cartography. Meiji-era maps, in fact, often clearly state that they are based on *Inō-zu*. To this day, the Historiographical Institute of the University of Tokyo holds a number of as-yet-uncataloged maps from the early Meiji period. Among these works are sheets of paper used to carefully trace the needle holes of the *Inō-zu*. There are also many examples of map drafts that are essentially *Inō-zu* to which modern-style hachures have been added. Such items give us a glimpse of how influential the *Inō-zu* were in the mapmaking processes of the period.

(Translated by Bret Fisk)

Note

This work was supported by JSPS KAKENHI Grant Number 24520737.

Suggested Readings

Gardiner, R. A. "Inō-Tadataka and Philipp von Siebold." *Geographical Journal* 142 (1) (March 1976): 176–77.

Hoyanagi, Mutsumi, ed. *Inō Tadataka no kagakuteki gyōseki* [The scientific achievement of Inō Tadataka]. Tokyo: Kokon shoin, 1974. (Although the book is written in Japanese, the last thirty pages provide a summary in English.)

Ōtani, Ryōkichi. *Tadataka Inō, the Japanese Land-Surveyor*. Tokyo: Iwanami Shoten, 1932.

30 No Foreigners Allowed: The Shogunate's Hydrographic Chart of the "Holy" Ise Bay

Suzuki Junko 鈴木純子

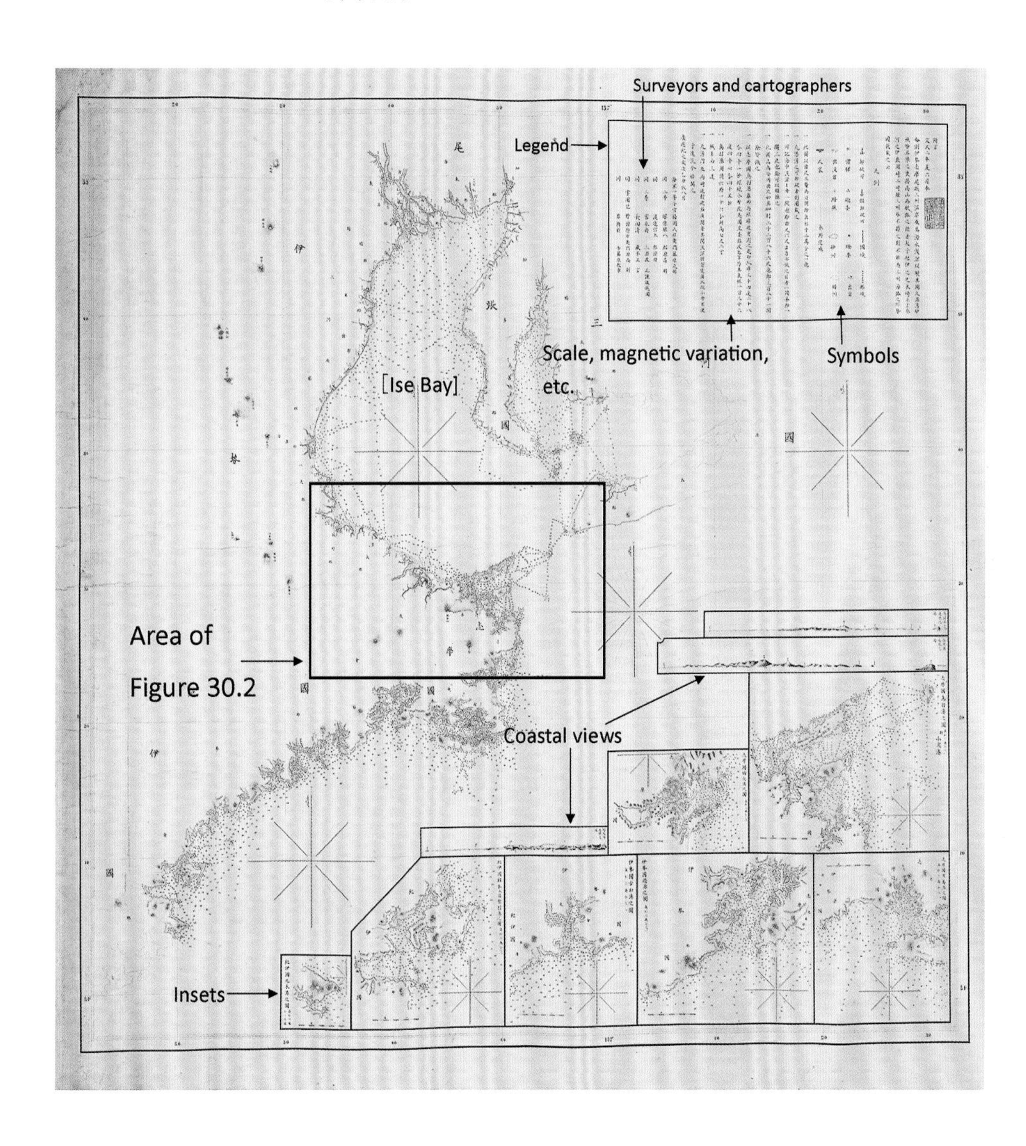

FIGURE 30.1 "Chart of the Biseishi [Owari, Ise, and Shima Province] Coast" [Biseishi kaigan jissokuzu 尾勢志海岸実測図], by Fukuoka Kyūemon 福岡久右衛門, Tsukahara Ginpachirō (塚原銀八郎), et al., 1865. Manuscript, 128 × 125 cm. Courtesy of the Historiographical Institute, the University of Tokyo. English labels added by author.

Ise Bay is located in the center portion of the south end of Japan's main island, Honshu. It is an enclosed bay that curves inland between the Atsumi and Shima Peninsulas. Near the entrance of the bay lies Ise Shrine, dedicated to the ancestors of the imperial family and long considered an especially holy area. Ise Bay, situated as it is in front of these sacred precincts, is the focal point of a hydrographic chart called the "Chart of the Biseishi Coast" (fig. 30.1). This chart was produced during the last days of the Tokugawa shogunate by officers from the shogunate's Naval Training School using modern Western techniques that they had only recently begun to study.

The name "Chart of the Biseishi Coast" is based on a combination of one character each from the names of the three provinces adjoining Ise Bay: Owari (Bi), Ise (Sei), and Shima (Shi). The map exists only as a handwritten copy. It was never printed and was never used for navigational purposes by either foreign or domestic vessels. In spite of this fact, eventual presentation to foreign powers, specifically Great Britain, was actually one purpose behind the map's creation. Below, I will first discuss the contents of the chart before moving on to a discussion of how foreign pressure and the appearance of Western vessels in the coastal waters of Japan prompted the chart's creation.

It is important to remember that the chart currently in existence is a reproduction created by the Geographical Bureau of the Meiji government based on the original, which was at that time kept by the Hydrographic Office of the Navy. It was long thought that this map had been completely lost to history, but the present handwritten copy was found several years ago among a collection of works that came into the possession of the Historiographical Institute of Tokyo University.[1] It has now been ascertained that, although the chart is a reproduction, it seems likely that a draftsman who had worked on the original, Iwahashi Shingo (also known as Iwahashi Noriaki), was subsequently employed by the Meiji government and was involved in its creation. We can therefore assume that the reproduction is accurate even though the original seems to have been lost in a fire caused by the Great Kanto Earthquake in 1923.

The hydrographic chart itself is a practical and objective representation of information regarding the coastal waters, seemingly devoid of other motives or intents. However, the fact that it was created for presentation to foreign powers becomes apparent when the chart is compared to others prepared by the same group of technicians during this period for Kanagawa Bay, Edo Bay, Osaka Bay, and the Ogasawara Islands. Put simply, the

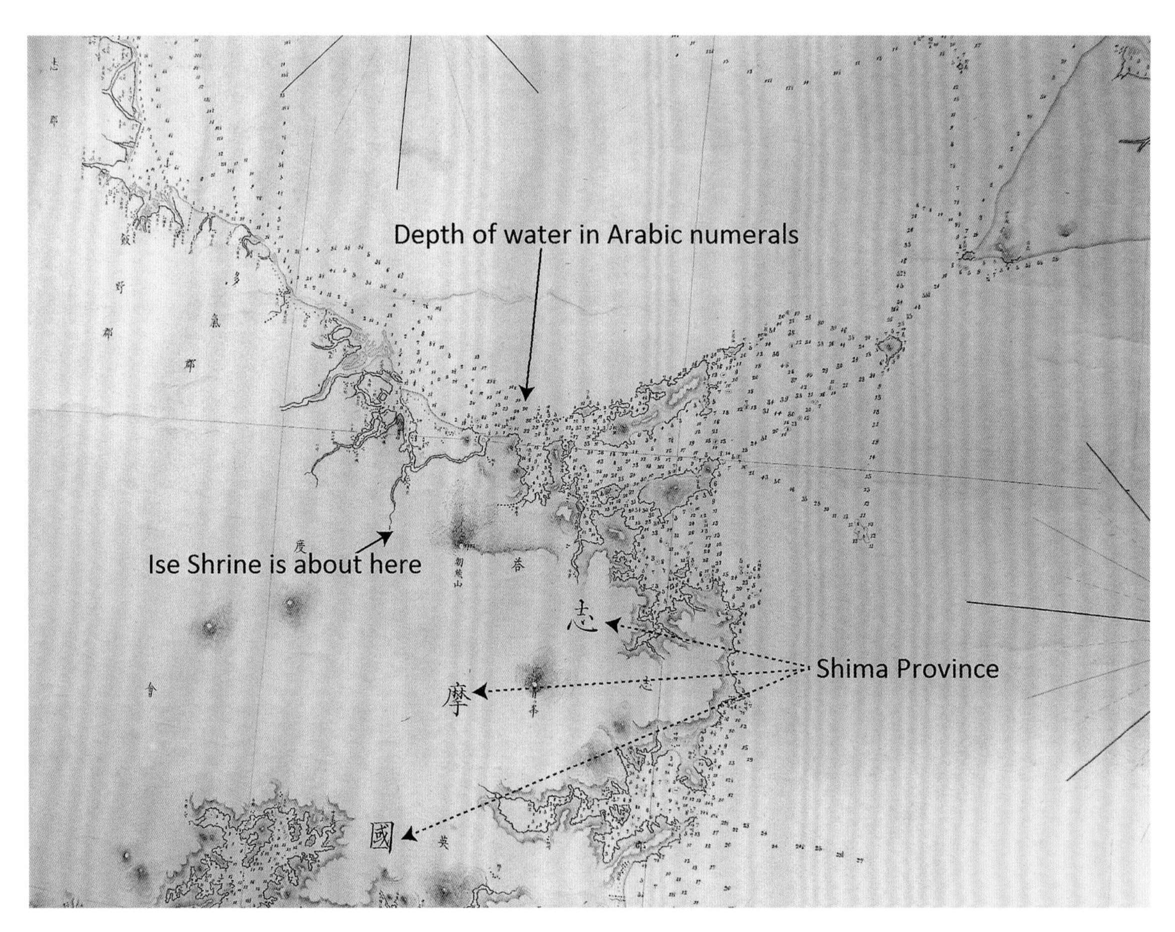

FIGURE 30.2 Detail from figure 30.1, showing the coastline in front of Ise Shrine, west of Irago Channel, at the entrance to Ise Bay. English labels added by author.

Ise Bay chart is produced in a Western fashion and much more in line with international standards.

As seen in figure 30.1, the general layout consists of two images: the main chart itself and an illustration showing the coastal view. The chart was constructed to serve well as a pasteup, and the coastal view was probably based on those found in Western charts. One difference from Western charts is the lack of a title inscription, but this is probably attributable to the fact that Japanese maps were folded for storage and were therefore generally labeled on the outside. The title "Chart of the Biseishi Coast" is written in brush and ink on the reverse side.

At the top right of the chart are explanatory notes regarding such elements of the map's composition as area displayed, symbols used, methods of displaying both depth of water and land elevations, and overall scale. In addition, information is provided regarding the time frame in which the chart was produced, the surveyors and draftsmen involved, locations where measurements were taken to provide data regarding latitude and longitude, tide times, and magnetic deviation. Five surveyors are listed, including Fukuoka Kyūemon, and the two draftsmen mentioned are Nomura Sōemon and Iwahashi Shingo. Although the date of the inscription is 1865, we know that surveying for the project began in 1862.

Elements that exhibit a Western influence include indication of scale, indication of water depth and land elevation, and reckoning of longitude from the prime meridian at Greenwich. Japanese units of measurement for depth and elevation on the chart, such as *ken* and *shaku*, are represented with Arabic numerals. This method of numerical indication is also a major difference from contemporary charts that provided the same information in Chinese numerals. It was still impossible at this point to make a complete change from using Japanese units of measurement to using internationally recognized units. Within the explanatory notes there is even an interesting passage directed toward Japanese readers that explains the Arabic numeral system. It should also be noted that the number of points used in the bathymetric survey were still very few, and the chart does not include depth contours. Apart from the direct area of the survey, the chart depicts coastlines as they were presented in the *Inō-zu* maps of Inō Tadataka.

As mentioned above, one reason for the creation of the chart was the pressure exerted by Western ships to open Japanese ports to commerce. From the end of the eighteenth century, ships from Britain and Russia began to approach Japan's shores, and the question of how to deal with them was one of the shogunate's main problems. Once European and American powers had gained the ability to establish absolute location by means of astronomical observation and the use of chronometers, they quickly began creating modern-style nautical charts in the beginning of the nineteenth century. Such advances in nautical cartography allowed the accurate determination of location even at sea and led to the attainment of better sea routes. Nautical charts thus became a necessity for safe seafaring, and their construction was overseen by the hydrographic offices of the British and other navies.

Once the opening of Japan's ports became a priority, the creation of hydrographic charts of its coastal waters became necessary. Efforts that had until then been focused on surveying the Chinese coast were redeployed to survey the coasts of Japan after the end of the Opium War in 1842. The arrival and surveying activities of the British ships *Samarang* (1845) and *Mariner* (1849) shocked the shogunate and led to its requiring feudal lords with lands that included coastal areas to create maps indicating water depth. Although indications of coastal lines and water depth are integral to nautical charts, there were no maps in that day that included water depth, and the only maps with accurate coastal lines were the *Inō-zu* produced in the early nineteenth century. The maps that were now produced at the instigation of the shogunate did eventually include measurements of water depth taken at six designated points from the shore, but these measurements were merely enumerated on the chart with often inconsistent labeling. While these were the first maps in Japan to include such important data, it is clear that their purpose was not to aid in navigation but merely to hint at how close to shore foreign ships would be able to approach.

Foreign surveying of Japan's coasts intensified from the 1850s onward. Fleets led by Matthew Perry (1854) and John Rodgers (1855) both conducted forced surveys, and the British also stepped up efforts during the 1860s. These were the conditions that led to the creation of the "Chart of the Biseishi Coast."

During this period steps were taken within Japan to improve nautical surveying and to establish a navy. The Astronomical Office of the Tokugawa government realized that establishing methods for astronomical observation at sea and determining position and course with reference to latitude and longitude were a genuine part of its own mission. As a first step in this direction, they began to read and translate Dutch works on navigation. In 1855, this basic knowledge proved useful when training exercises were undertaken in conjunction with the Dutch navy in Nagasaki. Staff from the Astronomical Office including Fukuoka Kyūemon learned much during this training regarding navigation and surveying techniques and went on to be instructors at the Naval Training School established in Edo. Fukuoka is listed at the head of surveyors for the "Chart of the Biseishi Coast" and also worked on the Kanagawa Bay chart.

The above-mentioned intensification of British surveying efforts during the 1860s is directly linked to the creation of this map. In 1861 the British surveying ships *Actaeon* and *Dove* requested permission to survey the Pacific coast. The shogunate initially intended to deny this request and instead provide charts of their own making. The British, however, rejected this, and further negotiations eventually allowed the British to conduct the survey as planned (including temporary landings at important points), albeit while flying the insignia flag of the shogunate and allowing its officers to accompany the surveyors aboard ship.

Among the six Japanese officers to accompany the British were two of the surveyors who would later create figure 30.1, Fukuoka and his colleague Tsukahara Ginpachirō. It must be assumed that much of the transfer of technological knowledge that took place at this time was later utilized in the creation of the "Chart of the Biseishi Coast."

Opposition quickly arose once reports of the British plans reached the Ise area. As mentioned above, Ise Bay was considered sacred. Not only did the local feudal lords protest, but the imperial court also ordered that foreign vessels be forbidden from entering the bay. The shogunate quickly renegotiated so that this area alone would be surveyed by the Japanese, with the results to be subsequently presented to the British. These events are the direct precursors to the creation of the "Chart of the Biseishi Coast." Only fifteen years had passed since the first maps to show coastal water depth had been produced domestically. During that short period, the Japanese had learned modern Western methods of nautical cartography. These rapid changes underscore the tension felt domestically to meet the challenge presented by the European and American powers.

During 1863 the British again requested permission to survey Ise Bay. At this time the proposal to present the "Chart of the Biseishi Coast" to the British was reconsidered within the shogunate, and the plan was finally abandoned. It was decided that presenting the chart to the British would be tantamount to letting them into the bay itself. As a result, the chart was never to fulfill its intended purpose. Nonetheless, the surviving map—a portion of which was reissued by the Meiji navy in 1873—serves as a symbol of the process of modernization and the development of modern nautical cartography in Japan.

(Translated by Bret Fisk)

Notes

This work was supported by JSPS KAKENHI grant number 24520737

1. Part of a collection of maps that the Historiographical Institute of Tokyo University received from the Ministry of the Interior. The collection was unorganized and housed in the Akamon Shoko vault until it was demolished in 2010. The collection is currently being organized.

31 Indigenous Knowledge in the Mapping of the Northern Frontier Regions

Tessa Morris-Suzuki

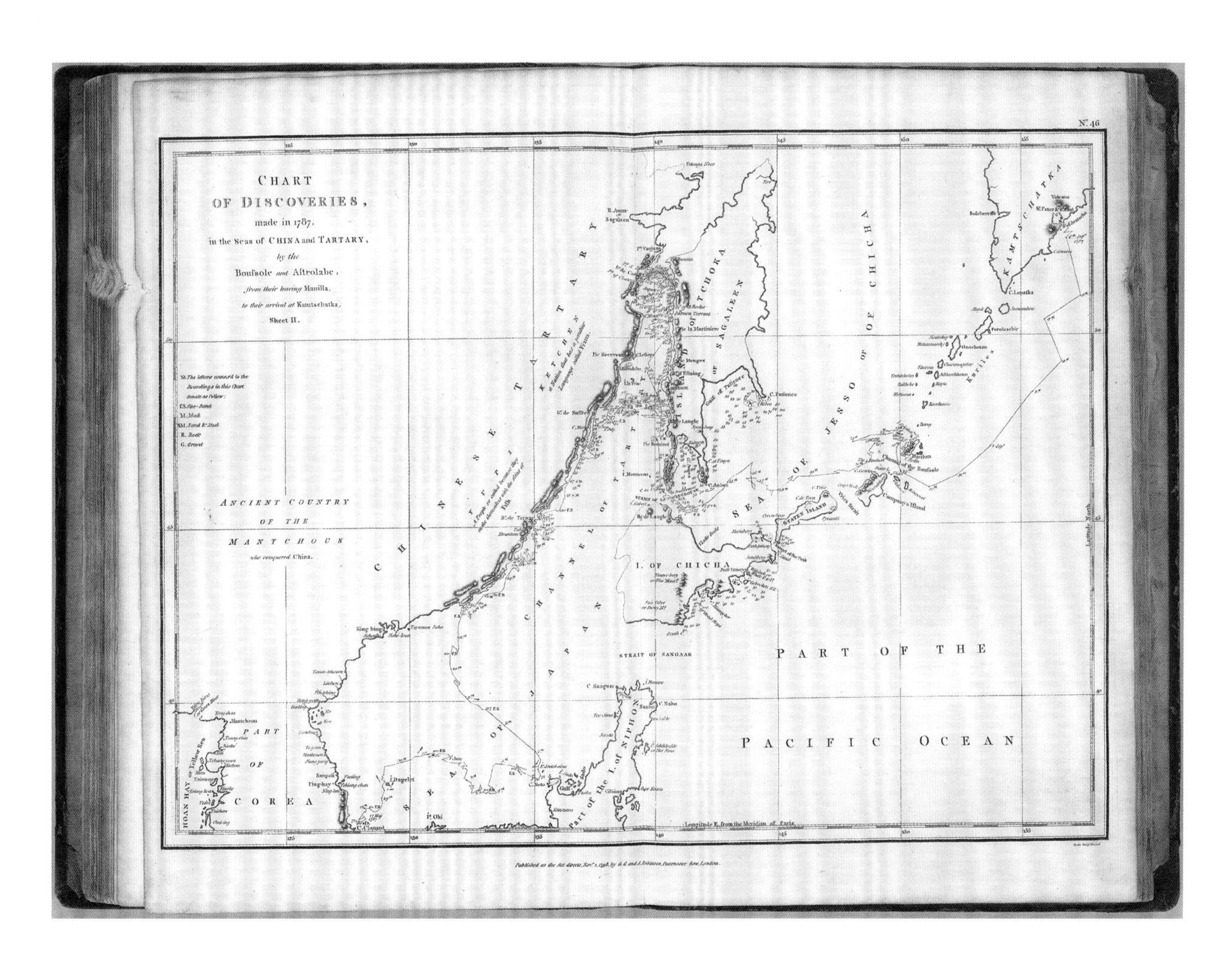

FIGURE 31.1 "Chart of the Discoveries made in 1787 in the Seas of China and Tartary by the Boussole and Astrolabe from their leaving Manilla and Arriving in Kamtschatka, Sheet II," by Jean-François de Galaup La Pérouse. Unpaginated illustration from *Charts and Plates to La Perouse's Voyage* (London: G. G. and J. Robinson, 1798). Copperplate engraving, 39 × 50 cm. Courtesy of David Rumsey Historical Map Collection.

On July 12, 1787, the French explorer Jean-François de Galaup de la Pérouse and the crew of his two ships, the *Boussole* and the *Astrolabe*, made landfall close to a village on a bay on the west coast of the island of Sakhalin, which La Pérouse named De Langle Bay after the captain of the *Astrolabe*. The Ainu inhabitants of the village called it Tomarioro, meaning (appropriately enough) "harbor" or "stopping place for boats."

La Pérouse's voyage was an attempt to map and describe many areas of the world still unknown to European geographers (fig. 31.1). Among these mysterious realms was the far east of "Tartary," stretching from the north of Japan to the eastern limits of the Chinese and Russian empires—the areas that we now know as Hokkaido, Sakhalin (or Karafuto), the Kurile (or Chishima) Islands, northeast China, and eastern Siberia. Hokkaido and its neighboring islands were at that time called Ezo in Japanese (spelled "Jesso" on the La Pérouse map), and dimly known Sakhalin was often referred to as Oku Ezo (Deepest Ezo).

The French expedition to the northern fringes of Japan provides fascinating insight into the fluid and shifting nature of Japan's boundaries and involved surprisingly complex exchanges of geographic knowledge among many different peoples. La Pérouse visited Sakhalin at a time when the northernmost Japanese domain of Matsumae was gradually extending its control, via coastal fishing and trading posts, to the northern limits of Hokkaido and into the southern parts of Sakhalin. But many Hokkaido Ainu communities still governed themselves and engaged in trade with their southern and northern neighbors: the Japanese to the south, the indigenous Nivkh and Uilta people who inhabited central and northern Sakhalin, and other indigenous communities on the eastern seaboard of the Asian mainland. These indigenous trading routes stretched up the western coast of Sakhalin Island and as far as the Amur River.

This was also the time when the Russian empire, which by now extended as far as the Kamchatka Peninsula, was rapidly extending southward into the Kurile Islands. Both the Japanese shogunate and the northernmost Japanese domain of Matsumae were becoming increasingly aware of the presence of foreigners ("Red Ezo," as they were known, after the color of their hair) on the northern fringes of Japan. Scholar-officials Ōishi Ippei and Mogami Tokunai were commissioned to explore the Kurile Islands and Sakhalin in 1786, and Mogami made many subsequent visits to the region. In 1801 two other samurai officials were sent to Sakhalin to find out more about the island and its connections to the lands beyond. Mamiya Rinzō and Matsuda Denjūrō in 1808 made an extensive survey of the coast of Sakhalin, and the following year Mamiya traveled via Sakhalin to the Asian mainland and as far as the Amur trading post of Deren (see chap. 32). These Japanese explorers were eager both to define the northern borders of Japan and to gather ethnographic information on the "Red Ezo."

Both Japanese and French explorers sought ethnographic as well as cartographic knowledge and produced detailed descriptions of the various peoples whom they encountered on their travels. The French explorers, though, were not content merely to describe. The expedition's doctor, Claude-Nicolas Rollin, brought the latest anthropometric techniques with him and conducted detailed measurements of the bodies of Sakhalin Ainu and of the indigenous people of the lower Amur. The maps produced from La Pérouse's voyage were also meticulously detailed, replete with numbers indicating latitude and longitude, dates of measurements, and estimates of sea depths and heights of mountains.

Modern science and indigenous knowledge intertwined. Both French and Japanese explorers made extensive use of the geographic knowledge of indigenous people in their mapping of the region, although in rather different ways. La Pérouse's crew could not speak any of the local languages and communicated with indigenous Ainu and Oroch people via sign language. He discovered that they had excellent knowledge of the geography of the west coast of Sakhalin and the mouth of the Amur and were able to sketch maps for him showing the outlines of the coast and the distances between various key points. Thanks to their mapping he was able to fulfill one of the main aims of his voyage by establishing that Sakhalin was an island separate from the Asian mainland.

But the La Pérouse map of the region embodies a crucial mistake, which tells us much about the limitations of communication between local residents and European explorers. When the Oroch people of the lower Amur sketched a map of the region for him, they pointed to the narrow straits between Sakhalin and the mainland and made gestures of pulling canoes across sand. La Pérouse interpreted this as meaning that the straits were blocked by sandbanks, which he carefully proceeded to inscribe on his maps, and which continued to be marked on many European maps until the mid-nineteenth century.

In the search for knowledge about the region, and about Russians and other "Red Ezo," Japanese explorers, like their French counterparts, also learned from the indigenous people. But, unlike La Pérouse, Japanese explorers like Mogami and Mamiya could make use of Ainu interpreters who spoke Japanese, other local languages, and in some cases also Russian. This enabled them to gain a much closer and more intimate knowledge of the local landscape and people. Mamiya, for example, although unable to travel to northern Sakhalin, provided a detailed description of its contours, and even an estimate of its population, on the basis of information given to him by the Ainu and Nivkh whom he met on his travels. From these descriptions and his own observations, Mamiya understood that the straits between Sakhalin and the mainland were passable, although subject to large tidal fluctuations that left broad sandbanks exposed on either side (the information that the Oroch people had clearly been trying to communicate to La Pérouse in sign language).

The accounts left by these Japanese travelers also enable us to see that, while Western explorers observed others, they themselves were also being observed. Mogami Tokunai, during a journey to Sakhalin in 1892, heard a fascinating account that offers a glimpse of an Ainu perspective on the scientific activities of La Pérouse's expedition. Ainu villagers told him that in the Tenmei era (1781–89), a ship of "Red-Haired People" had arrived at their village near the southern tip of Sakhalin. The foreigners sent a boat to the shore and summoned the local Ainu. "At last, many Ainu [Ezo] gathered, and they [the Red-Haired People] measured the Ainu from head to foot, and furthermore cut off a little of their hair with a small sword. The Ainu were absolutely astonished and alarmed, but they wanted to be helpful, so they stayed still." This is followed by a detailed Ainu description of a piece of equipment—possibly a tinderbox or sulfur matches—shown to them by members of La Pérouse's crew. "The Ainu people," says Mogami, "still tell the story of how fearful it was."[1]

At that time, Ainu from western Sakhalin traveled regularly to the Amur to pay tribute to Manchu officials, and there they met a multitude of peoples from as far away as the Korean borderlands and beyond. Chinese officials had conducted their own geographic and ethnographic surveys of the lower Amur region, describing and comparing different ethnic groups according to long-standing traditions of classification. The cartographic knowledge possessed by the indigenous people of Hokkaido and Sakhalin was not naive indigenous wisdom. It had been shaped by interactions with both the Chinese and the Japanese knowledge worlds and with numerous other ethnic groups.

On La Pérouse's map, national boundaries remain undefined. The activities of explorers like Mamiya were part of a process by which Japan would gradually come to claim full territorial control over Hokkaido and extend its claims into Sakhalin and the Kurile Islands. The mapping of this multiethnic region was a complex exchange in which Ainu, Frenchmen, Japanese, and others learned from one another. Knowledge flowed in multiple directions, along paths we are still retracing today.

Note

1. Mogami Tokunai and Yoshida Tsunekichi, eds., *Ezo sōshi* (Tokyo: Jiji Tsūshinsha, 1965), 187–88.

Suggested Readings

Bravo, Michael. "Ethnographic Navigation and the Geographical Gift." In *Geography and Enlightenment*, edited by David N. Livingston and Charles W. J. Withers, 199–235. Chicago: University of Chicago Press, 1999.

Dunmore, John, ed. and trans. *The Journal of Jean-François de Galaup de la Pérouse, 1785–1788*. Vol. 2. London: Hakluyt Society, 1995.

Latour, Bruno. *Science in Action: How to Follow Scientists and Engineers through Society*. Chap. 6. Cambridge, MA: Harvard University Press, 1987.

Walker, Brett J. "Mamiya Rinzo and the Japanese Exploration of Sakhalin Island: Cartography and Empire." *Journal of Historical Geography* 33 (2007): 283–313.

32 Mamiya Rinzō and the Cartography of Empire

Brett L. Walker

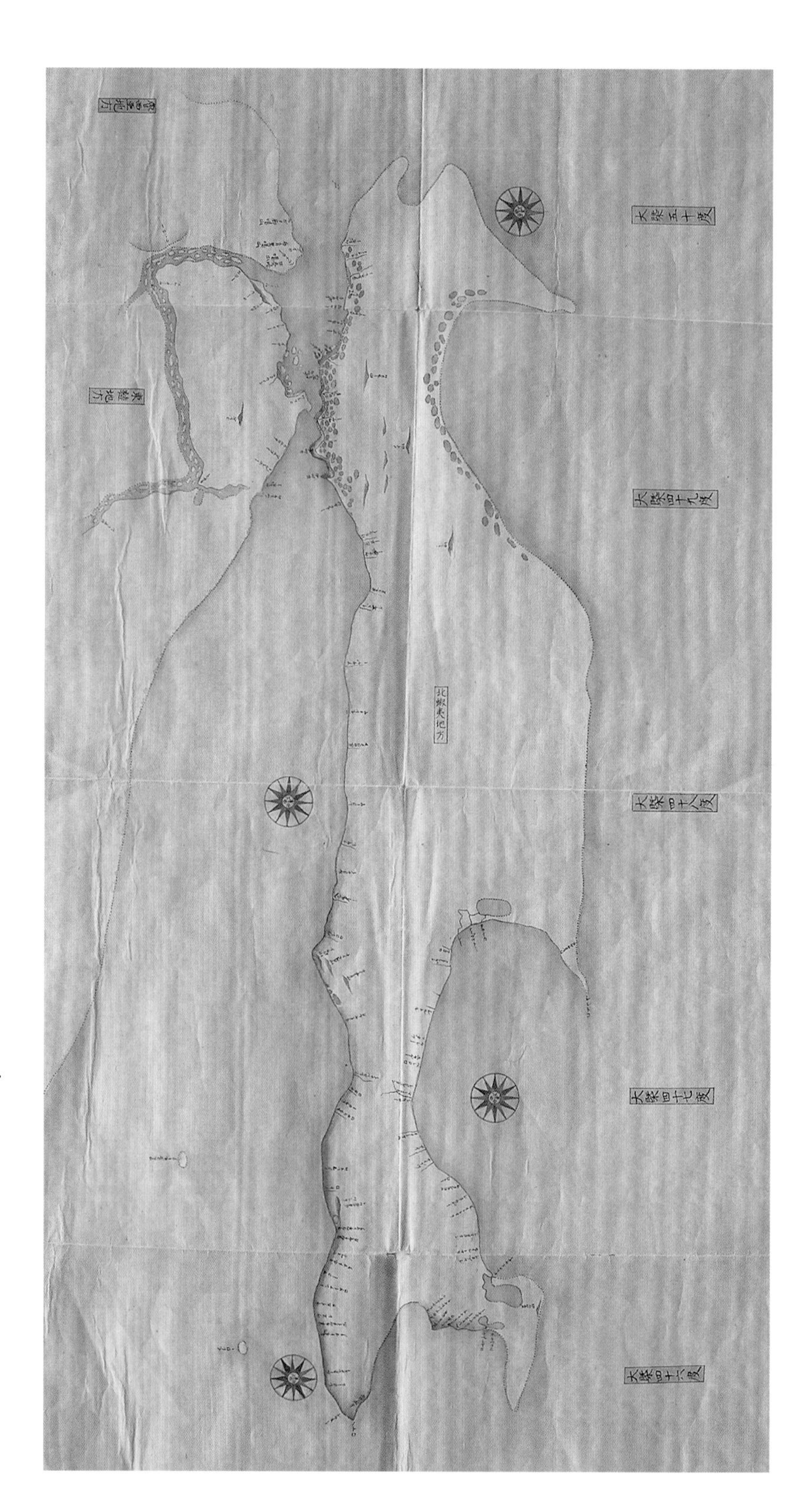

FIGURE 32.1 "Central Heilongjiang [Amur River] with Latitudinal Lines" [Kokuryūkō chūretsu narabini tendo 黒龍江中州并天度], by Mamiya Rinzō 間宮林蔵, 1810. Manuscript, 80 × 39 cm. Courtesy of the Resource Collection for Northern Studies, Hokkaido University Library.

In 1808, a Japanese explorer and cartographer named Mamiya Rinzō was sent on a secret expedition to probe the least-known margin of the Japanese archipelago. At the behest of the shogun, Mamiya and a companion sailed as far north as Sakhalin Island with a mandate to map the area and document its inhabitants, geographic features, and natural resources. During his travels, Mamiya drafted sketch maps of Sakhalin as well as of the central Heilongjiang, or Amur River, region, ultimately compiling the image shown here, which he called "Central Heilongjiang with Latitudinal Lines" (fig. 32.1). In the history of Japanese cartography, this map represents a significant departure from earlier mapping practices.

Mamiya's image of the far north effectively used Western cartographic science to reimagine Japan's sovereignty in relation to the Russian and Chinese empires. His surveying and mapping project contributed to envisioning the northernmost limit of the Japanese nation, as it might be imagined and understood through European eyes, as well as by the Qing empire. Prior to Mamiya's maps, variation in customs and clothing between Japanese and their Ainu neighbors—an early version of ethnic identity and cultural difference—had been taken as rough markers of national boundaries in the north. But where ethnic divisions between "Japanese land" and "barbarian land" had once sufficed to define the northern extent of the early modern Japanese state, by 1808 the complexities of the Qing tributary system and the steady progress of Russian exploration demanded a more precise demarcation of where Japan stood.

In a manner similar to that described by J. B. Harley, Mamiya's expedition to Sakhalin "anticipated empire" in the northern Pacific, and his maps played roles that were similar to those of European counterparts in empire building.[1] For starters, his map placed Sakhalin on a universally recognizable grid of latitude and longitude. At the same time, it symbolically emptied the land of its indigenous inhabitants, who were formally relegated to the pages of illustrated ethnographies. Both were radical moves. Mamiya's choice of cartographic idiom suggests an engagement with Western scientific techniques, at a time when policymakers around the world used maps as tools that not only constructed the nation but, increasingly, also imagined and reimagined the boundaries of sprawling empires.

Mamiya's northern foray and mapmaking enterprise came to symbolize Japan's age of northern exploration, which extended through the late eighteenth and early nineteenth centuries. These decades witnessed the journeys of Inō Tadataka and others as well, who, like Mamiya, plied the rivers and bays, and traversed the mountains and coastlines of Japan's northern latitudes. Unlike many of its mapmaking predecessors, this generation scrapped the traditional Japanese practice of writing extensive text or drawing pictures of native peoples on maps. Instead they favored longitudinal and latitudinal measurements, markings more common to the global early modern scientific and political community than to earlier Japanese map culture. Ethnography, in turn, was not only banished from the map but transformed as well, becoming characterized by empirical observations rather than secondhand information. The indigenous people that Mamiya met during his travels, such as the Ul'chi, Nivkh, and Sakhalin Ainu, replaced the "Cyclops" and other fictitious peoples covered in early Japanese and Chinese encyclopedias.

Mamiya and his travel partner Matsuda Denjirō spent months in Japan's northern borderlands in 1808. With the information they brought back from Sakhalin, Tokugawa officials and their teams of Edo cartographers, working in one of the global "centers of calculation," compiled a "Newly Revised Map of the World" (1810), which was designed to serve as a reference work for those in charge of Japan's increasingly delicate diplomatic affairs. Six years later, the shogunate issued an official copperplate engraving of the map, with revisions drawn from a much earlier Chinese compilation (the so-called Kangxi Jesuit Atlas). In 1810, Mamiya completed a draft of the extravagant "Map of Northern Ezo Island," a monumental work divided into seven large sheets with a scale of 1:36,000. Eventually, he reduced this unwieldy creation to one sheet and gave it the title it bears here,

"Central Heilongjiang with Latitudinal Lines." This one-sheet map proved critical to future Japanese interests in the northern Pacific. As Bruno Latour has observed elsewhere, "there is nothing you can *dominate* as easily as a flat surface of a few square meters."[2]

To gather geographic information, Mamiya and his companion relied on Ainu guides to journey into Sakhalin and the Amur River region. In the western Sakhalin town of Nayoro, Mamiya learned that Ainu traveled to the Qing outpost at Deren, on the Eurasian continent, where they paid tribute to Qing officials and, in return, received elaborate titles. On southern Sakhalin he discovered that a Japanese protoindustrial landscape, dominated by large-scale fisheries, overlapped with Qing imperial space in the form of the tributary order. One way that the Qing sustained that order in this remote outpost was through the dissemination of official titles and tributary duties. However, as he traveled through native villages, Mamiya was less struck by the fixity of economic zones or political allegiances than by the pervasive ethnic fluidity he saw. The northern Sakhalin Ainu had recently converted to the customs of the Ul'chi people, and a surprising number of merchants and adventurers were constantly coming and going between the island and the Asian mainland, exchanging material artifacts and more. In many regards, what he discovered was a cosmopolitan geography—and one that was far more engaged with northern Asia than it was with Japan, where the shoguns continued to enforce the restrictive maritime prohibitions that had been put in place in the 1630s.

In an age of heightened geopolitical competition, "Central Heilongjiang with Latitudinal Lines" was more useful than earlier Sakhalin maps because it aided Japanese visions of expansion. With Mamiya's maps and ethnographies in hand, later Japanese were able to travel through Sakhalin and the Amur River region as if for a "*second* time."[3] By keeping notebooks, writing ethnographies, and drafting scientific maps, Mamiya and his like-minded contemporaries constructed valuable knowledge about the edges of the Japanese world, far-flung terrains that were later incorporated into the expanding Japanese empire.

The one mathematical discrepancy in this map was a product of Mamiya's inability to calculate latitude based on celestial surveys. Mamiya relied on Inō Tadataka's calculation that a one-degree arc of the earth's surface equaled 3.93 kilometers and then conducted linear measurements to determine latitudes as he traveled. According to Japanese experts, when Mamiya calculated the distance between the villages of Shiranushi and Natsuko, he ended up being off by about 4.9 degrees. Despite this error, however, Mamiya had fixed Sakhalin on a globally recognizable and easily transferable grid, something no other cartographer had done. He also contributed to resolving a long-standing scientific debate by confirming, with his own eyes, that Sakhalin was an island rather than a peninsula (see Tessa Morris-Suzuki's essay in this volume).

"Central Heilongjiang with Latitudinal Lines" is thus a landmark document in more ways than one. The area that it mapped became disassociated from its people, who were documented separately. By separating the human from the natural as distinct categories of scientific knowledge, the land was symbolically emptied, placed on a grid for all cartographically literate people to read and then made accessible to policymakers in Edo. "Central Heilongjiang with Latitudinal Lines" provided a visual representation of Japan's future empire well before the Japanese formally claimed sovereignty over southern Sakhalin, nearly a century later. Nor has its usefulness expired in our day. By placing Tokugawa Japan within a global cartographic order, a map like this serves to undermine the notion that early modernity (and its accompanying mapmaking projects) was an exclusive domain of European nations. Western cartographers may have invented many of the cartographic tools, but policymakers in China, Japan, and elsewhere around the world quickly seized upon them and found ways to deploy them toward their own national and imperial ends.

Notes

1. J. B. Harley, *The New Nature of Maps: Essays in the History of Cartography* (Baltimore: Johns Hopkins University Press, 2001), 57.
2. Bruno Latour, "Drawing Things Together," in *Representation in Scientific Practice*, ed. Michael Lynch and Steve Woolgar (Cambridge, MA: MIT Press, 1990), 45.
3. As Bruno Latour has argued, "We start knowing something when it is at least the *second* time we encounter it." Bruno Latour, *Science in Action: How to Follow Scientists and Engineers through Society* (Cambridge, MA: Harvard University Press, 1987), 217, 219.

Suggested Readings

Walker, Brett L. *The Conquest of Ainu Lands: Ecology and Culture in Japanese Expansion, 1590–1800*. Berkeley: University of California Press, 2001.

———. "Mamiya Rinzō and the Japanese Exploration of Sakhalin Island: Cartography, Ethnography, and Empire." *Journal of Historical Geography* 33 (2) (April 2007): 283–313.

33 Outcastes and Peasants on the Edge of Modernity

Daniel Botsman

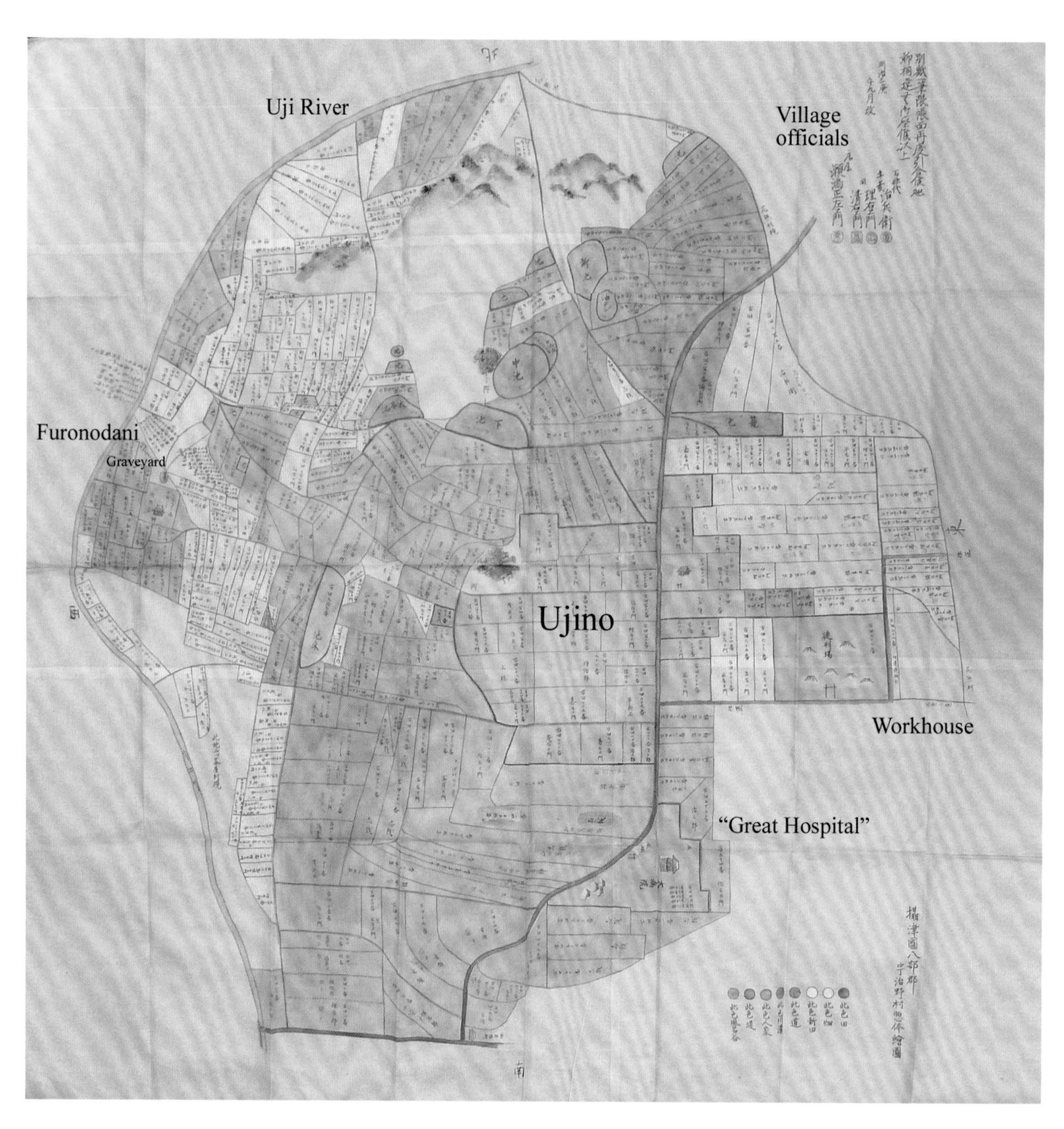

FIGURE 33.1 "A Complete Map of Ujino Village, Yatabe County, Settsu Province" [Settsu no kuni Yatabe gun Ujino mura sōtai ezu 摂津国八部郡宇治野村惣体絵図], 1870. Approved with the seals of the village headman, Sejima Shōzaemon, and three other officials. Manuscript, 135.4 × 133.2 cm. Courtesy of the Kobe City Museum, Hyōgo, Japan. English labels added by author.

In 1968 a department store in the city of Kobe sparked a national controversy when it began selling reproductions of historical maps as part of a special exhibit marking the centenary of the Meiji Restoration. Two maps in particular, depicting the area around the city at the beginning of the Meiji period, showed the locations of outcaste communities marked with the words *eta mura* (literally "great filth village")—a label that, in terms of its emotive power and historical significance, might well be compared to the English expression "nigger town." As in other parts of the world, Japan in the 1960s saw a growing concern with social justice. Led by the Buraku Liberation League (BLL)—a group founded to fight discrimination against those believed to be descendants of Tokugawa period outcastes—local activists decried the sale of these "discriminatory maps" as "completely unforgivable." The store's management quickly responded by abandoning the exhibit and offering to buy back all copies of the maps that had been sold.[1]

In the wake of this widely publicized incident, and others that followed, publishers, museums, and libraries in Japan became increasingly cautious about their handling of historical maps.[2] It soon became standard practice to erase or cover up all references to the existence of outcaste communities when republishing maps, and institutions that held the originals increasingly began to limit public access to any that might be viewed as problematic. The concern was not just with the use of hateful language, but also the possibility that old maps might be used by unscrupulous individuals to "out" the residents of particular neighborhoods as "former outcastes," thereby reinscribing old patterns of discrimination and exclusion with regard to employment and marriage. (Because Japan's "Burakumin," like the Dalits of India, are physically indistinguishable from the majority population, local knowledge of where "they" live has long been critical for identifying their communities.) This concern has not completely disappeared and, indeed, the introduction of online services such as Google Earth has rekindled debate about the possible abuse of historical maps.[3] In general, however, there has been a profound shift in thinking about the links between historical maps and discrimination since the late 1980s. Groups such as the BLL now openly criticize the practices of altering and suppressing maps and have come instead to view the responsible publication and display of uncensored maps, properly explained and contextualized, as a valuable method for raising awareness of the problem of discrimination.[4]

Like the maps that first sparked controversy in 1968, the one reproduced here depicts the Kobe area in the immediate wake of the Meiji Restoration (fig. 33.1). The whereabouts of the hand-drawn original is unknown, but several carefully traced copies have survived in local museums and archives. Such maps provide an excellent example of how cartography can be useful for understanding how outcaste communities lived, how such communities were integrated into the larger society, and how outcaste geographies helped shape other aspects of social development and change.

The map shows the lands of a village called Ujino in 1870, the third year of the Meiji era.[5] In the top right-hand corner, the seals of the headman and three other village officials attest that its remarkably detailed record of local land use is accurate. The village lands are divided up into a total of 420 plots, each numbered and marked with the name of an individual owner. Colors have been used to indicate different kinds of land use: rice paddies are shown in light blue, dry fields in yellow, and newly reclaimed paddy land (*shinden*) in off-white. The Uji River, which formed the village's western border, is shown in blue. Following standard practice, irrigation ponds are marked in a darker blue, embankments in brown, and roads in red. The large blank space at the top of the map represents a hilly area too steep for cultivation, while another area left blank at lower left indicates land controlled by a neighboring village. Ujino itself, which is to say the houses of village residents, is shown close to the middle of the map, marked in orange. Two red *torii* gates indicate the presence of local shrines; a Buddhist temple can also be seen near the heart of the main settlement. At the western edge of the map, however, we see a second Buddhist temple, in the middle of an area marked in purple. The map key indicates only that this area was called Furonodani. This was a "branch village," within the territorial limits of Ujino but inhabited by persons of outcaste status.

One immediate indicator of the disadvantages faced by Furonodani's residents is its location on the banks of the Uji River, which made it prone to flooding. It is also instructive to note how small it is in comparison with the "main" peasant village of Ujino. From the records of a local census we know that in 1871 there were some 1,093 people living in this tiny space. In Ujino, by contrast, an area perhaps three or four times as large, the total population, including servants, was around 400. The overall pattern of inequality is also clear from a consideration of cultivated lands. The compilers of the map used a derogatory prefix to indicate the names of outcaste landowners, and from this we can confirm that residents of Furonodani were mainly limited to farming a narrow strip of land around the edge of the community. In one area, just north of Furonodani and adjacent to the village graveyard, we see a particularly dense concentration of plots, too small to be properly mapped. The compilers here have simply listed the names of the outcaste owners.

With access to land so limited, it is hardly surprising that, in the 1871 census, only a small number of Furonodani households listed farming as their primary occupation. Like many outcaste communities, Furonodani had been responsible throughout the Tokugawa period for collecting the carcasses of dead animals (particularly cattle) from surrounding peasant communities. This helped ensure a strong connection to various kinds of leather work, but other occupational groups (day laborers, hair dressers, flower sellers, and the like) were also represented. The map also makes it clear that only a very small portion of the land cultivated by Furonodani residents was used to grow rice. This may have been because Furonodani, like many outcaste communities, was exempt from the payment of annual rice taxes. Instead, its residents had been required to perform special duties (*yaku*) in the nearby port town of Hyōgo: cleaning the area around the town's administrative headquarters, clearing away the bodies of sick travelers and beggars found dead in the town streets, and carrying out judicial punishments (such as displaying and guarding the severed heads of executed criminals). This long-standing connection to the management of poor, diseased, and criminal bodies helps explain one other striking aspect of the 1870 map of Ujino: the presence of both a hospital and a workhouse.

Under the commercial treaty concluded between the United States and Japan in 1858, the Japanese had promised to open Hyōgo to "trade and residence for American citizens." Ten years later, when final preparations were being made for the arrival of the first Westerners, Japanese officials began building a variety of new institutions in the area: a "Western-style" hospital and workhouse, as well as a new jail. There seems little doubt that the concentration in one small area of these new institutions, in many ways symbolic of the onset of modernity, was closely linked to an older geography of social inequality.

Notes

1. "Tonda Meiji hyakunen kinen: 'Sabetsu no kochizu' o sokubai," *Asahi Shimbun* (evening edition), November 13, 1968, 10. I am grateful to Timothy Amos for sharing this article and his knowledge of the subject with me.

2. For a thoughtful overview of this history in Japanese, see Yoshimura Tomohiro, "Ezu no tenji kōkai to sono igi," in the landmark exhibition catalog *Ezu ni egakareta hisabetsumin* (Osaka: Ōsaka Jinken Hakubutsukan, 2001), 93–97.

3. On the debates about Google Earth sparked by UC Berkeley's impressive efforts to digitize their collections of Japanese historical maps in partnership with Mr. David Rumsey, see Ishimatsu Hisamitsu, "Kariforunia Daigaku Bākurī-kō ni okeru Nihon kochizu no dejitaru-ka purojekuto ni tsuite," *Jōhō no kagaku to gijutsu* 59 (11) (November 2009): 557–62; and Kondō Toshikazu, "Gūguru Asu no kochizu shōgō sābisu to buraku sabetsu," *Buraku kaihō* 642 (February 2011): 188–94.

4. The current BLL position on historical maps (dating from October 2003) can be found online at http://www.bll.gr.jp/siryosi-tu/siryo-syutyo2003/guide-seimei-20031110.html.

5. The existence of this map was first brought to my attention by Usui Hisamitsu, who includes it in his *Hyōgo no burakushi 3: Bakumatsu-Ishin no senmin-sei* (Kōbe: Kōbe Shimbun Sōgō Shuppan Sentā, 1991), 144–45. For an overview of the history of Furonodani see the same author's *Hyōgo no burakushi 2: Tenkan-ki no senmin kōzō* (Kōbe: Shinbun Sōgō Shuppan Sentā, 1992), 158–73; and Mae Keiichi, "Settsu Furonodani no rekishi," in *Buraku mondai ronkyū 6*, ed. Hyōgo Buraku Mondai Kenkyūsho (Kobe: Hyōgo Buraku Mondai Kenkyūsho, 1981), 88–98.

Suggested Readings

Amos, Timothy D. "Portrait of a Tokugawa Outcaste Community." *East Asian History* 32/33 (June 2006/December 2007): 83–103.

———. *Embodying Difference: The Making of Burakumin in Modern*

Japan. Honolulu: University of Hawai'i Press, 2011.
Bayliss, Jeffrey P. *On the Margins of Empire: Buraku and Korean Identity in Prewar and Wartime Japan*. Cambridge, MA: Harvard University Press, 2013.
Howell, David L. *Geographies of Identity in Nineteenth Century Japan*. Berkeley: University of California Press, 2005.
Ooms, Herman. *Tokugawa Village Practice: Class, Status, Power, Law*. Berkeley: University of California Press, 1996.
Ōsaka Jinken Hakubutsukan, ed. *Ezu ni egakareta hisabetsumin*. Osaka: Ōsaka Jinken Hakubutsukan, 2001.

34 Converging Lines: Yamakawa Kenjirō's Fire Map of Tokyo

Steven Wills

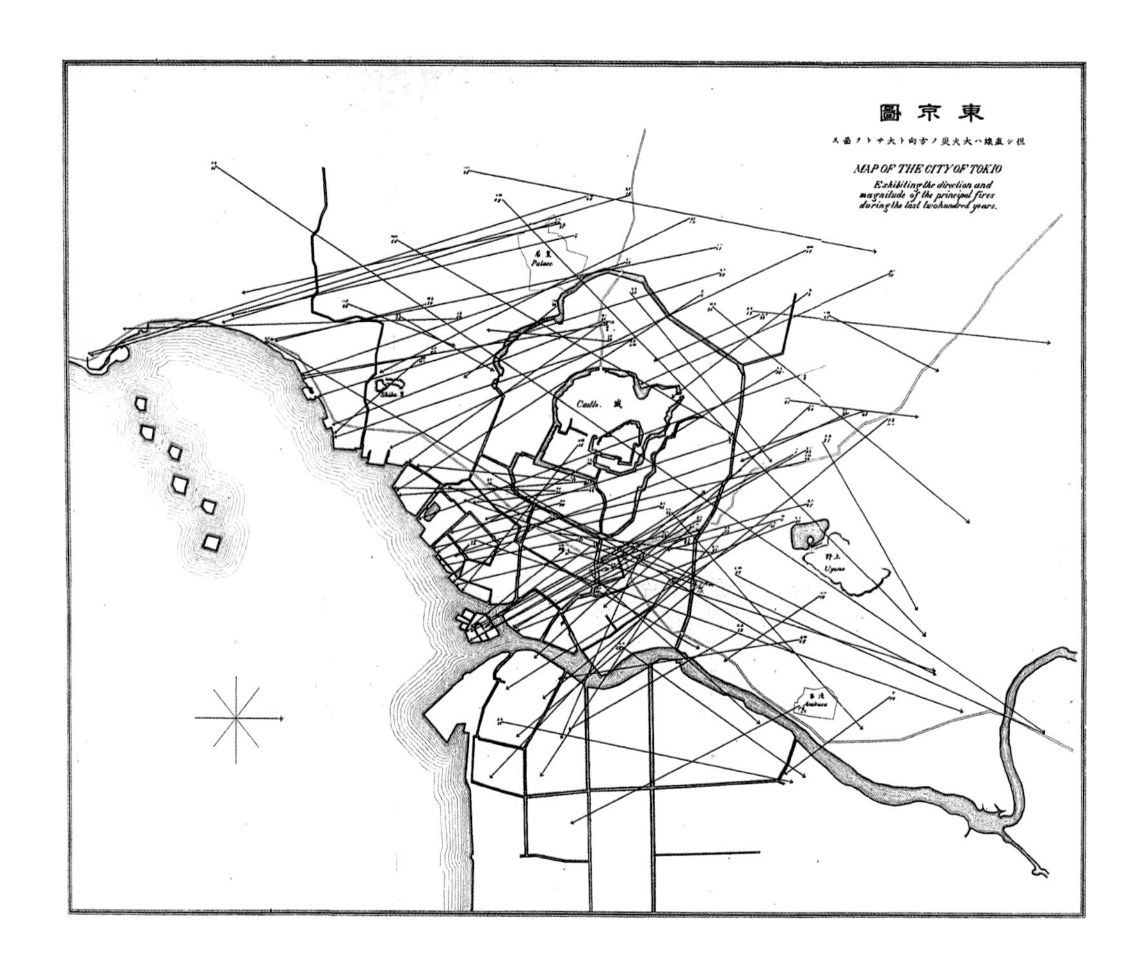

FIGURE 34.1 "Map of the City of Tokio" [Tōkyō zu 東京図], by Yamakawa Kenjirō 山川健次郎. Reprinted from *Fires in Tokio* [Tōkyō fuka kasairoku 東京府下火災録], in Thomas C. Mendenhall, *Report on the Meteorology of Tokio* [Tōkyō kishō hen 東京気象編], vol. 3 of *Memoirs of the Science Department, University of Tokyo* [Rika Kaisui 理科会粋] (Tokyo: Tokyo University, 1881), 126.

As one might expect in a city shaped profoundly by conflagration, the visual culture of fire imagery in Tokyo dates back to the city's very beginnings. Edo suffered its first catastrophic blaze as early as 1601, only eleven years after the establishment of the castle town by Tokugawa Ieyasu. In spite of the shogunate's repeated efforts to protect it from fire, the city continued to burn regularly throughout the nineteenth century. Fire imagery became an important cultural response. Monochrome pictures in illustrated books were common in the seventeenth century, and brightly colored scrolls and woodblock prints began to appear in the eighteenth century. Rather than mapping burned areas from above, these early images zoomed in to emphasize the pathos of individual suffering or personal heroism. In contrast, nineteenth-century broadsides (*kawaraban*) were more journalistic, with accounts of specific fires illustrated with hastily prepared maps showing the approximate extent of the damage.

Despite their differences, all of these early genres shared a preoccupation with isolated incidents. Yamakawa Kenjirō's 1881 "Map of the City of Tokio" (fig. 34.1) broke with that tradition by superimposing two centuries of fire data on a map of the city to highlight the areas of greatest vulnerability. In doing so, he contributed to a nineteenth-century trend of thematic map-making connected to urban reform efforts in Europe and the United States. Rejecting familiar images of flames, smoke, and panicked crowds, Yamakawa used modern cartography and meteorological observations to demonstrate deep historical patterns that were predictable, and thus at least potentially preventable.

As a two-term president of Tokyo University, Yamakawa is best remembered for his contributions to higher learning in modern Japan. Prior to his career as a high-ranking administrator, however, he led a rich and varied life. Yamakawa came from a prestigious samurai household in the northeastern domain of Aizu, and although he fought against the victorious imperial army during the Boshin War of 1868, he later received government support to travel to the United States as an exchange student. After distinguishing himself as the first Japanese student to earn an undergraduate degree from Yale, he returned to Japan and, in 1879, became the first Japanese professor of physics. His book-length study of the historical "Fires of Tokio"—including this map—was published a mere two years later.

Yamakawa produced his map by identifying ninety-three blazes between the years 1657 and 1880 that met his definition of a major fire: one that covered a distance of at least fifteen *chō* (roughly one mile) from its point of origin to its terminus. After plotting the burn lines of each fire on the map to measure the distance it traveled, he determined its directionality by measuring angles with a protractor (a novelty in Japan at the time). Once he had assigned distance and directionality to each blaze, he compiled all of his data in tables and charts at the end of the report. Although Yamakawa does not cite the source of his map, which took the form of a bird's-eye view of the city featuring rivers, canals, and a few bilingual labels of urban sites (Castle, Palace, Shiba, Uyeno, Asakusa), the fact that he relied on it to measure angles and distance suggests that he considered it to be an accurate representation of the city plan.

The main thrust of Yamakawa's quantitative analysis—which drew on both historical data and scientific observations recorded at the meteorological observatory of the new University of Tokyo—was to advance an argument about the nature of the city's fire problem. His report showed that two-thirds of major fires originated in the northern and northwestern areas of the city, and that the overwhelming majority of blazes occurred between the months of October and May, with a sharp peak in March. Charting those historical patterns against the observatory's data on wind for 1879 and 1880 demonstrates a remarkable correspondence between wind patterns and fire damage. Fires were much more likely to break out in the dry winter months, when strong northwesterly winds blow through the Kanto Plain, and fires that occurred during that peak season were typically more destructive. None of these correlations are apparent on the map, but with its rat's nest of crisscross lines converging in the downtown districts of Nihonbashi and Kanda, the map clearly indicates the areas of the city that were most vulnerable to conflagration. Yamakawa observes in his report that these areas were at great risk because of a deadly combination of population density, building practices, cultural attitudes toward fire, and seasonal variations in the regional climate. The policy implications were clear. Yamakawa concludes his report by calling on the central government to invest ¥100,000 annually toward fire prevention.

The timing of Yamakawa's study was hardly coincidental. In the early years of the Meiji period, when he published the report, leading officials were intent on transforming Tokyo into a capital befitting a state with world-power ambitions. Given the humiliating circumstances of Japan's opening to the West in the mid-nineteenth century, the fact that Tokyo suffered from devastating fires in the 1870s was a source of embarrassment and frustration for top officials. In the gendered language of high imperialism, vulnerability to fire was coded as feminine and weak, labels the Meiji government was eager to cast off. Yamakawa echoed this anxiety in the introduction to his report, pointing out that no European city burned as easily as Tokyo did, and damning Tokyoites' lax attitude toward fire safety as one of the "strange fruits of feudalism."

It is worth noting that Yamakawa conducted his research after the government's first effort to address the city's fire problem had ended in failure. After an 1872 blaze, Tokyo governor Yuri Kimimasa launched an ambitious project to turn the Ginza district, Tokyo's front door to the world, into a fireproof "bricktown." Ginza was indeed rebuilt to resemble a modern European boulevard, but at such great expense and in the face of such opposition from landowners that the initial plan to transform the entire city was abandoned by 1877. As Yamakawa documents in his report, over the next several years Tokyo suffered a series of conflagrations that highlighted the failure to expand the Ginza model. By the early 1880s, Tokyo elites had realized that piecemeal fire prevention was doomed to fail. Just as vaccinations require widespread adoption in order to protect an entire population, fire prevention only works when a sufficient segment of the built environment is transformed, and such thorough reforms could only be achieved by harnessing the authority of the central government. As it did in other public campaigns to combat "social problems," the Meiji government relied on the authority of knowledge produced by Western science to persuade recalcitrant communities to comply.

Armed with data and protractor, Yamakawa contributed to this effort by producing a novel image of Edo/Tokyo's fire hazard as a historical phenomenon. In doing so, he showed that the downtown area north of Nihonbashi Bridge was ground zero of the fire problem. Tokyoites had long known of the area's vulnerability and the connection between fire and the windy season, but Yamakawa's map presented that common knowledge in a way that demanded action. In 1882, the year after Yamakawa and his colleagues published their work, Governor Matsuda Michiyuki inaugurated a new Fire Prevention Ordinance that created fire breaks in the central districts and imposed new building codes and roofing regulations throughout the city. By targeting the sites of greatest vulnerability while imposing effective fire-safety measures on the greater part of the city, the ordinance effectively eliminated the devastating conflagrations Yamakawa had plotted on his map. Whether Matsuda or any government officials were aware of Yamakawa's work is unclear. But the effectiveness of the 1882 ordinance demonstrates that the combination of empiricism and historical inquiry Yamakawa employed in his study had enabled new ways of thinking about old problems that led to real solutions.

Suggested Readings

Churchill, Robert R. "Urban Cartography and the Mapping of Chicago." *Geographical Review* 94 (1) (January 2004): 1–22.

Karrow, Robert, and Ronald E. Grim. "Two Examples of Thematic Maps: Civil War and Fire Insurance Maps." In *From Sea Charts to Satellite Images: Interpreting North American History through Maps*, edited by D. Buisseret, 213–37. Chicago: University of Chicago Press, 1990.

Mendenhall, Thomas C. *An American Scientist in Early Meiji Japan: The Autobiographical Notes of Thomas C. Mendenhall.* Edited by Richard Rubinger. Honolulu: University of Hawai'i Press, 1989.

Yamakawa, Kenjirō. "Fires in Tokio." In *Report on the Meteorology of Tokio for the Year 2540 (1880)*, by Thomas C. Mendenhall, vol. 7 of *Memoirs of the Science Department, University of Tokio*, 71–88. Tokyo: Tokyo University, 1881.

———. "Tōkyō fuka kasairoku." In *Tōkyō kishō hen*, by Thomas C. Mendenhall, *Rika kaisui* vol. 3, pt. 2, 99–126. Tokyo: Tokyo University, 1881.

35 Mapping Death and Destruction in 1923

J. Charles Schencking

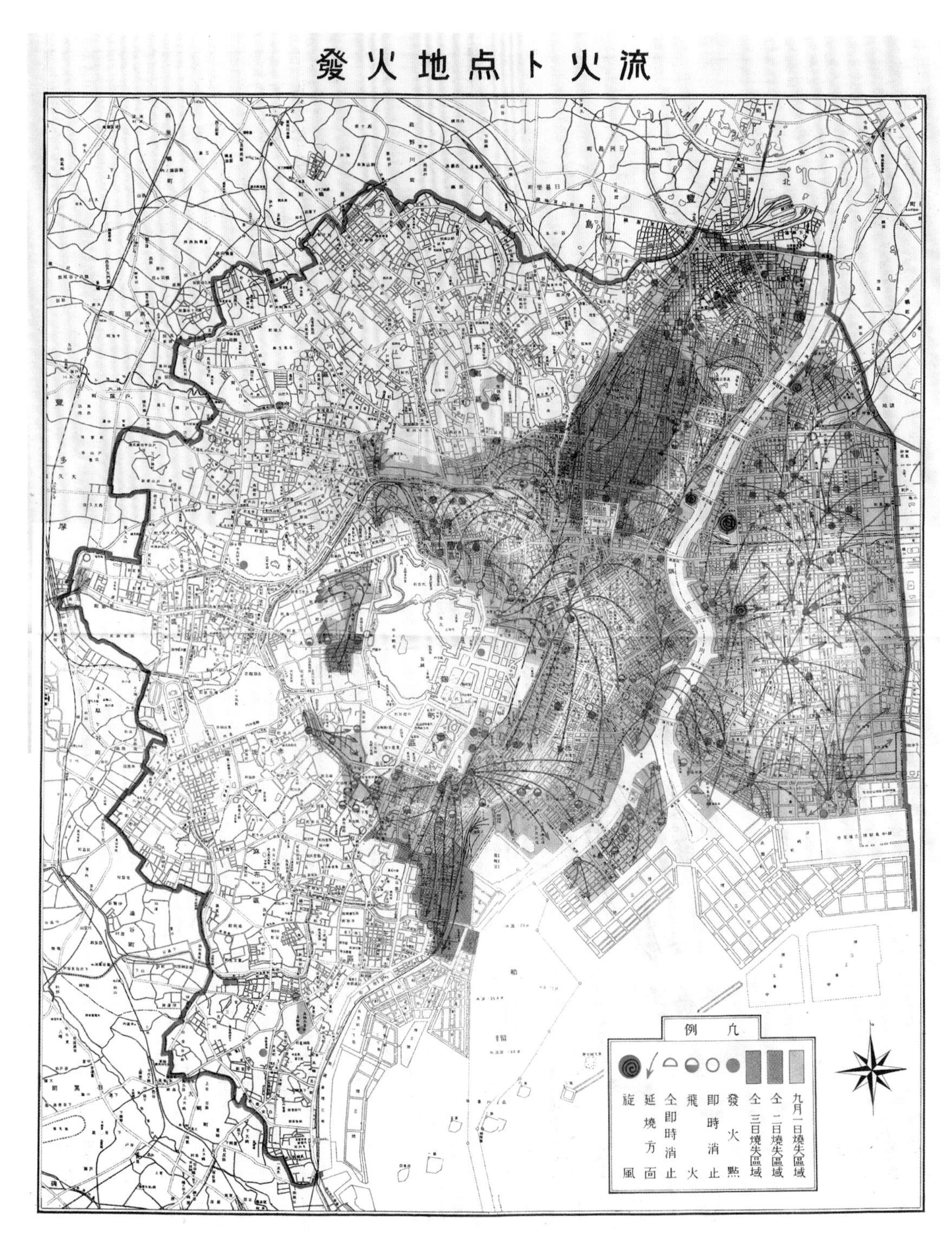

FIGURE 35.1 "Map Indicating the Origin Points of Fires and their Spread" [Hakka chiten to karyū 発火地点ト火流], from Nihon Tōkei Fukyūkai 日本統計普及会 ed., Survey of the Imperial Capital Reconstruction Project [Teito fukkō jigyō taikan 帝都復興事業大観], 2 vols. (Tokyo: Nihon Tōkei Fukyūkai, 1930), vol. 1, pt. 3, 6. Courtesy of Tokyo Metropolitan Archives.

Just before noon on September 1, 1923, the first of a series of tectonic upheavals rocked Tokyo. The destructive seismic waves triggered by the subduction of the Philippine Tectonic Plate beneath the Okhotsk Tectonic Plate south of Tokyo under Sagami Bay wrecked countless structures, ripped open great fissures in the land, and killed thousands in collapsed buildings. The fires that followed unleashed pandemonium, killed many more individuals, and burned over 45 percent of Tokyo and 90 percent of neighboring Yokohama to ash. The disaster also triggered a human-inspired catastrophe, as individuals and loosely organized bands of rabble—sometimes assisted by unscrupulous police and military officials—sought out Koreans and Japanese anarchists and murdered them out of fear, anger, hatred, or opportunism. Evincing the extraordinary destruction and death meted out by the calamity, one chronicler who scoured the postdisaster remains of Tokyo asked, "If this were not Hell, where would Hell be?"[1]

When order was restored to Japan's capital and relief work initiated, survey teams mapped the hellscape of Tokyo. Assisted by army engineers, surveyors conducted forensic investigations throughout the charred city by charting where fires had ignited and recording the paths they burned through the capital. Bureaucrats also mapped something far more ghastly: the locations where dead bodies had been collected and held for cremation. By documenting in vivid form that the poorer, more susceptible parts of eastern Tokyo bore the brunt of destruction and death, the resulting maps revealed the uneven social and economic landscapes of Japan's capital in the early twentieth century. In particular, they highlighted the vulnerabilities of eastern Tokyo's "low city," which contributed significantly to making the Great Kanto Earthquake one of the most deadly natural disasters of the twentieth century.

"Map Indicating the Origin Points of Fires and Their Spread" (fig. 35.1) provides a bird's-eye view of the burned-out remains of Tokyo, showing that 33 million square meters of the city was reduced to ash following the initial earthquake. The key at the bottom of the map indicates that fires raged for three days in the capital, with the areas that burned on the first (sienna orange), second (brown), and third (bright orange) days of September constituting virtually all of eastern Tokyo. Orange circles identify the combustion points of 110 individual fires; orange arrows emanating from the points track the directions that the fires spread after ignition. Large orange circles with black counterclockwise spirals symbolize the locations from which five firestorms erupted. These infernal whirlwinds burned virtually everything in their path. The firestorm that overwhelmed the site of the former Honjo Clothing Depot on the banks of the Sumida River—the largest whirlwind depicted on this map—was itself responsible for the deaths of nearly forty thousand people who had taken refuge in the area's only significant open space.

Why were virtually all of the fires concentrated in the eastern half of Tokyo? And why did they spread so quickly? The answer lies in large part with vulnerability. Vulnerability is often used to express the diminished capacity of a society to cope with, respond to, and recover from a natural hazard such as an earthquake, fire, or tsunami. Social, economic, political, and environmental vulnerabilities made the eastern wards of Honjo, Fukagawa, Kyōbashi, Asakusa, and Nihonbashi increasingly susceptible to fire as Tokyo grew into a crowded industrial city in the twenty-five years prior to the 1923 disaster. Between 1900 and 1920, the population of Tokyo nearly doubled, from 1.12 million to 2.17 million. In just twelve years, factories that employed five or more workers increased almost tenfold, with Tokyo hosting over 7,200 such worksites by 1919. Factory payrolls increased from 55,944 to 188,786 over the same period.

Many new factories and their employees settled in eastern Tokyo, an area already characterized by high concentrations of industry and cheap labor. The industrial expansion that coincided with the First World War further increased population densities. Honjo Ward, for example, hosted more factories, 1,113, than any other ward and possessed a population density of 112,118 persons per square mile in 1920. The eastern wards of Fukagawa, Kyōbashi, Nihonbashi, and Asakusa likewise possessed some of Tokyo's highest population and factory densities.

The cramped and congested residential and industrial conditions of eastern Tokyo allowed fires to spread quickly once ignited. Wood was the building material of choice for most homes, small businesses, and light industrial enterprises. Open spaces that might have served as firebreaks were virtually nonexistent. On the eve of calamity, Tokyo officials calculated that parks and open green space made up less than 2 percent of the overall urban environment of Japan's capital, compared with 9 percent for London, 14 percent for Washington DC, and 20 percent for Paris. A close inspection of the map also reveals the sizable number of narrow roads and alleyways that defined much of Tokyo. This byzantine network, a holdover from the Tokugawa shogunate, whose leaders sought to restrict the movement of enemy forces into and within the city, both enabled fires to spread quickly and hindered the rapid evacuation of residents.

Survivor accounts from the 1923 tragedy lend evidence to what this map of incinerated Tokyo reveals: that eastern Tokyo had become a hell on earth. Novelist Uno Kōji lamented, "As far as the eye could see, Tokyo had been reduced to ash."[2] "Watching the whirlpools of fire and undulating waves of flame" consume Tokyo, Kawatake Shigetoshi wrote, "was utterly beyond description."[3] He confessed that he would never use the phrase "burning hell" in a casual way again. Survivors also commented extensively on the dead. The "burned, bubbling, decomposing bodies" interspersed with "reddish brown electric wires . . . and blackened rubble" led

FIGURE 35.2 "Map Indicating Where Dead Bodies Were Accommodated" [Shinsai shitai shūyō zu 震災死体収容図], from Nihon Tōkei Fukyūkai 日本統計普及会 ed., *Survey of the Imperial Capital Reconstruction Project* [Teito fukkō jigyō taikan 帝都復興事業大観], 2 vols. (Tokyo: Nihon Tōkei Fukyūkai, 1930), vol. 1, pt. 13, 26–27. Courtesy of Tokyo Metropolitan Archives.

writer Koizumi Tomi to conclude that Japan's capital resembled a landscape "controlled whimsically by devils."[4] The dead, like the burned and fractured landscape itself, became an inescapable feature of postdisaster Tokyo.

Highlighting the disproportionate nature of casualties meted out by this disaster, "Map Indicating Where Dead Bodies Were Accommodated" (fig. 35.2) provides a ward-by-ward breakdown of where corpses were collected and held prior to cremation. Each collection center is represented by a number from one (一) to thirteen (十三), while the size of the number displayed on the map corresponds to the number of bodies accommodated. The collection center in Honjo, represented by the number ten (十), is larger than all others, for more than fifty thousand bodies were collected and cremated in this ward alone. In striking visual form, this map reveals that the wards of eastern Tokyo suffered far more casualties than the wealthier western wards.

This map relays two other significant pieces of information through two circular graphs positioned in the lower right corner. The one on the left documents that of the 58,656 bodies collected throughout the city, more than one in six were removed from Tokyo's waterways. This graph gives validity to a popular expression used in the weeks after the calamity to describe the fateful choice that many victims in eastern Tokyo faced: "death by burning or death by drowning."[5] The other graph reveals that attempts to classify the dead by gender were virtually impossible given the condition of many of the bodies collected.

Commentators repeatedly referenced the pitiful state of the dead bodies that littered Tokyo. As the 1923 "Song of the Great Taishō Earthquake Disaster" put it,

> Look at the ruins of the fires.
> The mountains of dead bodies of tens of thousands of our countrymen—
> The sight is too terrible to look at.
> It is more hellish than hell.
>
> It is impossible to tell from their faces
> If they are brothers or parents and children.
> Or, if they are men or women.
> Many of them are inflamed, and many are swollen.
>
> Heads are crushed, bellies are torn;
> Bones are broken; innards are pouring out.
> The strong stench assails our nostrils.
> The bodies are piled up as they are found . . .
>
> Ah, what a terrible disaster,
> Ah, what a terrible disaster.[6]

Though lacking the emotive punch of survivor accounts, maps document the human and physical traumas associated with the Great Kanto Earthquake at the macro level. Importantly, they encourage us to question what social, economic, and environmental vulnerabilities contributed to such disproportionate death and destruction meted out by Japan's most deadly natural disaster of the modern era.

Notes

1. Anonymous, "Shikai no toshi o meguru: saigai yokuyokujitsu no dai Tokyo," in *Taishō daishinsai daikasai*, ed. Dainihon Yūbenkai Kōdansha (Tokyo: Dainihon Yūbenkai Kōdansha, 1923), 37.
2. Uno Kōji, "Sanbyaku nen no yume," in Kaizōsha, *Taishō daishinkasai shi*, 38.
3. Kawatake Shigetoshi, "Sōnan ki," in Kaizōsha, *Taishō daishinkasai shi*, 26–27.
4. Koizumi Tomi, "Hifukushō seki sōnan no ki," in Kaizōsha, *Taishō daishinkasai shi*, 22.
5. Funaki Yoshi, "Hi de shinauka, mizu o erabauka," in *Kantō daishinsai: Dokyumento*, ed. Gendaishi no Kai (Tokyo: Sōfūkan, 1996), 53–57.
6. Soeda Azenbō, "Taishō daishinsai no uta," in Gendaishi no Kai, *Kantō daishinsai*, 210–12.

Suggested Readings

Hastings, Sally. *Neighborhood and Nation in Tokyo: 1905–1937*. Pittsburgh: University of Pittsburgh Press, 1995.

Schencking, J. Charles. *The Great Kanto Earthquake and the Chimera of National Reconstruction in Japan*. New York: Columbia University Press, 2013.

Seidensticker, Edward. *Low City, High City: Tokyo from Edo to the Earthquake*. New York: Knopf, 1983.

Sorensen, André. *The Making of Urban Japan: Cities and Planning from Edo to the Twenty-First Century*. London: Routledge, 2002.

Weisenfeld, Gennifer. *Imaging Disaster: Tokyo and the Visual Culture of Japan's Great Earthquake of 1923*. Berkeley: University of California Press, 2012.

36 Rebuilding Tokyo after the Great Kanto Earthquake

André Sorensen

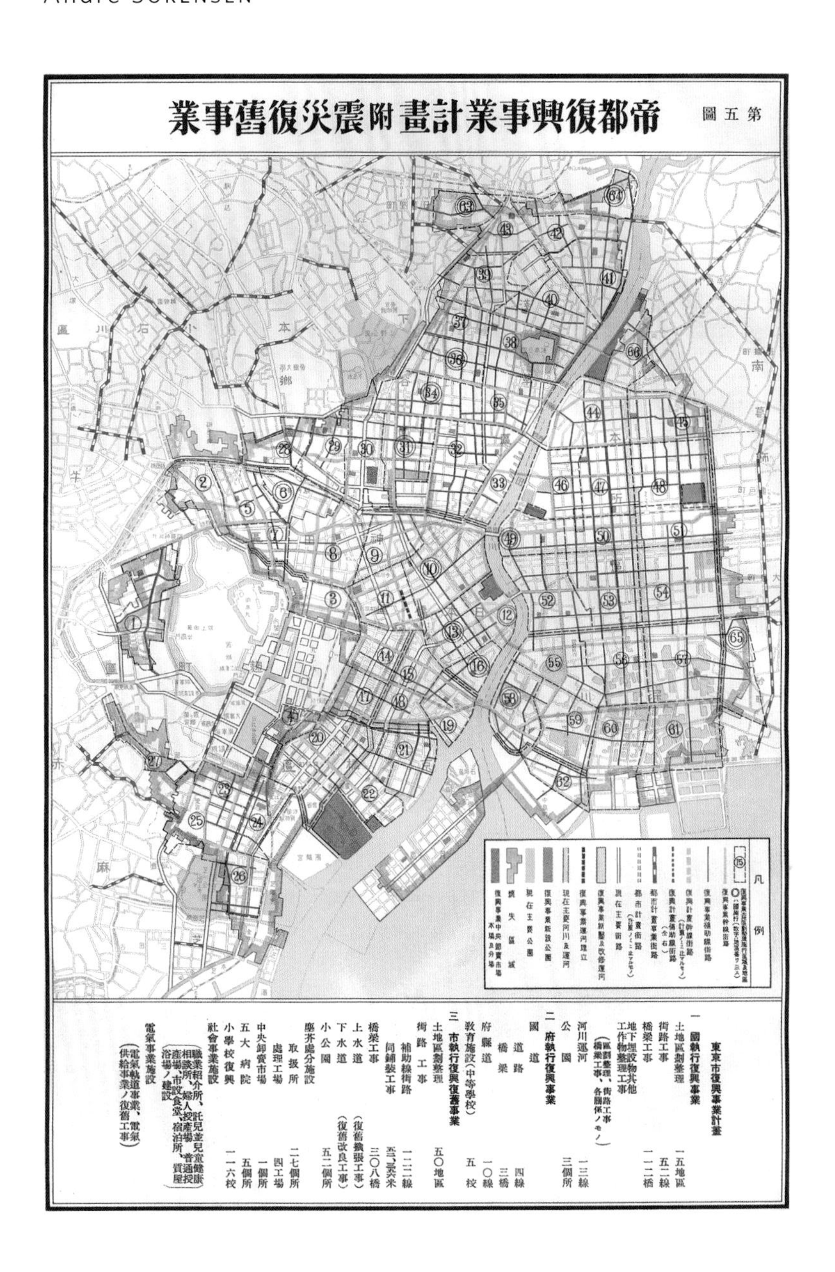

FIGURE 36.1 "Imperial Capital Reconstruction Project Plan: Disaster Reconstruction Projects, Map Five" [Teito fukkō jigyō keikaku: Shinsai fukkyō jigyō, dai go zu 帝都復興事業計画　震災復旧事業 第五図]. From *Teito fukkō jigyō zuhyō* (Tokyo: Tokyo Shi, 1930).

The map reproduced here (fig. 36.1) is one of many published in 1930 by the Imperial Capital Reconstruction Survey Commission as part of a massive (1,874-page) two-volume set documenting the reconstruction of Tokyo after the Great Kanto Earthquake of 1923.[1] Such commemorative volumes are certainly not unique to Japan, but Japanese government agencies, corporations, commissions, and neighborhoods are certainly among the most diligent and enthusiastic practitioners of the art.

The map shown here depicts the area of central Tokyo destroyed by fire after the earthquake (indicated with a heavy shaded red border) and comprehensively rebuilt over the next eight years with a new road system and new parks, bridges, fireproof schools, markets, and government buildings. The Great Kanto Earthquake, which destroyed the last vestiges of old Edo, has been viewed as a decisive event in the emergence of modern Tokyo.[2] While the case has been made that the reconstruction project failed to achieve the transformative renewal of Tokyo for which some had hoped, this map nevertheless documents a fundamental restructuring of the urban form of central Tokyo, a rare achievement in postdisaster reconstruction.[3]

It is important first to recognize the immense scale of the disaster, one of the worst ever to befall a Japanese city. Some one hundred forty thousand people were killed or missing, and over 44 percent of the urban area of Tokyo was destroyed by the earthquake or the fires it set off. Three-quarters of all households were affected by loss of a family member, a business, or a home.[4] Virtually all the old, densely populated central areas of Tokyo were burned to the ground. Destruction included the main commercial centers of Tokyo, the Kasumigaseki central government district (which housed most ministries), Tokyo City Hall, the major industrial areas east of the Sumida River, hundreds of bridges large and small, and huge numbers of schools and other public facilities.

A remarkable feature of this map is the way it brings together information about conditions before and after reconstruction. The commemorative volume of 1930 included dozens of maps and illustrations, but this one serves as a kind of overall guide to different elements of the effort. Following the legend at the bottom right of the map, from the left, the purple shading indicates wholesale markets created by the reconstruction project, including the new Tsukiji fish market completed in 1935. This was a much larger site than the previous fish market near Nihonbashi and would serve for nearly eighty years as the world's largest fish market (chap. 52). Two other new wholesale markets were created: one in Kanda and another to the east across the Sumida River on the site now occupied by the sumo wrestling arena.

Proceeding to the right, the next symbol is the red border that indicates the burned area. It is notable that a number of major roads were built outside this area. Next are parks, with a lighter green indicating predisaster parks, including the famous Hibiya Park (built as part of the Tokyo City Improvement projects from the 1890s), Ueno Park in the north, and Shiba Park in the south (each built on the site of a major premodern temple compound). In dark green are new parks created as part of the project. A few major parks rose from the ashes, such as Hamachō on the Sumida River and the Sumida riverbank parks near Asakusa in the north. The vast majority of new parks, however, are the dozens of small neighborhood parks, often associated with public schools, scattered throughout the central city. These open spaces created as part of the reconstruction project represent the majority of parks in central Tokyo at present.

Next are canals and rivers, featuring a clever cartographic device to show two time periods in a single map: predisaster canals and rivers are indicated with borders in grey, while the newly built are indicated in dark blue. Several major new canals and canal widenings are evident. For example, the canal bordering the Tsukiji fish market leading to the northeast was widened and lengthened at this time. The most revealing use of this before-and-after cartography, however, is the road system. The preexisting minor roads are indicated in light grey, preexisting arterial roads with solid red borders, and planned arterial roads with dashed red borders. The new arterial network created as part of the reconstruction project is in solid orange, with many entirely new major boulevards cut through what formerly had been densely built-up areas, such as at the mouth of the Kanda River in project district 10 (indicated with a blue number in a double circle). The reconstruction project created a comprehensive new network of major, intermediate, and minor roads, which has provided the fundamental structure of central Tokyo until the present day.

Certainly the most important and difficult aspect of reconstruction was the process of restructuring to create these new boulevards, canals, markets, and parks. Huge new areas of public space were created, virtually all of which had been private land before the disaster. This was accomplished with land readjustment projects, a process in which all the land within a project area, public and private, is pooled, property boundaries are erased, new streets and parcels are designed, and each private landowner receives back a slightly smaller parcel, preferably close to its former location. In this way all the landowners in a project area contribute equally to the expansion of public infrastructure, instead of just a few being expropriated entirely.[5] A special revision of the City Planning Act was passed in 1924 to require compulsory participation in land readjustment projects with no compensation for the first 10 percent of land area contributed. The map shows sixty-six such designated Reconstruction Land Readjustment Project Areas (*Fukkō jigyo tochi kukaku seiri shitei kuiki oyobi chiiki*), indicated with a dashed blue line and numbered one to sixty-six.

While land readjustment had been used to develop suburban areas in Japan since the 1890s and was a major element of the first City Planning Law of 1919, the Imperial Capital Reconstruction Project was its first major use in an existing built-up area. The use of land readjustment in this way was highly controversial. In rural areas land values typically rise rapidly with the coming of new roads and infrastructure. Within an existing urban area, however, the benefits are less clear, and many regarded compulsory contribution of a slice of their property to be expropriation without compensation, which was prohibited by the Japanese constitution.[6] Opposition movements mobilized when the plans were announced, and in 1925 the Tokyo City Assembly and the Lower House of the Diet passed motions opposing the compulsory contribution rule. The powerful Reconstruction Bureau, however, offered only minor changes in response and largely proceeded with the projects as planned.[7]

The Imperial Capital Reconstruction Project has had a lasting impact. First, the renovations that it set in motion modernized central Tokyo. Although the city would again prove vulnerable to fire during American bombing in 1945, the urban fabric created after the earthquake of 1923 endures today. Second, this project proved to be an incubator for urban planning techniques that would later spread throughout the Japanese empire. Over six thousand young planners and engineers involved in the project later formed the core cadre of experts who fanned out to cities throughout Japan and the colonies. New ideas about city planning that were developed and implemented in postdisaster Tokyo—especially in relation to parks, public markets, hierarchical street networks, modern bridges, and public housing projects—would later spread throughout the empire in the 1930s and beyond. Finally, the powerful role of land readjustment, as a tool both for suburban expansion and for inner city redevelopment, was firmly established. In the postwar period land readjustment was a key tool enabling large-scale land development projects, both in the suburbs and inner cities. While the use of the method has declined with slowing economic growth since the 1990s, it is again being used for many of the post-tsunami reconstruction projects in Tohoku.

Notes

1. Fukkō Chōsa Kyōkai, ed., *Teito fukkō shi* (Tokyo: Kōbundō Shoin, 1930).

2. Edward Seidensticker, *Tokyo Rising: The City since the Great Earthquake* (Cambridge, MA: Harvard University Press, 1990).

3. J. Charles Schencking, *The Great Kanto Earthquake and the Chimera of National Reconstruction in Japan* (New York: Columbia University Press, 2013).

4. Ishizuka H. and Ishida Y., "Tokyo, the Metropolis of Japan and Its Urban Development," in *Tokyo: Urban Growth and Planning 1868–1988*, ed. Ishizuka H. and Ishida Y. (Tokyo: Center for Urban Studies, 1988), 19.

5. André Sorensen, "Land Readjustment and Metropolitan Growth: An Examination of Land Development and Urban Sprawl in the Tokyo Metropolitan Area," *Progress in Planning* 53 (4) (2000): 1–113; Schencking, *The Great Kanto Earthquake*, chap. 8.

6. André Sorensen, "Land, Property Rights and Planning in Japan: Institutional Design and Institutional Change in Land Management," *Planning Perspectives* 25 (3) (2010): 279–302.

7. André Sorensen, *The Making of Urban Japan: Cities and Planning from Edo to the 21st Century* (London: Routledge, 2002), 128.

Suggested Readings

Fukkō Chōsa Kyōkai, ed. *Teito fukkō shi*. Tokyo: Kōbundō Shoin, 1930.

Ishizuka Hiromichi and Ishida Yorifusa. "Tokyo, the Metropolis of Japan and Its Urban Development." In *Tokyo: Urban Growth and Planning 1868–1988*, edited by Ishizuka Hiromichi and Ishida Yorifusa (Tokyo: Center for Urban Studies, 1988), 3–35.

Schencking, J. Charles. *The Great Kanto Earthquake and the Chimera of National Reconstruction in Japan*. New York: Columbia University Press, 2013.

Seidensticker, Edward. *Tokyo Rising: The City since the Great Earthquake*. Cambridge, MA: Harvard University Press, 1990.

Sorensen, André. "Land Readjustment and Metropolitan Growth: An Examination of Land Development and Urban Sprawl in the Tokyo Metropolitan Area." *Progress in Planning* 53 (4) (2000): 1–113.

———. *The Making of Urban Japan: Cities and Planning from Edo to the 21st Century*. London: Routledge, 2002.

———. "Land, Property Rights and Planning in Japan: Institutional Design and Institutional Change in Land Management," *Planning Perspectives* 25 (3) (2010): 279–302.

37 Shinjuku 1931: A New Type of Urban Space

Henry D. Smith II

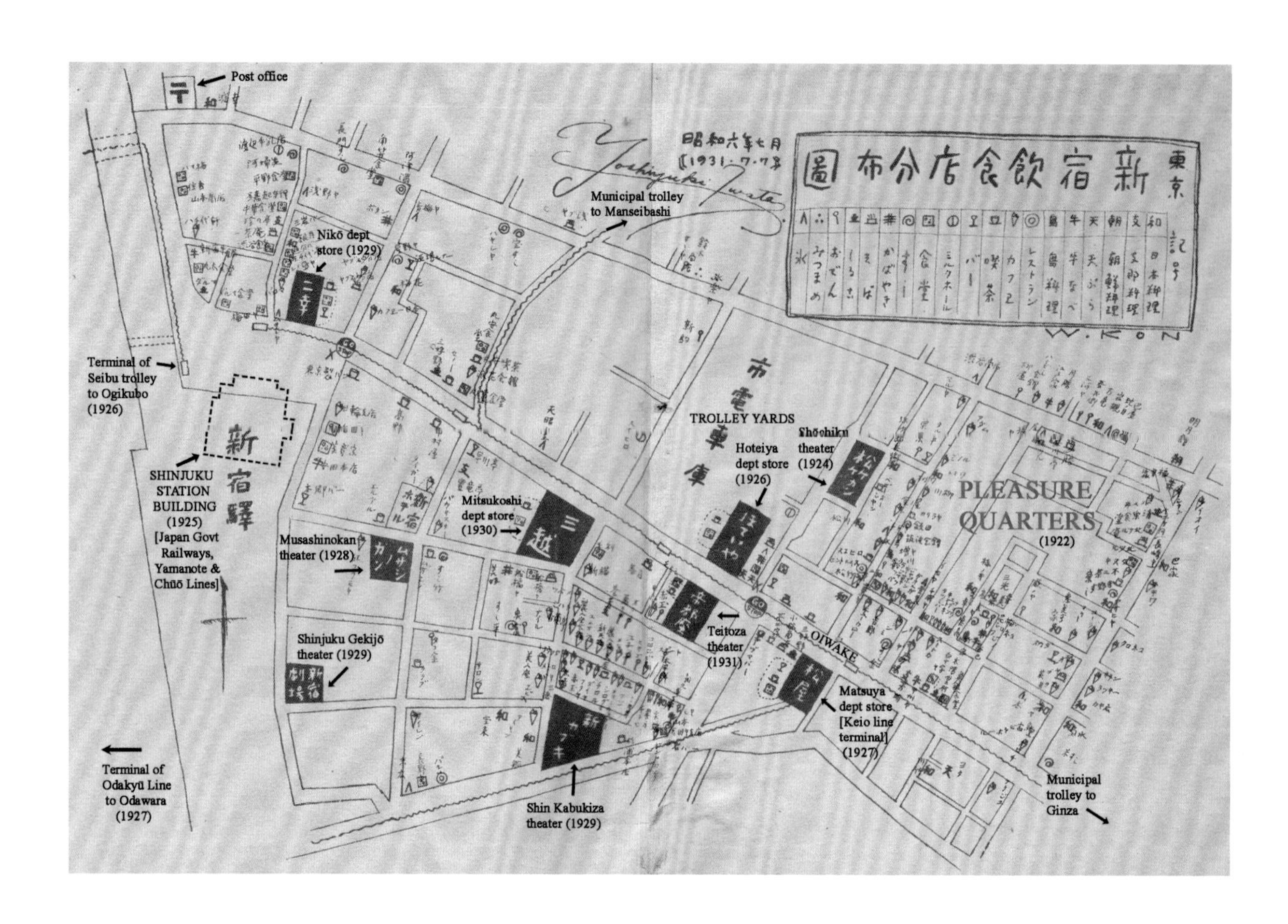

FIGURE 37.1 "Distribution of Drinking and Eating Places in Shinjuku, Tokyo" [Tōkyō Shinjuku inshokuten bunpu zu 東京新宿飲食店分布図], by Iwata Yoshiyuki 岩田義之 and Kon Wajirō 今和次郎, reprinted from endpapers of Kon Wajirō and Yoshida Kenkichi 吉田謙吉, eds., *Modernology Collecting* [Kōgengaku saishū (Moderunorojio) 考現学採集（モデルノロヂオ）] (Tokyo: Kensetsusha, 1931). 25.6 × 38.8 cm. English labels added by author.

The title of this map, "Distribution of Eating and Drinking Places in Shinjuku" (fig. 37.1), reveals its documentary purpose: to record the name and type of every such establishment in the Shinjuku area of Tokyo on a summer evening in 1931. It would be hard to overstate the historical importance of the geographic area that it depicts. Located some seven kilometers west of the center of Tokyo, Shinjuku was a wholly new type of urban space, a dense concentration of entertainment and consumption venues generated at the intersection of multiple rail lines. Although still a work in progress, Shinjuku in 1931 had already taken on its basic character as the "distillation" of the twentieth-century Japanese metropolis.[1]

Shinjuku began in 1698, as its Japanese name indicates, as a "new post town." It served as the first stop out of Edo along the Kōshū Highway, the main road to the west and gateway to the broad Musashi Plain and beyond. Few travelers actually stayed there, and it developed rather into

No.	Symbol	Explanation of symbol	種類	Type	Description	Count	% total (294)
1	 和	和 WA	日本料理	*Nihon ryōri*	Japanese cuisine	26	8.8%
2	 支	支 SHI	支那料理	*Shina ryōri*	Chinese restaurant, "*soba*" (noodles) in text—then "Shina-soba," now "ramen"	4	1.4%
3	 朝	朝 CHŌ	朝鮮料理	*Chōsen ryōri*	Korean restaurant, only one example; nature of food served is unclear	1	0.3%
4	 天	天 TEN	天ぷら	*tenpura*	tempura	3	1.0%
5	 牛	牛 GYŪ	牛なべ	*gyū-nabe*	beef pot, probably sukiyaki, text includes "beef bowl" (*gyūmeshi*)	7	2.4%
6	 鳥	鳥 TORI	鳥料理	*tori ryōri*	chicken dishes	2	0.7%
7		plate	レストラン	*resutoran*	restaurant, Western food	5	1.7%
8		cocktail napkin	カフエ	*kafee*	literally "café," but more like a bar with hostesses	92	31.3%
9		coffee cup	喫茶	*kissa*	coffee shop	23	7.8%
10		cocktail glass	バー	*baa*	bar	16	5.4%
11		lid of a metal milk can (from which milk was ladled in the milk halls)	ミルクホール	*miruku hōru*	"milk hall," serving milk, coffee and tea, and western-style sweets	4	1.4%
12		tray with items of a standard meal--rice bowl, soup, and two side dishes.	食堂	*shokudō*	"dining hall," but here a cheap eatery	36	12.2%
13		spiral roll of *makizushi*	すし	*sushi*	sushi	13	4.4%
14		metal grill (*ami*)	かばやき	*kabayaki*	grilled eel	5	1.7%
15		round wicker tray (*seirō*) for serving cold soba, with strands of noodles and/or chopsticks above	そば	*soba*	buckwheat noodles	14	4.8%
16		lacquer bowl and saucer with lid	しるこ	*shiruko*	*mochi* balls with sweet bean paste	5	1.7%
17		skewer with some item (probably an egg, or ball of ground fish)	おでん	*oden*	miscellaneous items steeped in soy sauce broth and served hot	21	7.1%
18	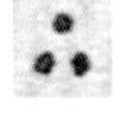	*azuki* beans (*mame*) used in *mitsumame*	みつまめ	*mitsumame*	mixture of beans, tokoroten, fruits; modern form from Asakusa Funawa 1903	2	0.7%
19		cone of shaved ice (in reality more rounded)	氷	*kōri*	shaved ice with syrup poured over	16	5.4%

FIGURE 37.2 Key to figure 37.1.

an unlicensed pleasure quarter, a settlement of dozens of "inns" lining both sides of the road. Shinjuku continued as such for fully half a century after Edo turned to Tokyo in 1868. But by 1918, the government had come to consider its brothels a public eyesore in the expanding modern city and ordered the removal of fifty-three such establishments to a confined area set back from the main road. On the 1931 map, this area appears to the upper right as a blank space, but its outline is clearly indicated by the many small drinking places and eateries that surround it. The new red-light district was completed by 1922 and operated continuously thereafter until legal prostitution in Japan was banned in 1955.

The old brothel settlement lay east of the modern Shinjuku we see here, ending before the Kōshū Highway forked at Oiwake (literally "branching"). At this point, the main highway led off to the southwest, while the secondary Ōme Road branched off toward the northwest.

The fork can be clearly seen on the map, at the Matsuya department store to the lower right. The basic triangular shape of modern Shinjuku was determined in 1885 by the siting of a station on a new railway line designed to provide north-south service through Tokyo. Steam locomotives were not welcome in the center of the city, so the line skirted settled areas, with the new station intentionally laid out five hundred meters west of the fork at Oiwake. This fortuitously created a triangle of sparsely settled land bounded by two highways and a railroad, into which the new urban configuration would expand. The north-south line, which would eventually evolve into the Yamanote loop line, was joined in 1889 by the east-west Kōbu Line, today the Chūō Line. Originally private, both the Yamanote and Chūō Lines became part of the nationalized rail system after 1906, and their electrification and double-tracking followed soon after, greatly increasing capacity and enabling easy entry to the city center. Meanwhile, Shinjuku was linked by electric streetcar from Oiwake to central Tokyo in 1903, while a second line from 1914 provided a more northerly route; both are seen on the map as wavy lines.

The truly dramatic expansion of Shinjuku came after the Great Kanto Earthquake of September 1, 1923, which set off fires that devastated most of central Tokyo and the low-lying areas to the east. Although the higher ground of the Yamanote area to the west was largely spared, the Shinjuku station area itself was consumed by a single fire set off when earthquake tremors upset a stove. But this also proved an opportunity, enabling a surge of modern rebuilding. Even more of a boost was the accelerated growth of the already-developing western suburbs, as thousands of families sought new homes on more solid ground. New private rail lines radiated westward from their new terminals at Shinjuku. The map shows the Keiō and Seibu Lines, while the Odakyū terminal lay on the still undeveloped western side of the station, the start of an eighty-two-kilometer line to Odawara from 1927.

Shinjuku was not only the gateway for new suburban commuters to downtown jobs, but also a growing market for their shopping needs. Along the main street were the four department stores seen on the map in blocks of solid blue, all occupying tall ferroconcrete buildings that dominated the landscape of the area. Nikō was the site of a Mitsukoshi branch from 1925 until Mitsukoshi moved to new building site seen here, eight stories above ground and three below. Hoteiya of 1926 would be joined immediately to the west by Isetan in 1933, which in turn absorbed Hoteiya shortly thereafter. Matsuya (no kin to the famous Matsuya on the Ginza), located above the Keiō terminal, folded soon after this map was made, unable to compete.

The other key large-scale institutions were the theaters, of which five are shown on the map (also in blocks of solid blue). All but one were devoted to the booming medium of film; the Shin Kabukiza alone offered more traditional performances. Note that all are located away from the main street, which was lined mostly with retail shops and the aforementioned department stores.

Let us now home in more closely on the "eating and drinking places" recorded on the map. The survey of these establishments was conducted single-handedly on the evening of July 7, 1931, by a Waseda University student named Iwata Yoshiyuki (whose flowery signature appears at top center). It was intended as a contribution to the "Modernology" movement created in the wake of the 1923 earthquake by Iwata's mentor, Waseda design professor Kon Wajirō. Kon argued for studying the urban present through the observation and documentation of the human activity they saw around them.

Iwata's map, pinpointing the name and location of 294 eating and drinking places, first appeared in the October 1931 issue of *Crime Science*—a venue hinting at the frisson of vice and transgression that Shinjuku then conjured up in the popular press. Kon wrote the accompanying text, analyzing the spatial distribution of the places recorded. The map was then slightly revised, and printed in two colors as the endpapers of Kon and Yoshida's second anthology of Modernology studies in December 1931.

The eating and drinking places are classified into nineteen categories, each shown with its symbol in the legend to the upper right (see fig. 37.2 for translation). In six cases, the symbols are Chinese characters (indicating such ethnic cuisines as Japanese, Chinese, or Korean); the rest show a characteristic serving utensil. The remaining fifteen categories offer an intriguing snapshot of the complex and constantly shifting taxonomy of consumption in 1931 Shinjuku. Take the "milk hall," for example, a curious fad of late Meiji that served milk along with Western-style cakes, tea, and coffee. Overlapping in function with the coffee shop, only four milk halls appear on the map, and the genre was soon to disappear entirely.

FIGURE 37.3 Night view of café street in Shinjuku. Reprinted from Historical Photos [Rekishi Shashin 歴史写真], no. 257 (October 1933).

Most numerous on the map was the "café," the talk of the town in Tokyo from the mid-1920s. The term was Western, but the Japanese reality—let us refer to it as *kafé*—was quite distinctive. There already existed the durable category of "coffee shop" (*kissaten*), offering a quiet space to sip coffee while chatting or reading, but the *kafé* was a different beast, offering mainly alcoholic drinks served by flirtatious waitresses—hostesses, really, working for tips. Iwata's survey shows fully ninety-two *kafé*, close to one-third of all eating and drinking places in Shinjuku, each indicated by the symbol of a folded paper napkin.

Looking back, we can see that Iwata was documenting the historical moment when the *kafé* of the late 1920s and early 1930s was at its peak. These establishments were not long lived and would in fact soon disappear from the scene entirely as the shadows of war darkened. The map does not capture such changes over time. Nor does it directly convey the frenetic pace of recent building in Shinjuku, or the press of the huge crowds that flowed through the station and into the streets and shops. But if we imagine what was obvious to any Japanese viewing the map in 1931 (perhaps even using it as a practical guide to the area), this array of varied symbols, each with a name inscribed in bright red, was more than enough to conjure up the fast-paced tempo that was a pervasive leitmotif in writing about Shinjuku.

Kon's analysis singled out three areas, of which two were dominated by the *kafé*. The larger was south and east of Mitsukoshi, with twenty-four folded-napkin symbols, of which the single densest street was famed for its nighttime neon, often featured in photos of this era (fig. 37.3). The other, mentioned earlier, is the area ringing the pleasure quarters, more spread out but with even more *kafé* napkins. A third area, known as "Eat-Your-Fill Alley," lay behind Nikō department store and consisted of numerous cheap eateries.

Shinjuku has expanded manyfold since 1931. The station now has more passenger traffic than any other in the world. Its spatial arrangement is far more three-dimensionally complex, with the amoeba-like spread of a vast underground area and the construction from the 1970s of a huge new district of thirty-odd skyscrapers west of the station (many over fifty stories, far exceeding the prewar limit of eight), including the new Tokyo Metropolitan Government Buildings (1991), which have effectively made Shinjuku the center of the entire city. But although the number of "eating and drinking places" has expanded some tenfold, the soul of Shinjuku remains much as it was in 1931: chaotic, diverse in function and scale, and throbbing with the constant movement of people as they commute, shop, socialize, eat, and drink.

Note

1. Julian Worrall and Erez Golani Solomon, *21st Century Tokyo: A Guide to Contemporary Architecture* (Tokyo: Kodansha International, 2010), 174.

Suggested Readings

Freedman, Alisa. *Tokyo in Transit: Japanese Cultures on the Rails and Road.* Stanford: Stanford University Press, 2011.

Silverberg, Miriam. "Constructing the Japanese Ethnography of Modernity." *Journal of Asian Studies* 51 (1) (February 1992): 30–54.

Worrall, Julian, and Erez Golani Solomon. *21st Century Tokyo: A Guide to Contemporary Architecture.* Tokyo: Kodansha International, 2010.

38 Mapping the Hōjō Colliery Explosion of 1914

Brett L. Walker

FIGURES 38.1A-B Untitled illustrations from "Documents on the Hōjō Colliery Disaster of March 12, 1914" [Hōjō tankō hensai shorui 方城炭鉱変災書類], 1914. Reproduced in *Sekitan kenkyū shiryō sōsho*, Vol. 26 (Fukuoka: Kyūshū Daigaku Seitan Kenkyū Shiryō Sentā, 2005). Courtesy of Kyushu University.

Japan's shift to a fossil fuel economy represents one of the most important transitions in its history. Prior to the Meiji Restoration of 1868, the country largely depended on charcoal and wood, renewable resources extracted from carefully managed forests. Such resources heated homes, boiled water, and—along with the sinewy power of human and nonhuman muscles, as well as swift river currents—drove Japan's protoindustrial economy. After 1868, however, Japan began switching to coal in a concerted effort to industrialize. Implicit in this energy transition were revolutions in Japanese political structures and a radical reconfiguration of geographic space, as newly engineered vertical geographies tapped stored carbon energy underground. The map shown here represents one such subterranean geography where labor, both human and nonhuman, extracted and transported the coal that rapidly revolutionized Japan.

In Japan as around the world, the age of fossil fuels restructured mass politics. Coal extraction and transportation workers were among the first to organize into militant unions; through the threat of general strikes, these same workers slowly democratized political practices in one country after another. In Japan as elsewhere, coal mines became important sites for often-violent protests, the development of a unionized labor force and, in the long run, the broadening of political participation. Many prominent progressive activists cut their political teeth at Kyushu coal mines. For as much as Japan's adaptation of Western political concepts eventually democratized the country, it was the recruitment of fossil fuels that enabled mass political participation in the first place.

Reliance on stored nonrenewable solar energy has its limits, however. To begin with, coal and oil supplies on earth are finite, including the rich seams once mined around the Hōjō mine. The fossil fuel civilization is term limited by physical and geologic realities. Moreover, the climate change that has resulted from burning fossil fuels has transformed human societies into geologic agents, serving to align geologic and historic calendars.[1] Geology is no longer simply the result of natural plate tectonics, volcanism, and erosion, but value-driven human decisions as well.[2] The coal mines that thrust Japan into the age of fossil fuels developed with astonishing speed in the decades bracketing 1900. But such growth also brought new and terrible dangers in its train. The December 15, 1914, explosion at the Hōjō Colliery, which killed 687 people, stands as one marker of Japan's coming of age as an industrial power.

The maps shown here (figs. 38.1A and 38.1B) were drafted shortly after that explosion, to accompany a report entitled "Documents on the Hōjō Colliery Disaster." Officials called for subterranean maps to pinpoint the location of the bodies that were discovered underground—burned, dismembered, drowned, or suffocated—in what remains to this day the worst mine explosion ever on Japanese soil. As shown here, rescue workers discovered the corpses of carpenters, foremen, miners, officials, and other underground workers scattered through the numbered mine shafts. Each dead body was represented by a ghostly vermilion form, apparently stamped on the map near the place where it was found. The map also illustrates the location of water sprinklers (used to suppress coal dust), water lines and proposed water lines (strung throughout the shafts), air pumps, coking areas, and coal transportation routes.

Such underground complexes were a strikingly new feature of the Japanese industrial landscape. The Japanese had mined some coal prior to the Meiji Restoration, but in the early modern era, miners merely scratched "raccoon-dog" holes into surface deposits. Starting in the 1880s, however, sloping shafts, often called "inclines," were drilled much deeper into the earth's crust. For industrial powerhouses like Mitsubishi, the company that opened the Hōjō Colliery in 1908, these shafts represented valuable sources of revenue. The Hōjō mine, one of the most productive, was also one of the deepest mines of its day, reaching nearly three hundred feet below the surface. But it was also one of the most dangerous. With the vertical geography created by the sloping inclines of late Meiji mining, runaway carts, cave-ins, floods, and fires, as well as coal dust and gas explosions, came to define the dangers of Japan's new energy regime.

The Hōjō Colliery was the seventh mine opened in Mitsubishi's Chikuhō Coalfield on the southern island of

Kyushu. The main coal vein at Hōjō had been discovered in 1897. Over the course of the next decade, engineers laid critical railway tracks to the mine and installed the mine shaft cage that lowered miners hundreds of feet in a matter of ear-popping seconds. As the veins nearer to the surface were exhausted, engineers constructed increasingly deeper inclines to access underground seams, requiring more elaborate pumping systems, coal dust suppression sprinklers, and methane gas dispersal fans. Output grew apace. By 1913, the Hōjō mine alone produced some 230,000 tons of high-quality bituminous and anthracite coal annually, while the Chikuhō region as a whole yielded up 10 million tons of coal—nearly half of Japan's entire annual production. By helping to fuel the ships and trains that enabled Japanese overseas expansion, such mines became central to Japan's sprawling empire. The Hōjō Colliery disaster map thus charts a vertical world that was intimately connected to Japan's horizontal ambitions.

The Hōjō mine was a technologically complex subterranean landscape. Engineers designed it, miners dug it, carpenters reinforced it, horses never left it, and foremen brutally ran it. Giant steel head frames anchored the mine's underground geography to the surface. The cages hoisted by these head frames weighed well in excess of one ton each. Massive Riedler's water pumps cleared the inclines, while whirling Rateau fans circulated air throughout the shafts. Even with these pumps and fans, the mines were very dangerous places.

In the early twentieth century, coal-mining accidents in Fukuoka Prefecture dramatically increased, claiming hundreds of lives. In the Chikuhō region, the overwhelming bulk of accidents involved gas or coal dust explosions. Such explosions had killed 210 souls at the Hōkoku mine as early as 1899 and hundreds more, at the same mine, in 1907. They were similarly responsible for 256 miners' deaths at the Ōnoura mine in 1909 and then 365 at the same mine just eight years later. In a word, the disaster at the Hōjō Colliery was hardly isolated or anomalous.

The explosion that shook this fragile subterranean world on a cold December day of 1914 completely sealed the upper ventilation shaft with rock and loam; the lower shaft remained open, and ventilation machines there groaned as they struggled to clear the noxious air. Rescue workers started retrieving bodies from the mine that afternoon after the levels of poisonous gases had subsided enough for them to enter the shafts. Even then, four rescue workers died from afterdamp poisoning. Stacked "like cordwood" and "baked like sweet potatoes," their bodies were added to those of the other victims that were hauled to the surface in the one functioning cage.

The intrepid prefectural official Meguro Suenojō, following his official investigation, determined that the coal dust explosion had been ignited by a faulty safety lamp. After analyzing the trace patterns of the coke and other burned matter in the shafts, Meguro determined that the ignition occurred near the junction of the "7½ incline" and the "16th side," the spot shown in the upper left center of the map. From this point, violent explosive waves cascaded through the mine shafts. Most miners were burned alive by the flames, while others suffocated when the intense flames sucked every last bit of oxygen out of this subterranean environment. Meguro discovered small traces of coke dust on the inside of the gauze mesh of one safety lamp. The mesh was designed to let oxygen in, but not such combustibles as coal dust or methane gas. The faulty safety lamp belonged to Negoro Yōjirō, a Hiroshima native who labored at the mine with his wife, Shizu, and his eldest daughter, Hatsuyo. (Unlike hard rock mines, where superstitions regarding "mountain deities" kept women on the surface, coal mines were worked by both men and women, often as teams.) There are six corpses represented on the map at the intersection of the "7½ incline" and the "16th side," shown as a row of vermilion figures floating against the blue background. Two of them were almost certainly Yōjirō and his wife, who died at his side.

The Hōjō Colliery disaster map traces the tragic costs of Japan's entry into the age of fossil fuels. It charts a subterranean environment braided by steep inclines that tapped coal veins hundreds of feet below the surface. Such underground mines mattered simultaneously as sites of capitalist production and radical political organization, an underground geography that fueled both political conflict at home and imperialist expansion abroad. In effect, the Hōjō Colliery disaster map documents an otherwise invisible geography of violence that the move to fossil fuels entailed.

Notes

1. Dipesh Chakrabarty, "The Climate of History: Four Theses,"

Critical Inquiry 35 (Winter 2009): 197–222.

2. J. R. McNeill, *Something New under the Sun: An Environmental History of the Twentieth-Century World* (New York: W. W. Norton and Co., 2000), 10–11.

Suggested Readings

Mitchell, Timothy. *Carbon Democracy: Political Power in the Age of Oil*. London: Verso, 2011.

Walker, Brett L. *Toxic Archipelago: A History of Industrial Disease in Japan*. Seattle: University of Washington Press, 2010.

39 Cultivating Progress in Colonial Taiwan

Philip C. Brown

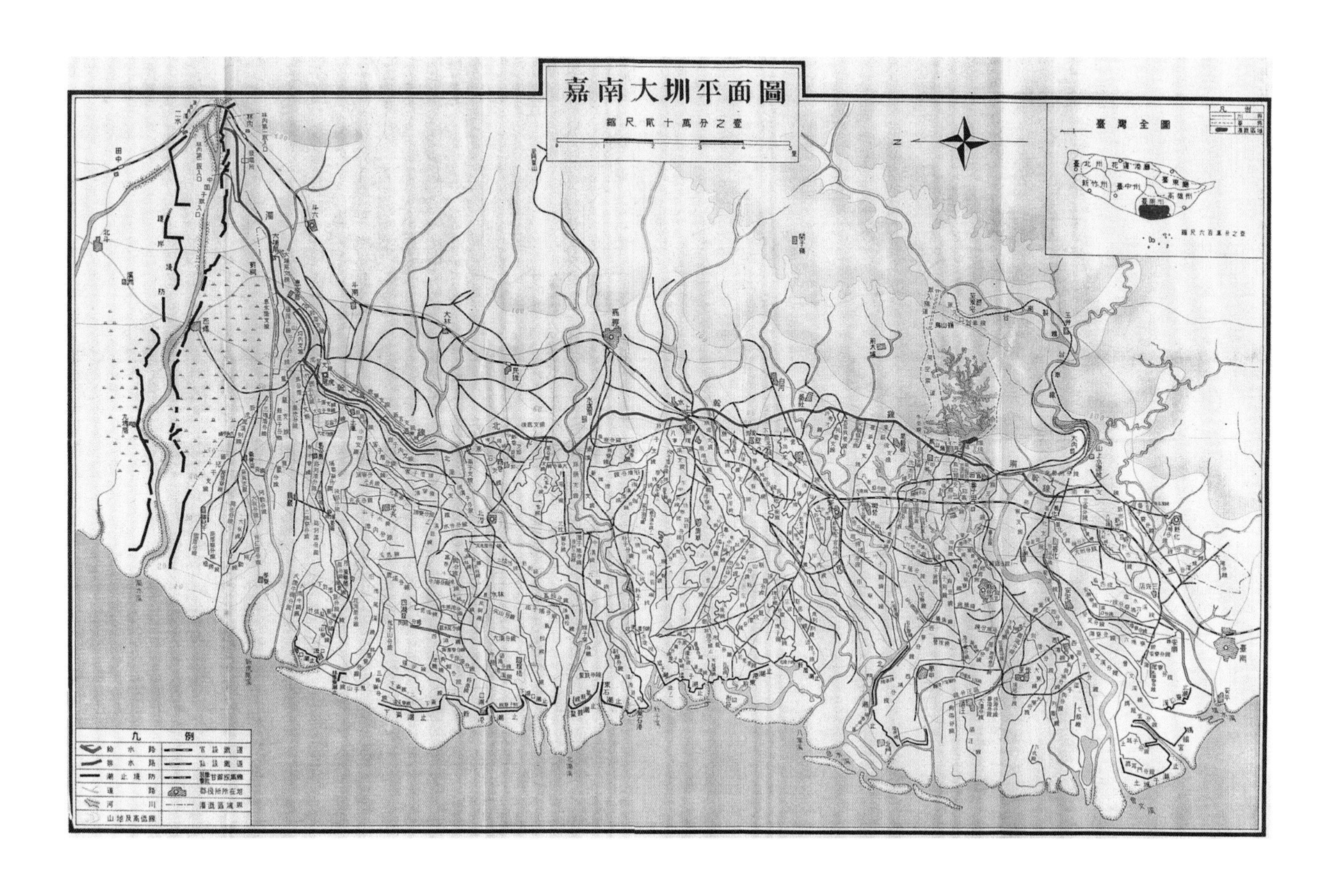

FIGURE 39.1 "Map of the Kanan/Jianan Water Control Project" [Kanan taishu heimen zu 嘉南大圳平面図], 1930. Produced for Eda Tokuji, *Kanan taishu shinsetsu jigyō gaiyō* ka 嘉南大圳嘉南大圳新設事業概要課 (Taipei: Kanan Taishu Kumiai 嘉南大圳組合, 1933). 56 × 42 cm.

"Map of the Kanan/Jianan Water Control Project" (fig. 39.1) celebrates the 1930 completion of water management and flood control structures that profoundly transformed agriculture and life in Japan's first formal colony, Taiwan.[1] When construction commenced in 1920, this enterprise constituted the largest single civil engineering project in East Asia. Its successful completion initiated an era of other similarly expansive projects undertaken by Japan on the Asian mainland, including in Korea and Manchuria. The map portrays a project impressive in scale and structural complexity. Upon completion, this enterprise developed more than sixteen thousand kilometers of irrigation and drainage lines that together transformed much of southern Taiwan by bringing some 150,000 hectares (about 370,000 acres) of land into agricultural production.

A project of this scope would not have been possible without Japan's victory in the Sino-Japanese War (1894–95), which brought Taiwan under Japan's control. Japan had achieved a military capacity sufficient not only to defeat the Qing dynasty but also to subdue all the indigenous peoples of Taiwan, a historic accomplishment. Such military prowess allowed the Japanese to undertake broad planning for the island's economic development and to carry out projects like the one depicted in this map. The war made an additional indirect contribution to the Kanan project: reparations from China funded major riparian projects on Japan's main islands that provided its engineers with relevant hydrological expertise.

A close look at figure 39.1 helps us understand the daunting challenge posed by the Kanan/Chia-nan (Jianan) project. Oriented with east at the top, the map's lower (western) half is visually busy. Numerous red water supply and blue drainage lines indicating newly constructed waterways crisscross this coastal plain, while roads are indicated in yellow and rail systems—many designed to transport agricultural products—in black-and-white-hatched lines. This part of the map depicts a belt to the sea that is about ten miles wide, gently dropping to sea level from an elevation of about seventy feet. By contrast, the top half of the map features a less complex pattern, representing an elevated region of lesser human activity. These uplands continue rising to more than twelve hundred feet over the course of about two miles, creating rapid runoff from Taiwan's abundant seasonal rains.

The Zhuoshui River, the single largest natural feature on the map (located to the extreme left), supplied most of the water for the project. The other source was man made: the Wushantou Dam, which created the large reservoir evident directly below the compass rose at top right. The planning and construction of this dam constituted a major technological achievement and constituted one of the two central components of the ten-year undertaking.

Considering that irrigation and flood control are typically considered distinct projects, the presence of both throughout the map is notable. The prominence of the red water supply lines reveals that much of the project was devoted to assuring an adequate supply of water to improve the productivity of existing farms and to expand the area of land under the plow. Yet the thick red line running left to right (north to south) suggests the challenge engineers faced. This transverse trunk line, not waterways running directly from the mountainous interior, supplied most of the new canals. With northern and southern lines branching from the Wushantou Dam, the trunk line traces the outline of highlands varying from 80 to 110 feet elevation. It is set back from the coast, skirting the foothills, with a gentle descent to the sea. Were the descent consistent and even, the project might not have required two other important features that the map details.

The many inland drainage channels constitute one such feature. Testifying to topography conducive to pooling of excess water and regular flooding, the drainage lines removed waters that could not be contained naturally within riverbeds and carried them to the sea or to irrigate other districts. Dikes also feature prominently on the map. Most clearly indicated by the long, black lines on either side of the Zhuoshui River, such dikes reveal a further attempt to engineer the landscape, whether to contain surplus water, improve drainage, control flood damage, or prevent the intrusion of saltwater into lands that would otherwise have the potential to support agriculture.

The cartographic techniques employed to create this map were just being perfected at about the time that engineers began to plan the Kanan project.[2] Unlike most of the maps discussed in parts I and II of this volume, this one was based on a land survey using modern instruments. It features all of the attributes we associate with state-of-the art maps of the early twentieth century, including a uniform scale and a diverse array of standardized symbols. The accurate measurement techniques that underlay this map provided the foundation for the territorial surveys that undergirded the plans and cost estimates of the irrigation channels, drainage canals, and flood control mechanisms of the Kanan project. Further, there were now sufficient numbers of well-trained surveyors to permit Hatta Yoichi, the project's originator, to conduct a detailed and reliable investigation of this extensive region in a matter of weeks.

For all the accomplishments it embodied, this map was selective in what it chose to represent. It did not reveal, for example, the opposition voiced by locals displaced or otherwise affected by the development.[3] In this instance, protests included opposition to fees to support the project and a new agricultural regime of crop rotation that ultimately became part of the pattern of agriculture made possible by the new system of water control.[4] Nor does the map show the sites where many Japanese and Taiwanese workers died in construction accidents. While deaths were common for large construction projects, the most lethal accident, in which more than fifty people died, occurred on a highly ambitious tunnel project that sought to overcome major challenges posed by the region's mountainous geography. (These deaths were memorialized in a tablet erected during the colonial era, on which the names of the Japanese and Taiwanese dead were intermingled.)

Despite its colonial origins and the controversies and challenges that marked the project's planning and construction, many Taiwanese and Japanese today would give this map a favorable, even triumphalist reading. As evidenced by websites, books, and television programs in both Taiwan and Japan, the project that the map represents is jointly celebrated as a part of Taiwan's historical trajectory, distinct from that of the mainland (as envisioned by either the Guomindang's Republic of China or the Communists' People's Republic of China) and embracing elements of the Japanese colonial era. A statue of Hatta Yoichi, the architect and manager of the Kanan project, was erected near the Wushantou Dam by his subordinates to celebrate the project's completion. Removed after the war, it was rediscovered and has become a focal point of commemorative events and a park named after Hatta in 2011. Every May Taiwanese and Japanese gather here to remember the sacrifices of those whose work transformed agriculture in southern Taiwan.[5]

Notes

1. Tokuji Eda, *Kanan daishun shinsetsu jigyō gaiyō* (Taihoku: Kōkyū hishin Kanan taishin kumiai, 1930).

2. Beginning with the major metropolitan areas of Tokyo, Osaka, and Kyoto, the Great Imperial Japan Land Survey Section (Dai Nihon teikoku rikuchi sokuryō bu) started by mapping urban areas, then expanded their efforts for a "rapid survey" of the Kanto area, and finally completed a national 1:50,000 topographical survey of all Japan in the first decade of the twentieth century. A revision of this effort was completed during World War I. On the rapid survey maps, see, for example, David S. Sprague, Iwasaki Nobusuke, and Takahashi Shin, "Measuring Rice Paddy Persistence Spanning a Century with Japan's Oldest Topographic Maps: Georeferencing the Rapid Survey Maps for GIS Analysis," *International Journal of Geographic Information Science* 21 (1) (2007).

3. For a broader perspective, consider that the World Commission on Large Dams indicates that some 40–80 million people were displaced by large dam projects alone in the century preceding its report. World Commission on Dams, *Dams and Development: A New Framework for Decision-Making* (London: Earthscan, 2000).

4. Misato Shimizu, "Taiwan shokuminchiki Taiwan suiri o meguru kenri no tōsō," *Nihongo/Nihon kenkyū* 1 (2011); "Suiri nettowaaku no saikōchiku: Kanan taishū kangai kuiki no kattō," *ISS Contemporary Chinese Studies* 9 (2012).

5. This fits one wave of late twentieth century reevaluation of Japan's colonial rule. Perhaps most well publicized within this new assessment, in 2003 former ROC president Lee Teng-hui delivered a series of lectures at Keio University which spoke favorably about the Taiwanese experience under Japanese colonial rule and Hatta Yoichi in particular.

Suggested Readings

Ho, Samuel Pao-san. "Colonialism and Development: Korea, Taiwan and Kwantung." In *The Pacific in the Age of Early Industrialization*, 273–324. Burlington: Ashgate, 2009.

Ka, Chih-ming. "Agrarian Development, Family Farms and Sugar Capital in Colonial Taiwan, 1895–1945." *Journal of Peasant Studies* 18 (2) (1991): 206–40.

———. *Japanese Colonialism in Taiwan: Land Tenure, Development, and Dependency, 1895–1945*. Boulder, CO: Westview Press, 1995.

40 Showcase Thoroughfares, Wretched Alleys: The Uneven Development of Colonial Seoul (Keijō)

Todd A. Henry

Before Land Readjustment		
Total Area:	147,650	100%
Residential:	123,944	84%
Parks:	3,333	2%
Roads:	20,373	14%

After Land Readjustment		
Total Area:	147,650	100%
Residential:	113,986	77%
Parks:	9,463	7%
Roads:	24,201	16%

FIGURE 40.1 Adapted from "Before and After Map of Land Readjustment for Zone Number Five (Pagoda Park Area)" [Dai go ku (Pagoda kōen fukin) kukaku seiri zengo no zu 第五区（パゴダ公園付近）区域整理前後ノ図], from *Report of Keijō City Planning* [Keijō toshi keikaku chōsasho 京城都市計画調査書], Keijō (Seoul), 1928, 271.

In 1910, nearly thirty years after settlers began living in its treaty ports, Korea was forcibly annexed to Japan's growing empire. With preferential treatment from the colonial state, the number of expatriates surged thereafter, increasing to nearly one million by 1945. Most settlers inhabited urban areas, where they dominated the ranks of business. Although generally supportive of settlers' commercial interests, Japan's government-general also tried to "assimilate" the colonized population, if only by enticing Koreans with the trappings of modernity. The wide and spectacular thoroughfares of Keijō (today Seoul) perhaps best embody the contradictions and limitations of transforming this city, once the capital of the Chosŏn dynasty (1392–1910), into a showcase of progress. Borrowing Western-inspired models used to update Japan's own cities, colonial officials embarked on an ambitious program of urban reforms, upgrading several axes that connected the Japanese-dominated "southern village" to the Korean-populated "northern village." However, their plans to create radial roads linking several plazas foundered (due to financial restrictions as well as opposition by both Japanese and Korean property holders), leaving most of the city unaltered.

It was in this context that in 1926, Sakai Kenjirō, a Japanese bureaucrat who then headed Keijō's temporary urban planning division, penned an instructive, if deceptive, description of the city's colonial infrastructure. As he wrote in the October issue of a government digest,

> In city planning, streets are the skeleton, transportation routes the arteries, and conveyers the blood. With these [elements], a city proceeds to grow, while buildings serve as the city's veritable muscles. Keijō's muscles are, in fact, extremely disordered and emaciated. Their organization does not form a fixed system or unity, but is an assemblage of muscular nodes which are poor and in ruins; with those winding roads and cul-de-sacs, it is barely living. The skeleton, heart, and blood of Keijō are complete, but no muscle adheres [to that body]. For this to happen, the city's residents must rely on their [own] awakening. What is that awakening? It is one thing: land readjustment.[1]

Likening the city to a human body, this bureaucrat drew on organic metaphors then circulating globally to capture both Keijō's current shortcomings and future solutions. Following fifteen years of urban reforms, Sakai praised fellow officials for remaking the royal capital into a showcase of Japanese modernity. He offered a remarkably positive evaluation of the colonial city, arguing that Keijō's skeletal streets, arterial routes, and sanguinary conveyers all operated productively. Doing so, however, clearly downplayed the city's unsanitary neighborhoods, especially those in the northern village, where most Koreans resided.[2] Although largely positive, Sakai's assessment did recognize some problems—namely, that the city's buildings, which he likened to its muscles, remained unhealthy. Rather than indicting the colonial state (of which he formed a part) for failing to upgrade the infrastructure around Keijō's allegedly diseased muscles, this colonialist blamed the city's Korean residents for failing to modernize their own living spaces.

Sakai's 1926 description captured only one "truth" about the complex and contradictory nature of this colonial city. The Japanese government-general, intent on justifying its authority, had poured inordinate resources into broadening the city's most visible boulevards. Perhaps the most important of these showcase thoroughfares was Taihei Boulevard. Often likened to Paris's Champs-Élysées, this impressive thoroughfare served as the city's main north-south axis, home to the Government-General Building at its north terminus and Keijō Train Station at its southern end. Other important monuments, such as City Hall and Korea Shrine, were also accessed from this central thoroughfare. Even as officials used Taihei Boulevard to impress residents and visitors alike, they largely neglected the capillary network of smaller streets that made up the bulk of Keijō's public infrastructure, especially in the northern village.[3] The resulting juxtaposition of showcase thoroughfares and wretched alleyways characterized the colonial city.

To overcome the gap between these two extremes, bureaucrats like Sakai proposed land readjustment. This German planning technique aimed to superimpose the city's rudimentary system of gridded roads onto the microspaces of dilapidated Korean neighborhoods—now recognized as a dangerous source of contagion for Japanese settlers. To finance this project, taxpaying landowners, a statistical minority in Keijō, were asked to relinquish a portion of their land in exchange for anticipated increases in property values and commercial profits.[4] The plan of 1928 provides an example of how this project of "cultural rule" was meant to play out in one district (fig. 40.1). According to detailed calculations, this district was composed of 84 percent residential lands, 14 percent roads, and only 2 percent parkland. To open up what they viewed as an unnecessarily cramped and unhealthy area, city planners sought to decrease the amount of land allocated to houses, making room to develop a system of gridded roads and to triple the size of Pagoda Park, a public site that served as the "lungs" of the northern village.[5]

To facilitate such sweeping projects, colonial officials had to rely on the civic-mindedness of Korean landowners. Lacking the power of eminent domain, they could only hope that local elites would come to see their common interests in urban modernization and relinquish a portion of their land. To encourage the necessary sacrifices, planners proposed creating landowner cooperatives.[6] In addition to providing financial support for government-led projects, these cooperatives would impose local sanctions, compelling all members to participate if at least two-thirds of residents owning at least two-thirds of land under discussion agreed.

Although some bureaucrats (including Sakai) were optimistic about property owners' commitments to land readjustment, more sober planners expressed serious concerns about ill-financed attempts to transform the microspaces of Korean neighborhoods. Naoki Rintarō, for one, remained pessimistic that Koreans would support these projects. The former head of the Tokyo Reconstruction Bureau, Naoki questioned the feasibility of carrying out land readjustment in the northern village, whose dense neighborhoods left little room for planners to maneuver. Even in Japan, the spirit of generosity and self-sacrifice he credited to Tokyoites after the 1923 Great Kanto Earthquake did not ensure popular support for land readjustment, which often had to be forced on reluctant landowners. In the colony, cooperation with officials was further complicated by the high percentage of mortgages on Korean properties and widespread suspicions that fees would accrue to their loans. Since the only other option, forced dislocation, threatened to disaffect the colonized population altogether, Naoki encouraged officials to focus their efforts on the rapidly developing suburbs, where some desperate Korean residents might relocate without having to expel poor Koreans from the dense city center.[7]

Given that the redevelopment of the northern village no longer appeared in the city plan of 1930, Naoki's pessimistic conclusion seems to have convinced Sakai and others to abandon this difficult project. As late as 1932, officials were, in fact, still complaining that Korean landowners refused to make even a 5 percent contribution of their land, a figure far lower than the 10–30 percent customarily required of their counterparts in the metropole.[8] Only after the passage of the Town Planning Act in 1934 did colonial officials begin to implement land readjustment, even then limiting these unpopular projects to the suburbs.[9]

Although specific in its aim of land readjustment, the 1928 proposal forms part of a much larger archive of unrealized modernization plans for historic cities like Seoul. Indeed, the history of colonial Korea was characterized by a series of bold plans that aimed to impress native inhabitants with the symbols of Japanese modernity, while allowing global capitalism to penetrate the infrastructure of formerly walled cities. To be sure, the urban reforms of the 1910s and early 1920s made significant inroads in this regard, creating a modest grid of boulevards that not only highlighted important monumental architecture but also smoothed the way for the circulation of people, goods, and ideas. Beyond this impressive and costly arterial infrastructure, however, the capillary network of Seoul's wretched alleyways remained largely unchanged throughout the period of Japanese rule. That proposals like this one failed to take root in urban centers like Seoul reminds us of both the ambitions and the frailties of modern colonialism.

Notes

1. *Keijō ihō* 59 (October 1926): 14.

2. On the city's precolonial transformations, see T'ae-jin Yi, "Seoul's Modern Development during the Eighteenth and Nineteenth Centuries" and "The Leaders and Objectives of the Seoul Urban Renovation Project of 1896–1904," in *The Dynamics of Confucianism and Modernization in Korean History* (Ithaca: Cornell University Press, 2007).

3. I borrow the arterial and capillary metaphors from Gyan Prakash, *Another Reason: Science and the Imagination of Modern India* (Princeton: Princeton University Press, 1999). On the city's post-1910 reconstruction, see my "Respatializing Chosŏn's Royal Capital: The Politics of Japanese Urban Reforms in Early Colonial Seoul, 1905–19," in *Sitings: Critical Approaches to Korean Geography*, ed. Timothy Tangherlini and Sallie Yea (Honolulu: University of Hawai'i Press, 2007), 15–38.

4. According to one estimate, as many as 80 percent of Keijō's Korean families lived in residences held under collateral security, while only 20 percent owned their own homes. *Tonga ilbo*, June 24, 1926.

5. *Keijō toshi keikaku chōsasho* (1928), plates between pp. 270 and 271.

6. *Keijō toshi keikaku chōsasho* (1928), 254–55. For Sakai's hope that Koreans would accept and implement land readjustment, see *Keijō ihō* 66 (March 1927): 34.

7. *Chōsen doboku kenchiku kyōkai kaihō* 133 (March 1929): 11–12.

8. *Chōsen to kenchiku*, July 1926, 4; and June 1932, 55.

9. On these wartime projects, see Son Chŏng-mok, *Ilche kangjŏmgi tosi kyehoek yŏn'gu* (Seoul: Iljisa, 1990), 281–300; and Ishida Junichiro and Kim Jooya, "Colonial Modernity and Urban Space: Seoul and the 1930s Land Readjustment Project," in *Constructing the Colonized Land: Entwined Perspectives of East Asia Around WWII*, ed. Kuroishi Izumi (Farnham: Ashgate, 2014), 171–92.

Suggested Readings

Henry, Todd A. *Assimilating Seoul: Japanese Rule and the Politics of Public Space in Colonial Korea, 1910–1945*. Berkeley: University of California Press, 2014.

Ishida Junichiro and Kim Jooya. "Colonial Modernity and Urban Space: Seoul and the 1930s Land Readjustment Project." In *Constructing the Colonized Land: Entwined Perspectives of East Asia around WWII*, edited by Izumi Kuroishi, 171–92. Farnham: Ashgate, 2014.

Korea Journal. Special issue, "Colonial Modernity and the Making of Modern Korean Cities" 48 (3) (Autumn 2008).

Uchida, Jun. *Brokers of Empire: Japanese Settler Colonialism in Korea, 1876–1945*. Cambridge, MA: Harvard University Asia Center, 2011.

Yi, T'ae-jin. *The Dynamics of Confucianism and Modernization in Korean History*. Ithaca: Cornell University Press, 2007.

41 Imperial Expansion and City Planning: Visions for Datong in the 1930s

Carola HEIN

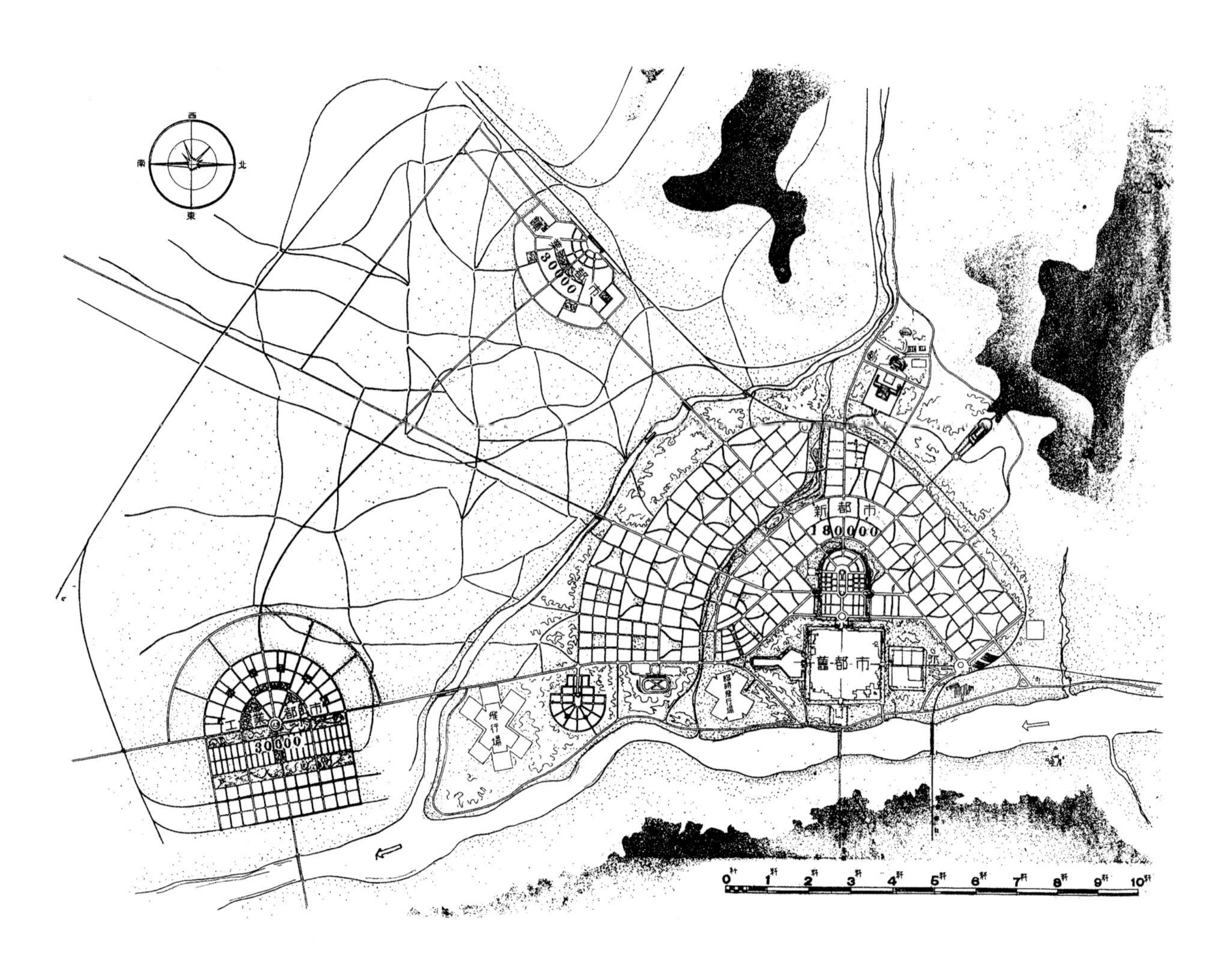

FIGURE 41.1 "Regional Plan for Datong, 1938" [Daidō no toshi kenkakuan 1938 大同の都市計画], from Uchida Yoshikazu 内田祥三, "Daidō no toshi kenkakuan ni tsuite 1," *Kenchiku zasshi* 53 (1939): 1292.

In 1938, a group of Japanese planners drew up a visionary proposal for a colonial settlement next to the historic walled city of Datong, some three hundred kilometers west of Beijing. Following the Japanese invasion and occupation of the city, the puppet Northern Shanxi Autonomous Government invited Uchida Yoshikazu (Yoshizō), professor at Tokyo University, to develop a plan for Datong's expansion. Uchida was only too happy to comply. Together with his son and a student, Uchida spearheaded a group that developed a radically new urban master plan for Datong. Although it was never realized on the ground, this proposal—with its accompanying close-ups of a citywide master plan and a forward-looking neighborhood design—encapsulates the state of urban planning in Japan in the late 1930s.

The Datong project built upon popular urban planning ideas of the time while paying homage to local forms and traditions and including references to specific Japanese colonial ideas and needs. But it owed more to global planning concepts than to any in-depth familiarity with urban life in north China. The entire design was created after a three-week study visit to Datong in September and October 1938. The professor's son prepared the drawings after the group returned to Tokyo, while one of his students, Takayama Eika, focused on construction plans; Uchida himself wrote a narrative report. The three presented their findings together the following year at a meeting of the Architectural Institute. Uchida published two articles in *Architectural Journal* in late 1939, and Takayama Eika, together with Uchida and his son Yoshifumi, wrote a jointly authored piece in *Contemporary Architecture* in 1940.[1]

The group drew up a carefully researched set of technical drawings for a main city of 180,000 inhabitants and two satellite cities of 30,000 inhabitants each (figs. 41.1 and 41.2). They also designed a close-up of the city plan and a typical neighborhood (fig. 41.3). Detailed drawings accompanying these designs proposed three phases of growth around the historic city, with different street profiles to be applied to roads and roundabouts, and the specific types of courtyard housing to be built.

Uchida's first article provided both historical context and contemporary reasons for investing in the expansion of Datong. Under the name Pingcheng, he noted, the city of Datong had served as the capital of the Northern Wei in the fifth century CE. This period saw the construction of the Yungang Grottoes, whose ancient carved Buddha images had been studied and celebrated by Japanese architect Itō Chūta. Uchida's Datong was thus an important historical site with potential as a tourist destination. But it was also a place of great industrial potential, located close to important mines and major transportation routes coveted by the Japanese state.

Uchida's proposal made space for both of these dimensions of Datong. On the one hand, it called for preserving the historic walled city center west of the Yuhe River; on the other hand, it targeted areas for rapid growth. The planners suggested creating a greenbelt along the river that would enclose both the historic city and proposed large-scale structures, including (from north to south) a railway station, airport, and sports arena. To accommodate projected growth, the planners also proposed an entirely new satellite town adjacent to the old one. Extending the old city to the west, they designed a new town center with administrative buildings embedded in greenery and surrounded by a half-moon-shaped new development—with a population density

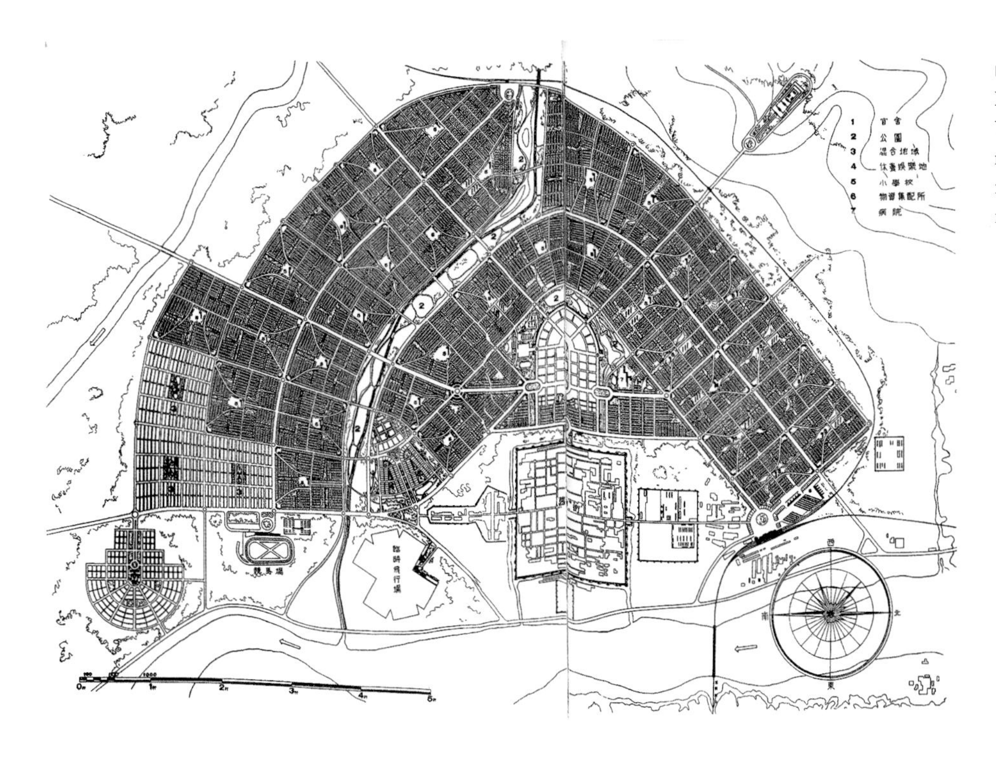

FIGURE 41.2 "Citywide Plan for Datong, 1938," from Uchida Yoshikazu, "Daidō no toshi kenkakuan ni tsuite 1," *Kenchiku zasshi* 53 (1939): 1290–91.

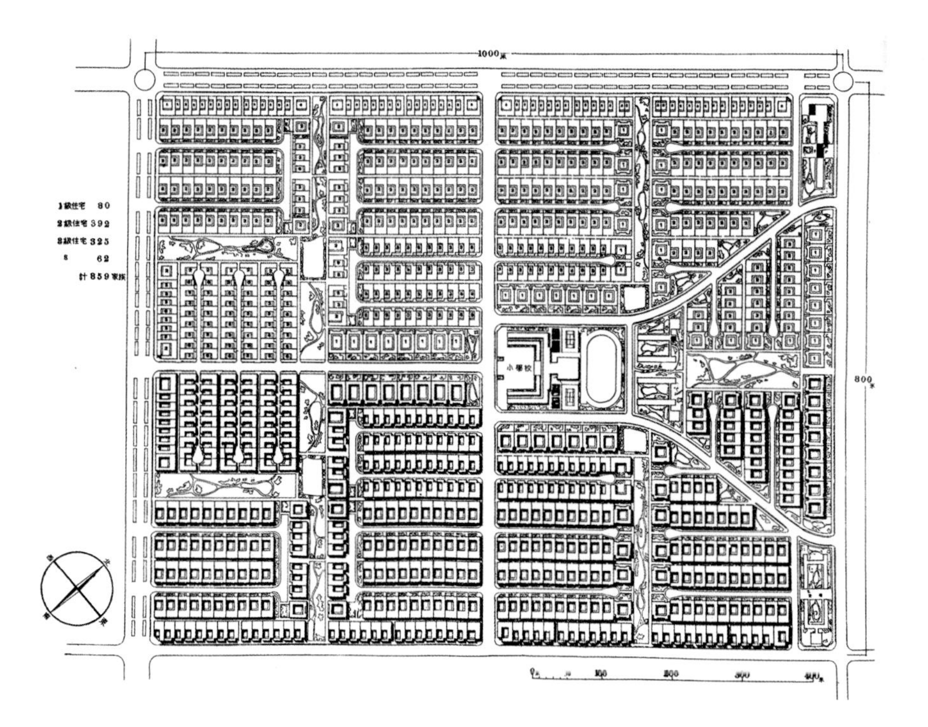

FIGURE 41.3 "Neighborhood Plan for Datong, 1938," from Uchida Yoshikazu, "Daidō no toshi kenkakuan ni tsuite 1," *Kenchiku zasshi* 53 (1939): 1293.

lower than that of Japanese cities—capable of accommodating 180,000 residents. Green areas and a railway line were suggested for the development's northern and northwestern side, while the Shili River to the south and southwest would complete the external contours of the new city. A smaller river running through the city—separating two housing areas—was framed by another greenbelt: a multipurpose space that could be used for both recreation and aerial defense if required. The planners took a regional approach to development, connecting Datong with other existing cities and providing for two additional cities nearby to house up to 30,000 inhabitants each. The first would be an industrial center (separated by a green zone and major roadway from the housing developments), while the second would be a coal-mining town.

Uchida's approach shows the influence of European and American design principles on Japanese planners. To be sure, some elements of the Datong plan were based on Japanese traditions. This is most evident in the Shinto shrine proposed for the mountainous area on the northeastern side of the city and linked to the center by a large road.[2] But the main design elements of the Datong plan emanated from Europe and the United States. Throughout the 1930s, many Japanese planners embraced ideas that had been developed since the 1920s by such European planners as Paul Wolf, Walter Christaller, and Raymond Unwin. Prior to his trip to Manchuria, Takayama Eika published a study of Western housing schemes that had a great influence on planners in Japan.[3] As the Japanese planners studied foreign practices, they implemented those ideas in colonial plans, thus simultaneously importing and exporting planning concepts.

This was particularly true in the case of the Uchida team's design for residential housing, where clearly defined neighborhood units reflect the direct influence of Western planning principles. The authors specifically reference the planned community of Radburn, New Jersey, a town for the motor age designed by Clarence Stein and Henry Wright in 1928. That proposal in turn was informed by British garden-city concepts that separated different modes of traffic and called for both self-contained pedestrian networks and housing blocks with cul-de-sacs. While acknowledging the negative connotations of this system within Japan, Uchida argued that residents of Datong would find its hierarchy of streets efficient and convenient.[4]

The rectangular housing areas they envisioned—each covering nearly a square kilometer—would collectively form a crescent surrounding the administrative and commercial city center. About five thousand people were to be housed in each unit, which would have its own school, park, and sports facilities. No through traffic would be allowed in these residential areas, most of which would be accessible only via dead-end streets that connected to pedestrian greenways leading to a central park. A closer look at Takayama's collection of housing districts shows this plan to be a nearly identical copy of an American garden-city neighborhood realized in Detroit in 1931. The planners of Datong made only minor transformations to the original design, such as changing the orientation of some buildings, replacing apartment houses with courtyard buildings, and eliminating churches.[5] The Uchida report insisted on the necessity of building large houses for metropolitan immigrants, as Western countries had done in their colonies. In regard to the housing design itself, however, the planners developed several model buildings based on the area's typical Chinese courthouse buildings, which were deemed most appropriate for the climate.

Colonial planning in the West is generally discussed together with the problem of ideology.[6] In the case of Japan, plans like these were part of a broad attempt to "overcome modernity" while preserving Japanese identity.[7] This made the prospect of borrowing urban forms from the West a fraught issue. Referring to Le Corbusier's Algiers project, Takayama took the view that the importation of urban planning concepts from abroad (or their export to one's own colonies) was problematic only when intended as a vehicle for the transmission of culture.[8] The diffusion of pragmatic planning techniques, by contrast, ought to be seen as the nonideological work of a craftsman. This kind of deliberate attempt to imitate, adapt, and "overcome" Western urban planning experiences was developed to an unusual degree in Japan's colonial city plans. While a few innovative projects were proposed for the Japanese archipelago during these years as well, the weight of existing political, social, and physical contexts sharply limited planners' scope for maneuvering at home. Japan's colonies in Asia, by contrast, afforded planners opportunities to envision (and sometimes impose) more radical urban change. The same was true of other imperial states operating in Asia, as one can see in the radical interventions by Russian planners in Dalian or their German counterparts in Tsingtao.[9]

As part of their own modernization since the mid-nineteenth century, the Japanese had made a careful study of European and American practice in multiple arenas for many years. Colonialism was a critical part of the example set by the European powers, and the Japanese eagerly followed their lead, adopting Western colonial techniques even while searching for locally adapted solutions. Although it was never realized, the proposal for Datong is particularly interesting in this regard. It provides a snapshot of the global knowledge Japanese planners could bring to bear in the late 1930s, as well as a glimpse of how they sought to deploy that knowledge in an attempt to further expand and entrench the Japanese empire in Asia.

Notes

1. Uchida Yoshizō, "Daidō no toshi kenkakuan ni tsuite 1," *Kenchiku zasshi* 53 (1939): 1281–94; Uchida Yoshizō, "Daidō no toshi kenkakuan ni tsuite 2," *Kenchiku zasshi* 53 (1939): 1354–68; Takayama Eika, "Datong toshikeikakuan," *Gendai kenchiku* 8 (1940): 48–57.

2. Koshizawa Akira, "Shinto keikaku: Jingū kankei shisetsu seibi jigyō no tokushoku to igi," *City Planning*, "Special Review Issue," 32 (1997): 73–78.

3. Kishida Hideto and Takayama Eika, *Gaikoku ni okeru jūtaku shikiji wari ruireishū* (Tokyo: Dōjunkai, 1936); Takayama Eika, *Gaikoku ni okeru jūtaku shikiji wari ruireishū, zoku* (Tokyo: Dōjunkai, 1938).

4. Uchida Yoshizō, "Daidō no toshi keikakuan ni tsuite 1," *Kenchiku zasshi* 53 (1939): 1281–94, 1294.

5. Takayama Eika, *Gaikoku ni okeru jūtaku shikiji wari ruireishū* (Tokyo: Dōjunkai, 1936), 188.

6. See for example: Jean-Louis Cohen and Monique Eleb, *Casablanca: Colonial Myths and Architectural Ventures* (New York: Monacelli Press, 1989); Paul Rabinow, *French Modern: Norms and Forms of the Social Environment* (Chicago: University of Chicago Press, 1976); Anthony D. King, *Colonial Urban Development: Culture, Social Power and Environment* (London: Routledge, 1991); Gwendolyn Wright, *The Politics of Design in French Colonial Urbanism* (Chicago: University of Chicago Press, 1991).

7. Richard Calichman, *Overcoming Modernity: Cultural Identity in Wartime Japan* (New York: Columbia University, 2008).

8. In an interview with Izosaki Arata, in *Tokushū kindai Nihon toshikeikaku shi, hito to shisō, jōkyō*, Toshi Jūtaku 7604 (4/1976).

9. Joe Nasr and Mercedes Volait, eds., *Urbanism—Imported or Exported? Foreign Plans and Native Aspirations* (Chichester: Wiley, 2003); S. V. Ward, "Re-examining the International Diffusion of Planning," in *Urban Planning in a Changing World: The Twentieth Century Experience*, ed. Robert Freestone (London: Routledge, 2000).

Suggested Readings

Hein, Carola. "The Transformation of Planning Ideas in Japan and Its Colonies." In *Urbanism—Imported or Exported? Foreign Plans and Native Aspirations,* edited by Joe Nasr and Mercedes Volait, 51–82. Chichester: Wiley, 2003.

Myers, Ramon H., and Mark R. Peattie. *The Japanese Colonial Empire, 1895–1945*. Princeton: Princeton University Press, 1984.

42 A Two-Timing Map

Catherine L. Phipps

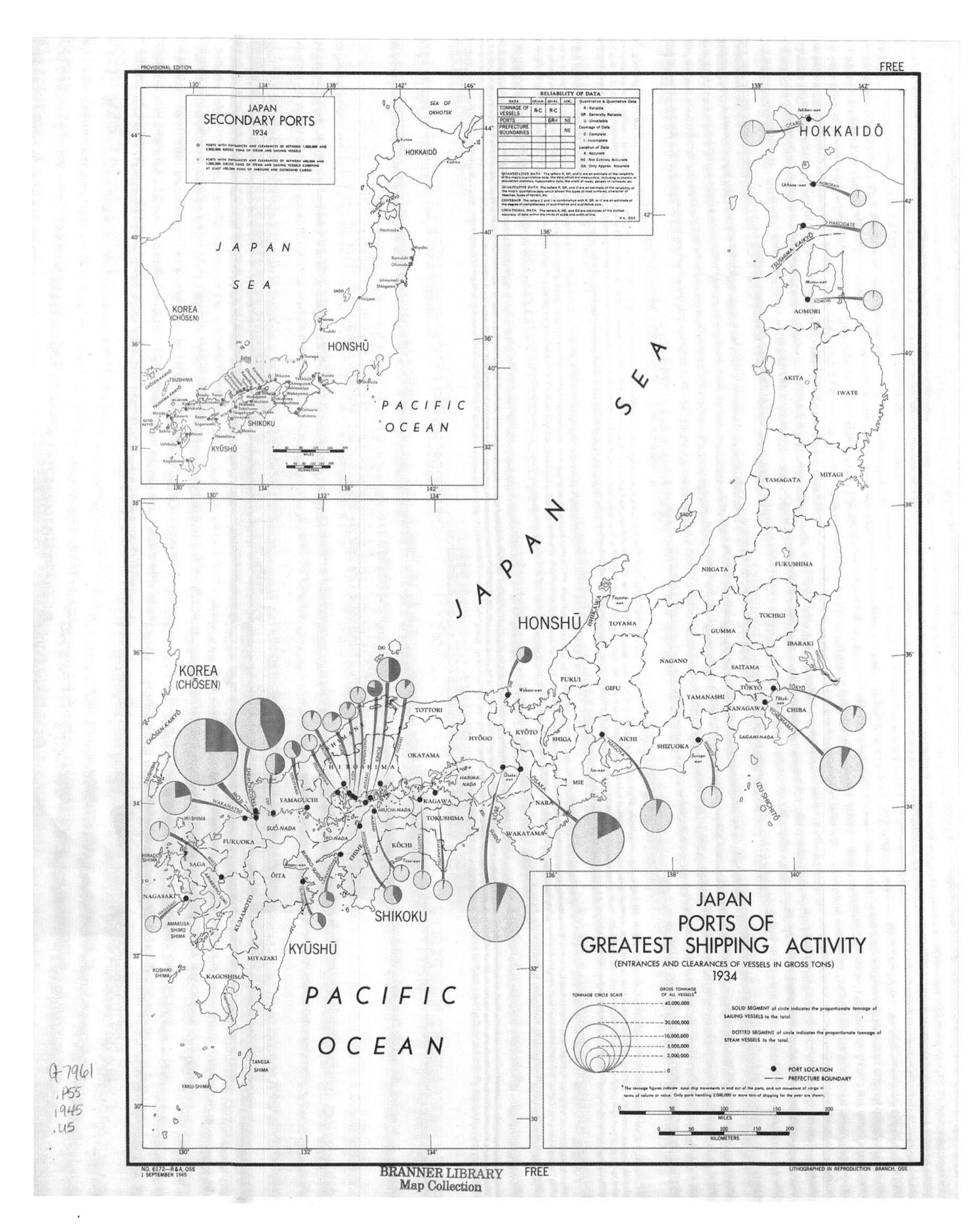

FIGURE 42.1 "Japan Ports of Greatest Shipping Activity 1934." OSS, Research and Analysis Branch, record group 226, September 1, 1945, map no. 6172. 42 × 31 cm. Courtesy of Branner Earth Sciences Library, Stanford University.

This seemingly straightforward map of Japan's four main islands offers a quick view of the nation's port capacity as of 1934 (fig. 42.1). Two elements catch the eye: the circular pie charts and their dramatic concentration along the western half of the Inland Sea. A third but less obvious feature is one of omission—namely, the absence of Japan's empire. The 1930s marked an era of aggressive imperial expansion and growing militarism as the Japanese tightened control over their colonial holdings, particularly Korea, Taiwan, and Manchuria, while mobilizing the nation for "total empire."[1] A 1930s image that isolates Japan from its colonies—especially a map of ports, which are by nature connective—appears at odds with history and makes its main features more difficult to comprehend. In order to make sense of this portrayal, then, we need to know more about its data, the information it implies and omits, and the mapmaker's intentions.

The pie charts, as the fine print clarifies, depict the gross shipping tonnage handled by each port. Most maritime nations adopted the convention of using gross tonnage in shipping statistics in the mid-nineteenth century, and the British Moorsom system, which calculates a ship's volume (rather than weight), provided the basis to standardize shipping on an international level.[2] This map's data, including the percentages of both sailing ships and steamers, suggest the general robustness of marine transportation in Japan. It offers a sense of how much traffic and what size vessels each port could accommodate. Shipping tonnage was also used to assess such things as tariffs and fees, further offering a potential gauge of a port's facilities and services. Still, tonnage is an imperfect measure of a port and does not capture such things as trade partners, cargo amounts and values, or the numbers, points of origin, or nationalities of passengers handled by a given site. Knowing such things is crucial to decoding the map's circles and their distribution.

According to this map, the five ports of greatest shipping activity in 1934 were, in order: Moji, Kobe, Osaka and Shimonoseki (tied), and Yokohama. The inclusion of Kobe, Osaka, and Yokohama is expected, since all three have been among Japan's top transportation hubs from the age of the treaty ports (1858–99) to the present. Moji and Shimonoseki are less familiar today but served as key ports from the late nineteenth century until shortly after World War II. Nonetheless, it is surprising to see Moji in first place and Yokohama—often called the country's front gate—in fifth. Pitting this map against trade statistics for the same year, however, arranges the ports in a more familiar hierarchy. Osaka ranks highest in total trade, followed by Yokohama and Kobe. Osaka outperformed Moji more than twofold in terms of trade amounts (tonnage) and more than fourfold in terms of value (yen), while Shimonoseki came in at half Moji's amounts. Nagoya and Wakamatsu, which rate in a tier below the top five in shipping capacity, each carried a considerably higher volume of trade than Shimonoseki.[3]

The disparity between shipping capacity and trade volume, particularly in the case of Shimonoseki, demonstrates that empire is constitutive of the map even if this dimension is concealed. Looking more closely, we find that nearly half of Shimonoseki's shipping tonnage came from sailing ships. Such vessels generally operated in coastal waters rather than on the high seas, and a good portion of Shimonoseki's traffic ran just within East Asia. In fact, the proximity of Shimonoseki and Moji to Japan's continental colonies played a major role in the two cities' development. The modern fortunes of both ports were based on their location on major domestic and international shipping routes and their coincident roles as regional trade and transportation centers. Moji exported Kyushu's vast reserves of coal throughout Asia, while Shimonoseki became the first designated site linking Honshu with Korea in the early 1880s. By 1934, the Korea trade formed a full 25 percent of its activities, with domestic coastal trade making up most of the rest. Both ports also served a high volume of passengers, who in making the short trip between Kyushu and Honshu, from one domestic port to another, or between the homeland and its colonies, help account for the twin ports' tremendous activity.[4]

The heavy concentration of major ports along the Inland Sea is also explained by a continental orientation.

The Inland Sea waterway connected Yokohama, Kobe, and Osaka with coastal seas along the Asian mainland, providing a well-traveled maritime thoroughfare used by domestic and foreign vessels alike. The traffic pattern from the continent to these ports, established by the Western powers that subordinated both China and Japan during the mid-nineteenth century, overlapped with older domestic shipping routes, prompting a push for greater port development along this vital waterway. By contrast, the Tohoku region suffered, at least in part, from its distance from major shipping lanes. The long, exposed Pacific coastline reaching northward from Tokyo to Hokkaido was almost devoid of major maritime entrepôts, and the western side of Tohoku was not much better served. The inclusion on the inset map of four minor ports along the Sea of Japan (in a region that was becoming known as the "backside of Japan") attests primarily to long-standing local efforts by residents there to trade with Korea, Manchukuo, and Russia.

Although its data were based on Japanese government statistics from 1934, the map before us was created a decade later by the Research and Analysis Branch of the United States Office of Strategic Services (OSS), predecessor to the Central Intelligence Agency. It is dated September 1, 1945—one day before Japan formally signed the instrument of surrender ending World War II. With the Occupation forces arriving and the emancipation of Japan's colonies already underway, why did the OSS publish a map of Japan's shipping capacity as it had stood a full decade earlier?

At the end of the war, many ports and shipping lanes were impaired; Yokohama, Osaka, and Tokyo were among those that sustained severe damage in air raids. Meanwhile, sunken vessels and mines blocked many harbors and navigation routes.[5] Upon taking command of the country and its devastated infrastructure, the General Headquarters, led by General Douglas MacArthur, halted all Japanese shipping, requisitioned Japan's major ports, and put the country's surviving ships under the control of the US Pacific Fleet. Additional Allied ships were also employed in carrying out the twin goals of demilitarizing and democratizing Japan.[6] Although there is no clear evidence that this particular map, with its dated information, was ever used by the General Headquarters, Japan's port system was essential to the Occupation.

From 1941 to 1945 the Research and Analysis Branch created a series of maps, including this one, to depict transportation routes and the locations of key industries. The date 1934 does not appear to have any singular significance and may well simply reflect the latest year of data available to intelligence units. The portfolio as it stands today in the US National Archives appears rather disjointed, reflecting the fact that geographers had piecemeal information and could provide only unbalanced coverage.[7] While their aims in presenting these specific findings may now be somewhat opaque, we can still read Japan's history in them.

In the end, this map belongs to two different times. On the one hand, it shows Japan in the year 1934, when it had a very active shipping industry and a complex, transnational system of ports serving the country's rapidly growing empire. On the other hand, the same map represents a view of the archipelago in mid-1945 as the Occupation forces were taking over Japan's transport network while also stripping it of imperial holdings as outlined by the Potsdam Declaration. In this context, the features of the empire and the activities of its harbors and ships throughout East Asia may well have seemed secondary, if not irrelevant, for the defeated land. The map was created at a time when Japan was suddenly reduced to a country without colonies and knowing past shipping capacity would have been useful in reconfiguring these ports to meet a new set of demands.

Notes

1. Louise Young, *Japan's Total Empire: Manchuria and the Culture of Wartime Imperialism* (Berkeley: University of California Press, 1998).

2. Yrjö Kaukiainen, "Tons and Tonnages: Ship Measurement and Shipping Statistics, c. 1870–1980," *International Journal of Maritime History* 7 (2) (June 1995): 29–32.

3. Naimushō Dobokukyoku Kōwanka, ed., *Dai Nihon Teikoku kōwan tōkei: Shōwa 9 nen* (Tokyo: Kōwan Kyōkai, 1936), 71–80.

4. Shimonoseki Shōkō Kaigisho, ed., *Shimonoseki keizai jijō* (Shimonoseki: Shimonoseki Shōkō Kaigisho, 1937), 143–44.

5. Ishimaru Yasuzo, "The Korean War and Japanese Ports: Support for the UN Forces and Its Influences," *NIDS Security Reports* 8 (December 2007), www.nids.go.jp/english/publication /kiyo/pdf/2007/bulletin_e2007_5.pdf: 56–57.

6. Yoshiya Ariyoshi, *Senryōka no Nihon kaiun* (Tokyo: Kokusai Kaiun Shimbunsha, 1961), 42–44.

7. Trevor J. Barnes, "Geographical Intelligence: American Geographers and Research and Analysis in the Office of Strategic Services, 1941–1945," *Journal of Historical Geography* 32 (2006): 160.

Suggested Readings

Barnes, Trevor J. "Geographical Intelligence: American Geographers and Research and Analysis in the Office of Strategic Services, 1941–1945." *Journal of Historical Geography* 32 (2006): 149–68.

Dower, John W. *Embracing Defeat: Japan in the Wake of World War II*. New York: W. W. Norton, 1999.

Kobayashi Shigeru. "Japanese Mapping of Asia-Pacific Areas, 1873–1945: An Overview." *Cross-Currents: East Asian History and Culture Review* 1 (1) (May 2012): 137–71.

Phipps, Catherine L. *Empires on the Waterfront: Japan's Ports and Power, 1858–1899*. Cambridge, MA: Harvard University Asia Center, 2015.

Young, Louise. *Japan's Total Empire: Manchuria and the Culture of Wartime Imperialism*. Berkeley: University of California Press, 1998.

43 Visions of a New Order in the Asia-Pacific

David FEDMAN

FIGURE 43.1 Untitled illustration from *Declaration of Greater East Asian Cooperation* [Daitōa kyōdō sengen, 大東亜共同宣言], by Dai Nippon Yūbenkai Kōdansha 大日本雄辯會講談社, 1942, 8–9. Bonner Fellers Papers, box 4, folder 13, Hoover Institution Archives.

FIGURE 43.2 Untitled illustration from *Declaration of Greater East Asian Cooperation* [Daitōa kyōdō sengen, 大東亜共同宣言], by Dai Nippon Yūbenkai Kōdansha 大日本雄辯會講談社, 1942, 10–11. Bonner Fellers Papers, box 4, folder 13, Hoover Institution Archives.

War, like empire, stretched the boundaries of the Japanese public's geographic imagination. Just as Japan's colonization of Taiwan, Korea, and Manchuria in the first half of the twentieth century had drawn the nation's gaze toward its northeast Asian neighbors, the imperial military's sudden "southern advance" in late 1941 drew attention to places scarcely familiar to many Japanese. Ceylon, Darwin, Surabaya, Luzon: these and other place-names gained currency across the homeland as the imperial navy pushed swiftly into the far reaches of the Asia-Pacific with the goal of establishing a "new order" of geopolitics in the region.

Maps served as a principal means through which the Japanese public learned about the progress of this diplomatic and military campaign. 1940s Japan accordingly saw a proliferation of maps of the Asia-Pacific region and the world. The Japan Broadcasting Corporation (NHK), for example, began selling inexpensive "news listening maps" designed to orient listeners during wartime radio broadcasts. Illustrated guides to the progress of the Greater East Asia War also made their way into the burgeoning cartographic marketplace, as Japanese of all ages strove to keep pace with wartime developments.[1]

Of the many geographic images to gain expression within this growing body of wartime cartography, few appeared as regularly as the Greater East Asia Co-prosperity Sphere, an imperial concept of East Asian solidarity in the face of Western colonialism. Cloaked in the rhetoric of racial harmony and economic autarky, the Co-prosperity Sphere emerged in Japanese policy circles as a powerful argument for Japan's expansion throughout the Asia-Pacific region. Figures 43.1 and 43.2, a pair of maps included in an illustrated children's booklet published in 1942 by the Great Japan Debate Society, evoke both the rosy dreams and the darker reality of Japan's imperium in the Asia-Pacific.

Figure 43.1 presents a nightmarish landscape of unhampered Western dominion. British, French, American, and Dutch forces freely lord over their spheres of influence in Asia—much to the delight of Winston Churchill and Franklin Roosevelt, who look on approvingly from above. Particularly pronounced in this map is the extractive nature of imperial rule: timber is felled at gunpoint in New Guinea; precious metals are mined by forced labor in Borneo; oil is extracted in Sumatra; sugar is exported from Java. Equally revealing is the map's depiction of Western military might. Bombers project American air power from bases in China while British, French, and American frigates, tanks, and artillery assert control over large swaths of Southeast Asia. "Look!" states the text in the right-hand corner of the map, "America, England, the Netherlands, and others have been keeping us down and doing bad things to us in Greater East Asia."

Figure 43.2, by contrast, depicts the dream of a Japanese-led Greater East Asia Co-prosperity Sphere. Gone are the fat-cat Western imperialists and the oppressive shackles of their rule in Asia. In their place stand the liberating Japanese military forces, which sail, fly, and march their way into occupied lands, where they are greeted with open arms and waving flags—as Churchill and Roosevelt shake in their boots above. The accompanying text narrates Asia's emancipation: "Japan stood up to take back Greater East Asia into our own hands. Its powerful army drove away the enemy from the Greater East Asia region." The latter claim registers visually with the British forces fleeing from Burma, American soldiers taken prisoner in the Philippines, and Japanese air raids destroying air bases in China. (Also noteworthy is the fact that Korea and Taiwan—Japan's longest-held colonial territories—are cast in red in both maps as extensions of the Japanese homeland.)

It goes without saying that figure 43.2 both overstates the achievements and oversimplifies the reality of the Co-prosperity Sphere. For one thing, the scope and stability of the geopolitical order suggested by this map was never realized. While strategists, diplomats, and military planners routinely invoked the Co-prosperity Sphere in the abstract, little in the way of a concrete political arrangement actually materialized until well into the war. Less a formal entity than a rhetorical tool, the Co-prosperity Sphere was always in flux; its contours were imagined as much as created, making it impossible to pin down.

The rhetorical underpinnings of the Co-prosperity Sphere were also rife with contradiction. Although Japanese leaders took great pains to espouse pan-Asian harmony, for example, these declarations were difficult to square with more strident claims to racial, cultural, and national superiority. These contradictions were not lost on many of the Asian peoples subjected to Japanese rule. In many cases the "liberators" proved to be as harsh as the Western powers they ousted, if not more so. The initial jubilation of liberation depicted in figure 43.2 quickly gave way to dismay and outrage as Asian people in these occupied territories came to terms with the harsh reality of Japan's heavy-handed rule.

In this connection, it is worth noting the maps' unmistakably martial nature. Infused with imagery of war waged in the air, on land, and on the sea, both images evocatively portray Japan's military posture during a time of intense hostilities in the region. Especially striking is their emphasis on the joint Anglo-American offensive, which Japanese strategists saw as the gravest threat to their homeland. Indeed, while concerns about the so-called "ABCD [American, British, Chinese, Dutch] encirclement" were widespread, it was the Anglo-American alliance that most viscerally unnerved Japanese leaders. This sentiment found its clearest expression in the Greater East Asia Joint Declaration of November 7, 1943:

> The United States and Great Britain, in seeking prosperity, have oppressed other nations and peoples. Particularly in Greater East Asia they have engaged in persistent invasion and exploitation, have given free rein to their ambition to enslave the region, and ultimately have undermined the roots of stability in the region. Herein lies the cause of the Greater East Asia War.[2]

The text accompanying figure 43.2 puts the point more bluntly: "We of Greater East Asia will combine our power to destroy America and England."

That this jingoistic message was intended for Japanese youth is a sobering point. Maps like these remind us that the mobilization for total war ensnared all levels of Japanese society. Circulated in an educational pamphlet meant to inform youth throughout Japan's colonies about the goals of the Co-prosperity Sphere, they present military aggression in an accessible and visually appealing form. In addition to illustrating the causes and consequences of Japan's righteous liberation of Asia, the map promoted the supposed affinity of all Asians, who can be seen warmly embracing the Japanese throughout the image. "The peoples of Greater East Asia," states the accompanying booklet, "have had honest hearts since ancient times. . . . Let us use our honest hearts to live in harmony like brothers and sisters." In this and other ways, the booklet and its maps obscured the ideological, political, and cultural struggles that shaped, often turbulently, the lived experience of Japan's imperial subjects throughout Asia.

Most tellingly of all, these maps sanitize war by turning it into a comic strip. In their cheerful cartoon aesthetic, battle zones are stripped of suffering and humanity. This is perhaps to be expected from a publication geared toward a young readership. Still, it highlights the ways in which propaganda in general—and maps in particular—dehumanized Japan's war abroad for readers at home. Sanitized portrayals of war zones were not particular to the Japanese, nor were they confined to children's fare. They were, and are, a hallmark of wartime cartography.[3]

Notes

1. "Taiheiyō sensō," *Shōwa nimannichi no zenkiroku* 6 (Tokyo: Kōdansha, 1989): 16.
2. This translation comes from Jeremy Yellen, "The Two Pacific Wars: Visions of Order and Independence in Japan, Burma, and the Philippines, 1940–1945" (PhD diss., Harvard University, 2012), 342.
3. For a detailed treatment of wartime cartography in the United States, see Susan Schulten, *The Geographical Imagination in America, 1880–1950* (Chicago: University of Chicago Press, 2001), chap. 9.

Suggested Readings

Aydin, Cemil. *The Politics of Anti-Westernism in Asia: Visions of World Order in Pan-Asian and Pan-Islamic Thought.* New York: Columbia University Press, 2007.

Dower, John. *War without Mercy: Race and Power in the Pacific War.* New York: Pantheon Books, 1987.

Duus, Peter. "Imperialism without Colonies: The Vision of a Greater East Asia Co-prosperity Sphere." *Diplomacy and Statecraft* 7 (1) (1996): 54–72.

Hotta, Eri. *PanAsianism and Japan's War, 1931–1945.* New York:

Palgrave Macmillan, 2007.
Iriye, Akira. *Power and Culture: The Japanese-American War, 1941–1945*. Cambridge, MA: Harvard University Press, 1981.
Saaler, Sven, and J. Victor Koschmann, eds. *Pan-Asianism in Modern Japanese History: Colonialism, Regionalism, and Borders.* London: Routledge, 2007.

PART IV

Still under Construction

CARTOGRAPHY AND TECHNOLOGY SINCE 1945

Introduction to Part IV

Kären Wigen

The rosy promise of pan-Asian prosperity that lit up Japanese children's magazines in 1942 would soon be exposed as a sham. The very next year, the tide began to turn against Japan. The imperial navy's attack on the American fleet at Pearl Harbor had already brought the United States into the fight, unleashing the fury of war across the Pacific. Early victories for the Japanese gave way to a string of defeats; by 1944 the front had reached the home islands. Guided by modified Japanese maps, American bombers blew up Japan's factories and reduced its major cities to rubble in the space of half a year. The atomic bombs dropped on Hiroshima and Nagasaki in August of '45 dealt the final blow. Within days, the emperor's voice could be heard through the static on radios across the country announcing the unthinkable. Japan's imperial dream—East Asia's colonial nightmare—was over.

In the short term, defeat brought desperation to ordinary Japanese. Food, jobs, and shelter were all in short supply. The repatriation of six and a half million servicemen and colonists from Korea, Taiwan, Manchuria, China, and Southeast Asia only aggravated the crisis; for nearly five years, millions faced the very real prospect of starving. While the Allied Occupation could put democratic reforms in place by executive order, rebuilding homes and livelihoods proceeded at a crawl. But when war broke out on the nearby Korean Peninsula in 1950—on the eve of the treaty negotiations that would bring a formal end to the Occupation—Japanese industry came back to life. Factories that had once been slated for dismantling now went into high gear, making tanks and munitions for the US Army. Even after hostilities ground to a halt, Cold War strategists in Washington made sure that Tokyo retained access to the technology, capital, and markets the Japanese needed to rebuild their domestic infrastructure and manufacture consumer goods, for sale abroad as well as at home. With their new national constitution founded on the principles of human rights, democracy, and pacifism, the Japanese people were able to make a fresh start, initiating a period of unprecedented peace and prosperity.

So began the wide-ranging, long-running process of postwar landscape transformation that is the subject of part IV. For students of spatial history, the years after 1945 present a conundrum. On the one hand, documentation for this era is richer than for any previous time in the human past. In Japan as elsewhere, the cartographic archive grew exponentially in the later twentieth century; since the invention of the Internet, it has swelled yet again. On the other hand, the resulting embarrassment of riches gives us too much to take in. With millions of maps covering thousands of themes, a comprehensive overview recedes out of reach. Fundamental changes have been wrought, and mapped, in every sphere of Japanese life; only a few can be captured in a volume like this.

That said, the essays in part IV offer precious perspectives on an era of rapid change. While interested readers can readily turn elsewhere for studies of party politics, social movements, or international relations, the contributors to this final part of *Cartographic Japan* have elected to revisit some of the chief themes that have preoccupied Japan's mapmakers since the Edo era: urban life, people on the move, sacred landscapes, and hazardous events. Along the way, their contributions illuminate the broad arc of postwar Japanese history, from recovery to growth to newfound vulnerabilities.

We begin with a look at the cartographic tools of US Army bombers in the final phase of the war (discussed in the chapter by Cary Karacas and David Fedman). The maps that facilitated the American campaign of incendiary air raids, from planning to damage assessment, took advantage of earlier topographic surveys undertaken by the Japanese government. Yet they are also among the first images of Japanese cities ever made by a conquering power. In one of many paradoxes that characterize this tragic era, these maps embody Japan's role reversal from colonizer to colonized. Moving forward to the immediate aftermath of defeat, a map of occupied Tokyo from 1946 shows how the Supreme Command for the Allied Powers provided for thousands of foreign troops in the ruined capital (chapter by Cary Karacas).

Crucial though they were, food and housing were

not the only urgent issues addressed during the Occupation. The US and its allies also moved swiftly to reorganize the symbolic landscape, stripping Shinto of its special political status and repurposing some sacred sites. As a prime symbol of the emperor-centered religion that had helped fuel Japan's wartime ideology, Mount Fuji in particular became the subject of a protracted legal battle—one that began under the Occupation but continued for more than a decade afterward (chapter by Andrew Bernstein). Meanwhile, determined Japanese visionaries began making maps for the future. One plan drafted in the early postwar years redesigned Hiroshima around a commemorative peace park (chapter by Carola Hein); another laid out a whole new town on the rural outskirts of Osaka (chapter by André Sorensen).

By the time Senri New Town was built, Japan had entered a period of unprecedented double-digit economic growth, kicking off another wave of urban migration. Workers, students, and families from rural areas across the archipelago flocked to Japan's newly rebuilt cities, with Tokyo emerging as by far the most powerful magnet. The sustained influx transformed the capital—a process brought into focus by our next group of essays. Starting with a snapshot of Tokyo's overstrained road system on the eve of the 1964 Olympics (discussed by Bruce Suttmeier), successive authors highlight the proliferation of subway lines (Alisa Freedman), the struggle to sustain a traditional shopping district (Susan Paige Taylor), and the rise and demise of the city's famed waterfront fish market (Theodore C. Bestor).

As Tokyoites began bracing for the 2020 Olympics, Japan found itself in the throes of yet another round of creative destruction. Most heavy industries had long since moved offshore in search of cheaper labor, the loss of manufacturing jobs only partially offset by new opportunities in the knowledge economy. Precariousness was the watchword of the day, as millions found themselves stuck in low-wage, part-time service jobs.[1] Nor was fragility marked only in the economic sphere. Japan's continuing vulnerability to natural hazards was made terrifyingly clear in the triple disaster (earthquake, tsunami, and nuclear meltdown) that struck northeastern Honshu on March 11, 2011.

That telling moment forms the backdrop for the volume's concluding theme: new developments in the digital age. Our final section begins with a trio of essays on natural hazard cartography in the wake of 3/11, with individual authors probing the frontiers of earthquake forecasting (Gregory Smits), crowd-sourced mapping (Jilly Traganou), and a new genre of run-and-escape cartography (Satoh Ken'ichi). But just as the digital age has put powerful tools in the hands of citizens seeking to understand contemporary phenomena, so has it given new techniques to scholars seeking to understand the past. The mapmaking software known as geographic information systems (GIS) has opened new historical vistas, revealing spatial patterns in data that were collected during earlier periods but never before represented in cartographic form. One area where such techniques are being used to good effect is in the study of historical demography. In the case explored here by Fabian Drixler, plotting nineteenth-century stillbirth statistics reveals a lingering culture of infanticide that was essentially invisible until computer-assisted cartography came along. Reconstructions of this kind can be expected to illuminate many more corners of the Japanese past.

Meanwhile, complementing these data-driven applications of computer cartography are new forms of analysis for old maps. In our final pair of essays, Japanese scholars Nakamura Yūsuke and Arai Kei—pioneers in this line of research—recount recent breakthroughs in the study of the *kuniezu,* the massive manuscript maps of the provinces introduced in part I. Digital scanners have proven a boon to students of these fragile and unwieldy materials, allowing for unprecedented magnification and side-by-side comparison of widely scattered documents. Moreover, with painstaking care, these scholars and their students have launched a novel project to reconstruct provincial maps at full scale. Recounting the excitement of their discoveries, the final essays in the volume leave us at the cutting edge of Japanese historical cartography today—bringing us back full circle to the archival treasures with which *Cartographic Japan* began.

Note

1. Anne Allison, *Precarious Japan* (Durham: Duke University Press, 2013).

44 Blackened Cities, Blackened Maps

Cary Karacas and David Fedman

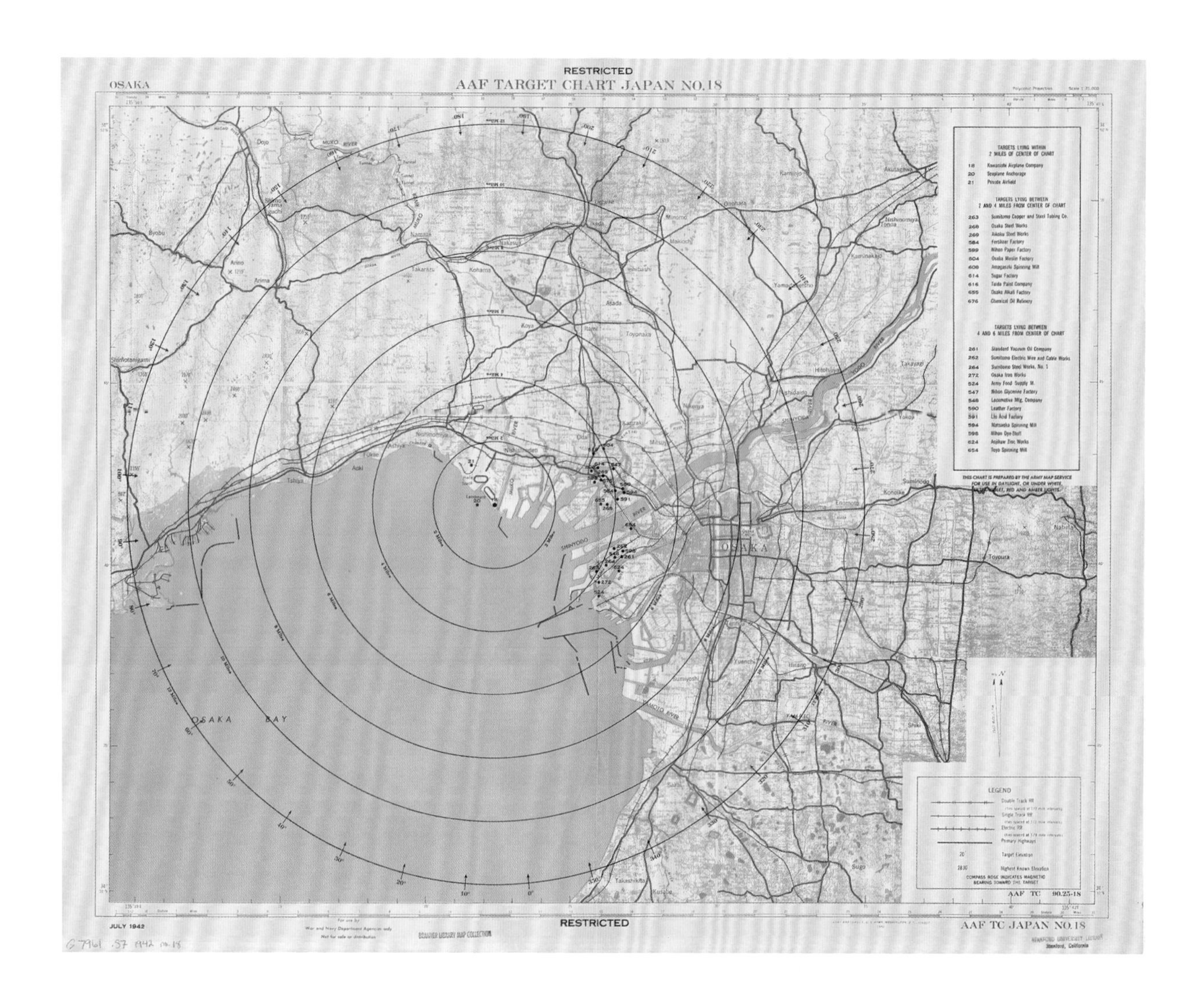

FIGURE 44.1 "AAF Target Chart Japan No. 18e Osaka," July 1942. Army Map Services, US Army. 57 × 68 cm. Courtesy of Branner Earth Sciences Library, Stanford University.

For many among the first wave of American soldiers and civilians to arrive in Tokyo following Japan's surrender in August 1945, words failed to convey the ruinscapes they encountered. "There is really no way to describe a bombed-out city," wrote photojournalist John Swope. "There is simply nothing left—that's all there is to it."[1] Perhaps no one painted a portrait of this destruction more vividly than the *New York Times* reporter George Jones, who wrote in a dispatch from Tokyo, "There should be no mistaken interpretation of the word ruins. There were blocks on which not a single building stood, where the construction and civilization wrought in the past centuries had been obliterated, leaving reddish soil—and nothing else."[2] Although observations of this sort focused principally on the capital and the two cities destroyed by nuclear bombs, urban wastelands stretched across the Japanese archipelago. By war's end, the US Army Air Forces (AAF) had targeted sixty-six cities for destruction by incendiary bombing.

Initially, military strategists did not imagine striking such a blow to the Japanese homeland. "Use of incendiaries against cities," Henry Arnold, the AAF's top commander, had professed, "is contrary to our national policy of attacking only military objectives."[3] Although as early as the 1930s some strategists had noted the vulnerability of Japanese cities to fire, the AAF in the Pacific prioritized the high-altitude precision bombing of Japan's aircraft engine plants, to be followed by "other war-making targets."[4] This is reflected in a substantial set of target maps of Japanese metropolitan areas created by the US military starting in 1942 as they planned the coming air war.

"AAF Target Japan No. 18e Osaka" (fig. 44.1), one of many such maps produced in 1942, visually conveys this strategic doctrine. The dots sprinkled throughout the concentric circles mark the location of military installations and industrial sites in the greater Osaka area, including munitions depots, billeting facilities, and chemical plants. This particular map centers on a target undeniably martial in nature: the Kawanishi aircraft plant, one of the region's largest manufacturers of airplane parts. In addition to conveying the AAF's focus on military sites, the map also hints at how new forms of spatial intelligence allowed for the production of a wide array of wartime cartography. In this instance, army cartographers relied heavily on Japanese cadastral survey maps, the likes of which cartographers and other specialists within the intelligence community—especially the Office of Strategic Services (OSS)—labored mightily to secure as they scrambled to address the paucity of geographic information about Japan.

Target maps of this sort, however, lost their utility as war planners attached to the AAF reimagined how America's expanding air power in the Asia-Pacific region might deliver a decisive blow against the Japanese empire. Based on research into the flammability of Japanese

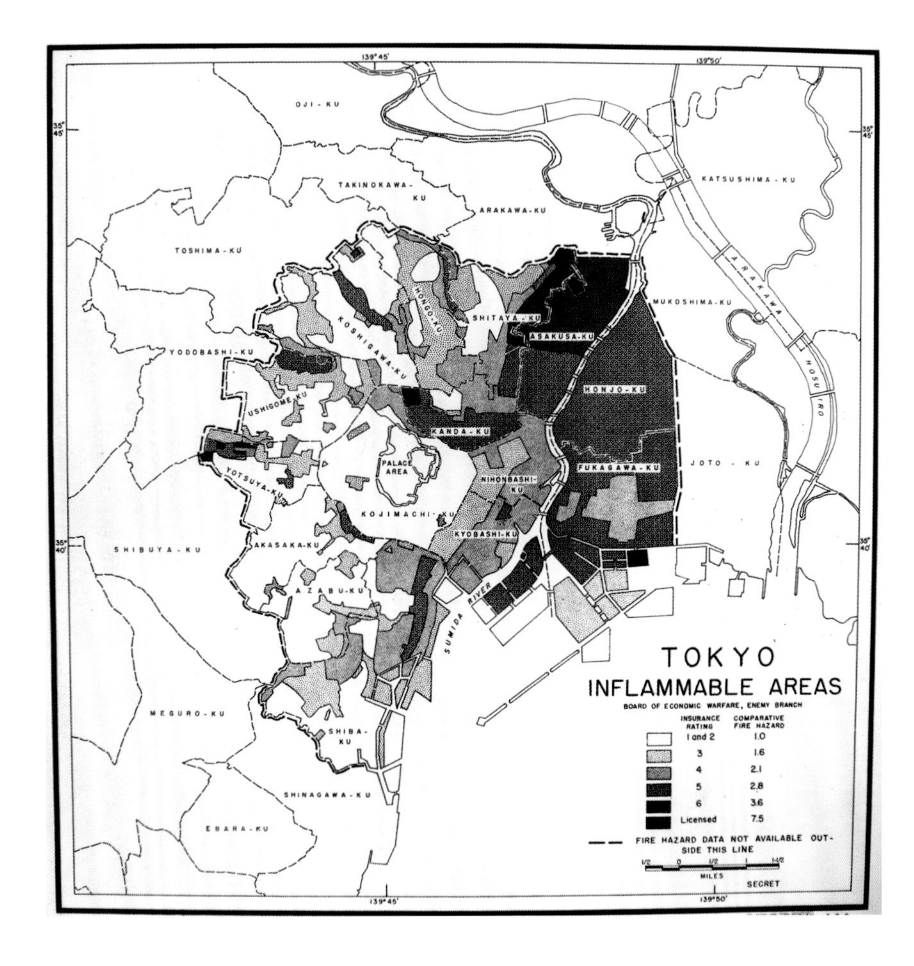

FIGURE 44.2 "Tokyo Inflammable Areas," November 1942. OSS map no. 877. 51 × 46 cm. US National Archives, Cartographic and Architectural Section, record group 226: 330/20/8.

traditional architecture, an awareness of the high concentration of Japan's war industries in urban areas, and a growing appreciation of the destructive potential of incendiary bombing (as demonstrated by Britain's attacks upon German cities, often with AAF support), planners in Washington began to include entire Japanese cities on their list of recommended targets. Maps made later in the war illustrate this shift. Cartographers working within the OSS made thematic choropleth maps of the largest cities, such as "Tokyo Inflammable Areas" (fig. 44.2), which detailed levels of population concentration and flammability for each ward. The most densely populated residential areas—also the most flammable—attracted the attention of the AAF's Committee of Operations Analysts. Successful large-scale incendiary attacks on such neighborhoods in Japan's six largest cities, the committee estimated, would destroy over two hundred square miles and kill a half million people, thereby denting Japan's ability to further prosecute the war.[5]

With the American seizure of the Mariana Islands in the summer of 1944 and the subsequent arrival of newly developed B-29 Superfortress bombers, the plan of the Committee of Operations Analysts to incinerate Japan's largest cities became a tactical reality. Given the insignificant results yielded by the first few months of precision-bombing attempts and the growing fear that the AAF would not play a central role in the defeat of Japan, it comes as no surprise that Henry Arnold and other high-ranking airmen enthusiastically welcomed this shift.[6] The US government, after all, had made a significant investment—costlier than the Manhattan Project—in the B-29 bomber, and a lack of compelling results would have doubtless deflated hopes for an independent air force once the war ended.

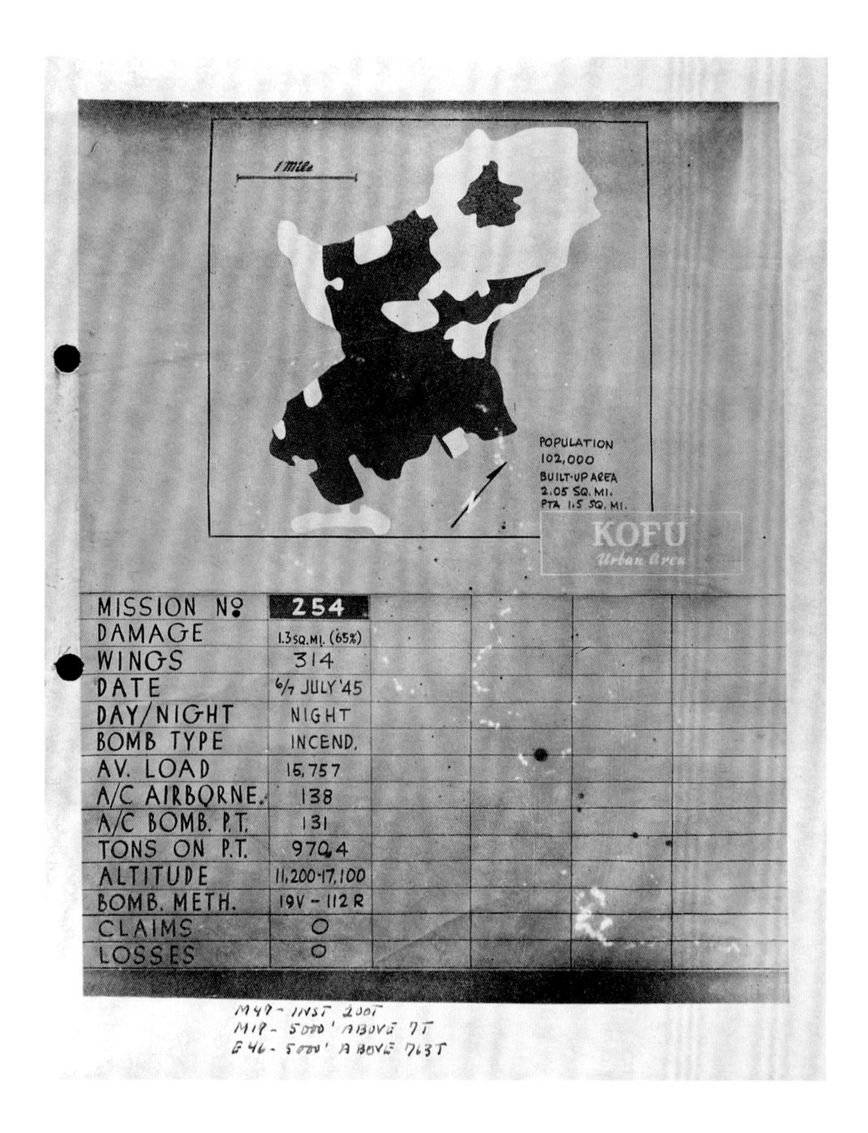

FIGURE 44.3 "Damage Report Map of Kofu City," July 1945. XXI Bomber Command. 20 × 27 cm. US National Archives, record group 243, series 59, box 5.

Maps produced by the Marianas-based XXI Bomber Command throughout the first half of 1945 clearly reveal the decisive shift toward firebombing Japan's cities. Following the first, weeklong assault on the six largest urban areas in March 1945, the command's mapping division created damage assessment maps. Together with aerial photographs, these visually conveyed to Washington the extent of destruction that each raid accomplished. As the air war expanded in scope and intensity, the swaths of black that demarcated destroyed neighborhoods grew on successive maps printed over successive months. Pleased with the results of targeting Tokyo, Yokohama, Kawasaki, Osaka, Nagoya, and Kobe, analysts then recommended attacking the country's medium-sized cities as well. Where concentric rings on target maps in 1942 had focused on military sites, those published toward the end of the war targeted the heart of the built urban environment.

Representative of the voluminous body of maps generated by this lethal project is the Kofu urban area damage assessment map (fig. 44.3). Produced after a July 1945 firebombing raid on a city that analysts themselves admitted had no military or industrial significance, maps such as this one reveal the American embrace of the incendiary destruction of Japanese cities as a legitimate wartime tactic. Tellingly, save for population figures, the maps contain little indication of the communities that resided within the targeted areas. This was new. While the 1942 target charts had also obscured the social dimensions of Japan's urban spaces, they did not reduce whole cities to target zones. These earlier maps included a variety of indicators of the city as a lived space, noting the locations of schools, community buildings, and railways. The dozens of damage assessment maps that the XXI Bomber Command produced, however, reduced representation of the Japanese city to but two values: black (the area destroyed by fire) and white (that which escaped the flames). Stripped of the professionalism and nuances of previous maps, the stark rendering of the Japanese city in these black-and-white categories conveys not only the extent of the destruction achieved by the AAF but also a troubling moment in the history of modern war: one in which a city and its inhabitants—including those too young, old, or enfeebled to contribute to the war effort—were reduced to target zones selected for indiscriminate destruction.

Notes

1. Carolyn Peter, John Dower, and John Swope, *A Letter from Japan: The Photographs of John Swope* (Los Angeles: Grunwald Center for the Graphic Arts, 2006), 220.

2. George Jones, "Writer over Tokyo Says Ruin Is Real," *New York Times*, August 29, 1945, 3.

3. As cited in William Ralph, "Improvised Destruction: Arnold, Lemay, and the Firebombing of Japan," *War in History* 13 (2006): 498.

4. Haywood Hansell, *Strategic Air War against Japan* (Maxwell Air Force Base, Alabama, 1981), 50.

5. Committee of Operations Analysts, *Economic Effects of Successful Area Attacks on Six Japanese Cities*, 4 September 1944. Source: US National Archives, record group 18, entry 57, box 8.

6. A detailed examination of the shift to incendiary tactics can be found in Thomas Searle, "'It Made a Lot of Sense to Kill Skilled Workers': The Firebombing of Tokyo in March 1945," *Journal of Military History* 66 (2002): 103–33.

Suggested Readings

Fedman, David, and Cary Karacas. "A Cartographic Fade to Black: Mapping the Destruction of Urban Japan during World War II." *Journal of Historical Geography* 38 (2) (July 2012): 306–28.

Henrikson, Alan. "The Map as an 'Idea': The Role of Cartographic Imagery during the Second World War." *American Cartographer* 2 (1975): 19–53.

Searle, Thomas. "'It Made a Lot of Sense to Kill Skilled Workers': The Firebombing of Tokyo in March 1945." *Journal of Military History* 66 (2002): 103–33.

Selden, Mark. "A Forgotten Holocaust: U.S. Bombing Strategy, the Destruction of Japanese Cities and the American Way of War from World War II to Iraq." In *Bombing Civilians: A Twentieth-Century History*, edited by Yuki Tanaka and Marilyn Young, 77–96. New York: New Press, 2007.

Sherry, Michael. *The Rise of American Air Power: The Creation of Armageddon*. New Haven: Yale University Press, 1987.

45 The Occupied City

Cary Karacas

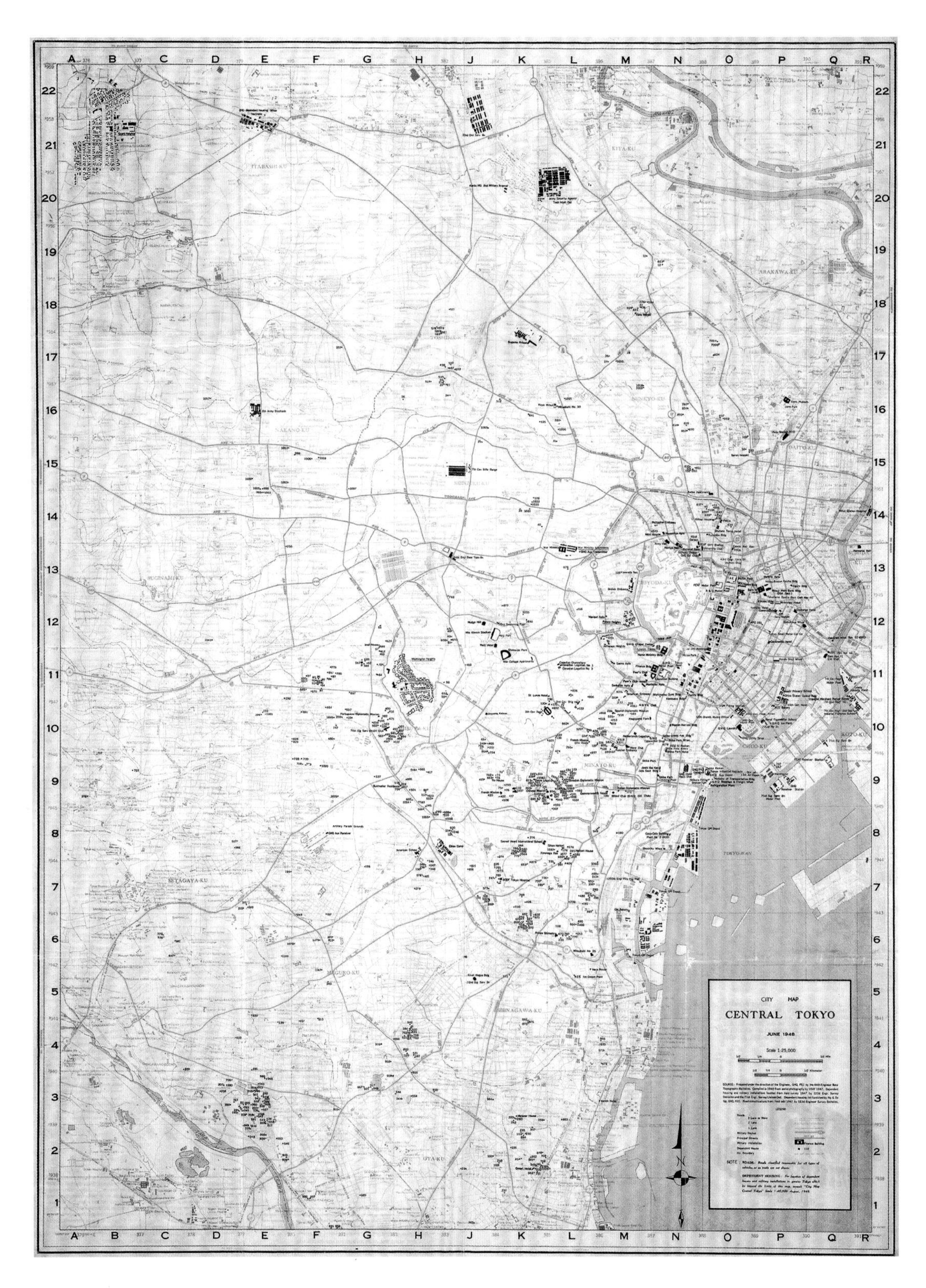

Among the myriad maps produced during Tokyo's four-hundred-year history, the one featured here (fig. 45.1) provides us with a unique snapshot of the Japanese capital under foreign rule. First published in August of 1946, eleven months into the seven-year Allied Occupation of Japan, it shows how America's temporary sovereignty over the country was spatially made manifest in Tokyo. The immediate purpose of this map, however, was practical. Cartographers attached to the Eighth Army composed "City Map Central Tokyo" primarily to orient soldiers and their dependents to the bewildering landscape that was postwar Tokyo. If the extensive firebombing was not enough to disorient newcomers, the circuitous network of the capital's main roads—few of which had designated names or street signs—was sure to confound them.

The map highlights three features: arterial avenues, requisitioned buildings, and structures that housed the Occupation forces. The roads appear as a skein of red veins spreading out in web-like fashion over the pale city background (onto which the Japanese names of neighborhoods and specific buildings are inscribed in light sepia). Avenues A to Z, designated as such by the new sovereigns, radiate outward from the Imperial Palace (located in the map's center right), while streets newly numbered 1 through 60 form concentric rings around the palace. Streets in the Marunouchi business district and adjoining areas to the east of the palace received particular attention because of their heavy use by Occupation forces, with many given personal names (fig. 45.2). A paramount example is Tokyo Boulevard (Avenue A), the street dividing the Imperial Palace grounds from the fourteen-story Dai Ichi Mutual Insurance Building, location of the Occupation's General Headquarters (GHQ), which was led by Douglas MacArthur, the Supreme Commander for the Allied Powers. Assigning such designations was a practical necessity for the occupiers, especially given that the main form of travel for most was by road.

FIGURE 45.1 "City Map Central Tokyo." GHQ, 64th Engineering Base Topographical Battalion, June, 1948. 90 × 66 cm. US National Archives, record group 331.

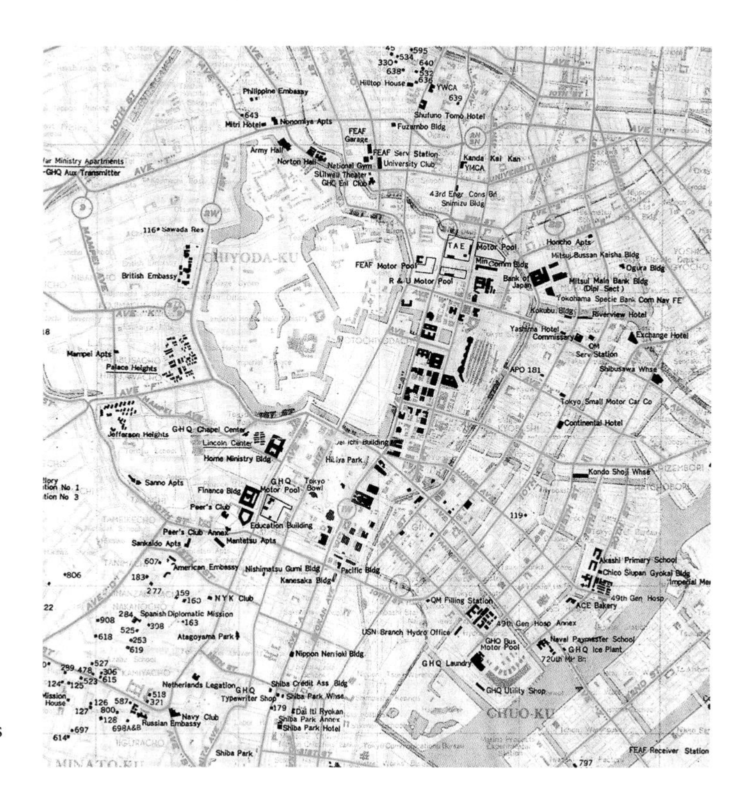

FIGURE 45.2 Detail from figure 45.1, showing requisitioned buildings and houses surrounding the Imperial Palace.

The rest of the map is given over to a few thousand black shapes that indicate buildings and houses under control of the occupying power. The requisition of almost all prime space began soon after Allied troops entered Tokyo in September 1945. The first directive to the Japanese government, in fact, demanded lodging for fifty-four hundred officers and twenty-two thousand enlisted men.[1] A subsequent stream of directives ordered the Japanese to turn over specific government buildings, cultural venues, hospitals, military quarters, piers and storage facilities, airfields, private residences, and even some public schools. Every Japanese Army base changed hands, as did Tokyo's two largest prisons. The requisition of a quarter of the Tsukiji Central Wholesale Market and 80 percent of the wharves, piers, and associated storage units lining Tokyo Bay allowed the Allies to store supplies arriving by ship from the West Coast of the United States, including many that would in turn be shipped to military units throughout the archipelago.[2]

"City Map Central Tokyo" also shows us where the occupiers played. In addition to the soldiers based in the capital, thousands of Americans stationed elsewhere in Japan and throughout Asia visited as soldier-tourists. So did the majority of the forty thousand British Commonwealth troops stationed on Shikoku and Kyushu. Seeking to create "wholesome venues" for these throngs of enlisted men, the commanding officer of the Eighth Army requisitioned Tokyo's best cinemas, clubs, theaters, and sports arenas. Meiji Stadium, renamed Nile Kinnick Stadium (fig. 45.3, *top right*) in honor of the famed University of Iowa halfback, hosted USO shows, rodeos, and annual year-end "Rose Bowl" football games.[3] If soldiers tired of the nearby Olympic-sized swimming pool, they could cool off across the Sumida River at the Kokugikan, the main site for sumo tournaments in Tokyo until Eighth Army engineers converted it into Japan's first ice-skating rink, named Memorial Hall (fig. 45.1, R13). Asia's largest theater, the Takarazuka—right across from the requisitioned Imperial Hotel designed by Frank Lloyd Wright—became the most popular cultural facility used by Occupation forces. Renamed after Ernie Pyle, the Pulitzer Prize–winning journalist killed on Okinawa's Iejima Island, the theater became a "must-see" for soldiers and dependents wanting to watch first-run movies, beauty pageants, and stage shows like *Arsenic and Old Lace* or Gilbert and Sullivan's *Mikado*.[4]

Finally, "City Map Central Tokyo" shows us where the occupiers lived. While barracks housed most forces, the higher-ranking officers set their sights on the houses of Tokyo's elite. In the fall of 1945, GHQ ordered the Tokyo Metropolitan Government to collect the locations and floor plans of all structurally sound Western-style residences with at least six rooms located within a forty-five-minute drive of the Imperial Palace. Every dwelling on the resulting list was requisitioned, with around one thousand of these "US Houses" represented on the map as numbered black squares. Residents were given one week to evacuate, with orders to leave all of their fur-

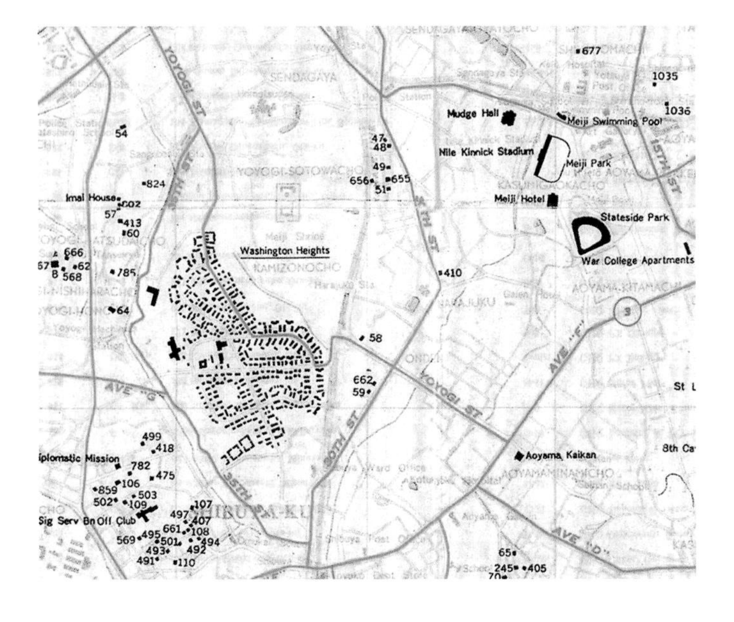

FIGURE 45.3 Detail from figure 45.1, showing the Washington Heights dependent housing complex and numbered requisitioned "US Houses" in Shibuya Ward.

nishings behind. Following a thorough cleaning, the Japanese government had to stock each house with linens, silverware, and "all other appointments and furnishings suitable for an officer." Additionally, each house was to be staffed with servants, including English-speaking stewards, experienced cooks, gardeners, laundresses, and furnace attendants.[5]

While the number ebbed and flowed, on average twenty-five thousand soldiers attached to either GHQ or the Eighth Army were stationed in the burned-out city at any given time.[6] To accommodate those with dependents, the Eighth Army converted hotels and office buildings into small family apartments. It also took advantage of bombed-out open spaces to erect Quonset huts on prime real estate near the Imperial Palace grounds (see Palace Heights in fig. 45.2). These stopgap measures bought time while the Eighth Army planned the largest housing projects ever to be constructed in Japan. In July 1946, its Engineering Division published guidelines for housing that reflected the "standards of American community life." Single and two-story duplex units ranging between 890 and 1,900 square feet were to accommodate soldiers ranking from lieutenant to brigadier general, along with their War Department civilian counterparts, who wished to bring their wives, children, pets, and even cars to Tokyo.[7]

After Japanese structures in the area were razed, construction began on a ninety-two-hectare parcel of land just south of Meiji Shrine. In a matter of months, Yoyogi Parade Ground, used since 1909 for military drills, had become Washington Heights (fig. 45.3), a gated community of several hundred housing units that resembled many an early postwar American suburb. In addition to the houses, the self-contained community featured a chapel, movie theater, recreation hall, library, fire and gas stations, schools, and a shopping center with a commissary, dispensary, and barber/beauty shops. An even bigger housing community called Grant Heights was planned for a tract of land that the Tokyo Metropolitan Government unsuccessfully lobbied to be used for housing some of the two million Tokyoites displaced by firebombing. A former airfield, Grant Heights (top left of fig. 45.1) featured 1,262 housing units accommodating fifty-six hundred Americans (with nine hundred school children) and twenty-four hundred Japanese servants.

One way to read "City Map Central Tokyo" is as a visual indictment of what some officials privately referred to as a "luxury occupation." The Japanese government was obliged not only to build whole communities but also to manufacture everything deemed needful in a middle-class home. For an estimated fifteen thousand living rooms, the government had to provide overstuffed armchairs and upholstered davenports, bookcases, magazine racks, Venetian blinds, carpets, telephones, and the stands on which to place them. The bathrooms required porcelain enamel sinks and tubs, water heaters, showerheads, and toilet paper holders. Mattresses, bed frames, and chests of drawers would go into the bedrooms, and the kitchens would be filled with toasters, percolators, dishes, rolling pins, and ice picks. No wonder analysts for GHQ found that the drain on raw materials just to house the Americans contributed to the country's rampant inflation.[8]

Finally, "City Map Central Tokyo" provides us with some clues regarding the more permanent effects of the Occupation on the city's landscape. While the 1952 US-Japan Security Treaty stipulated that, following the end of the Occupation, "as a rule, the Army and Air Force will be stationed outside of urban areas," Tokyo proved the exception, with the United States holding on to 208 military bases in the capital.[9] Despite repeated demands, the United States did not hand back the land accommodating the Washington Heights and Grants Heights dependent housing complexes until 1964 and 1974, respectively. During the long interval between requisition and return, Tokyo's population tripled, pressing hard on both open space and housing. After serving as the Athletes' Village for the 1964 Tokyo Olympiad, most housing units in the former Washington Heights were finally torn down to create Yoyogi Park, now one of the largest open spaces in central Tokyo.[10] A decade later, Grant Heights was likewise handed back to Japan. After being held by the United States for almost three decades, the area today hosts the largest housing complex in all of Tokyo, accommodating forty thousand people. As these examples illustrate, while "City Map Central Tokyo" visually conveys a brief, singular chapter in the spatial history of Japan's capital, it also hints at the specific ways, temporary and lasting, in which the occupation of the country affected Tokyo's landscape.

Notes

1. Takemae Eiji, ed., *GHQ shirei 'Scapin-A' sōshūsei zen 18 kan* (Tokyo: Emutei Shuppan, 1997).

2. United States Eighth Army, *Eighth Army Experience in Japan, 1945–1947* (1947), 49, 52.

3. *Pacific Stars and Stripes*, 16 October 1945; 5 November 1945. United States Eighth Army, *Occupational Monograph of the Eighth United States Army in Japan, August 1945—January 1946* (n.d.), 180.

4. *Pacific Stars and Stripes*, 28 February 1946.

5. Takemae, *GHQ shirei*.

6. US Department of the Army, *Reports of General MacArthur*, vol. 1, supp., *MacArthur in Japan: The Occupation: Military Phase* (Washington DC: US Department of the Army, 1966), 86.

7. US National Archives, RG 331, box 8108, folder 24, "GD, Subject: Assignment of Abodes for Dependent Housing in the Tokyo Area," AG 620 (24 Jun 46). RG 331, box 407, folder 75, "Construction Standards for Dependent Housing, Japan and Korea," AG 620 (15 July 46).

8. US National Archives, RG 331, box 8333, folder 10, "Impact of Occupation Requirements on the Japanese Economy."

9. Senryōgun Chōtatsushi Hensan Iinkai, *Senryōgun chōtatsushi, dai 1* (Tokyo: Chōtatsuchō Sōmubu Chōsaka, 1956).

10. Tokyo Metropolitan Government, *Tonai beigun kichi kankei shiryōshū* (Tokyo: Tokyo Metropolitan Government, 1964).

Suggested Readings

Crockett, Lucy Herndon. *Popcorn on the Ginza: An Informal Portrait of Postwar Japan*. New York: William Sloane, 1949.

Dower, John W. *Embracing Defeat: Japan in the Wake of World War II*. New York: W. W. Norton and Co., 1999.

Koizumi Kazuko, Takayabu Akira, and Uchida Seizō. *Senryōgun no jūtaku no kiroku*. Tokyo: Sumai no Toshokan Shuppankyoku, 1999.

46 Sacred Space on Postwar Fuji

Andrew Bernstein

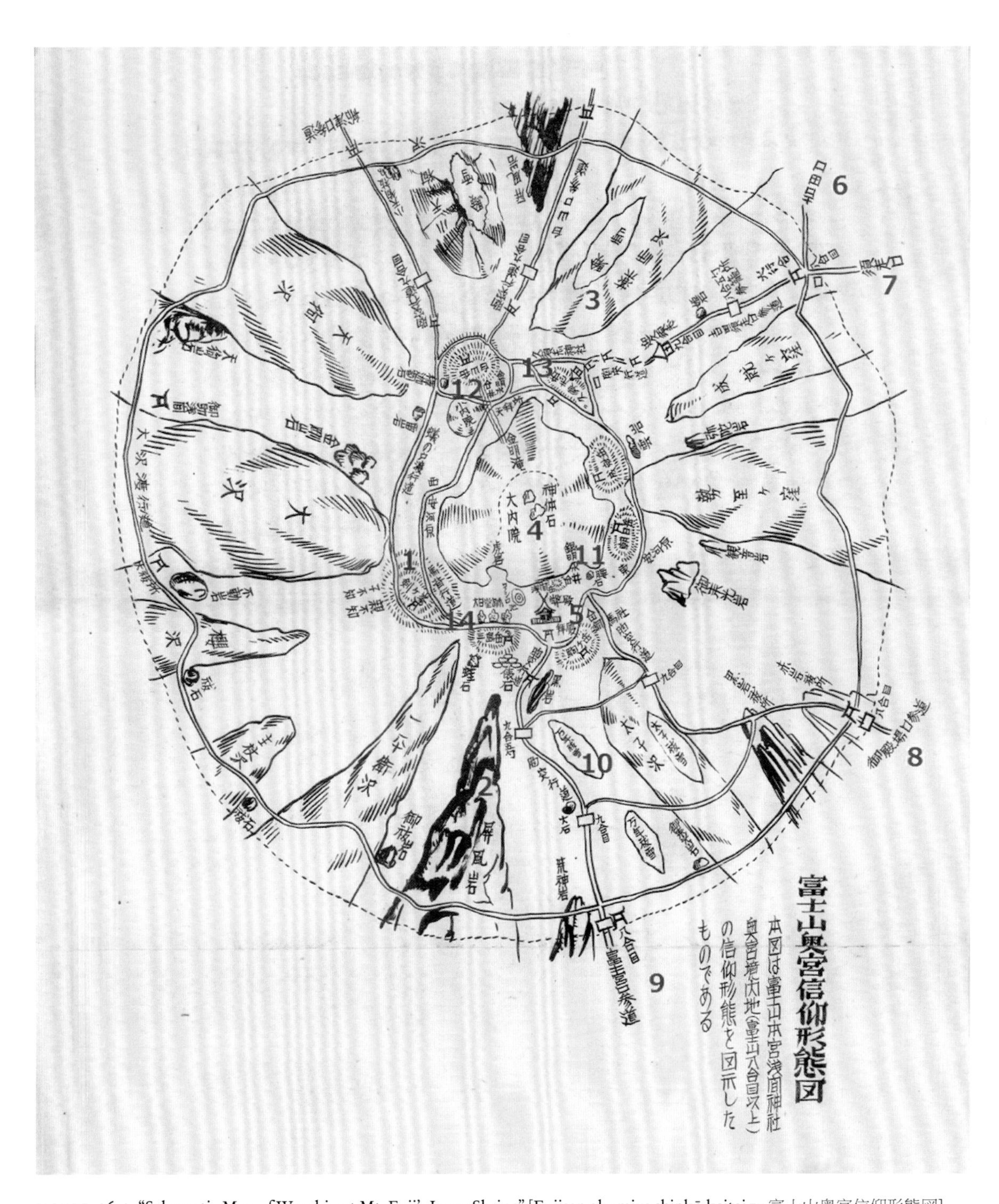

FIGURE 46.1 "Schematic Map of Worship at Mt. Fuji's Inner Shrine" [Fujisan okumiya shinkō keitai zu 富士山奥宮信仰形態図]. Fujisan Hongū Sengen Taisha 富士山本宮浅間大社, November 1951. Black and white print, 25.6 × 36 cm. Courtesy of Fujisan Hongū Sengen Taisha, Fujinomiya City, Japan. Numbers added by author.

The US military, assisted by Allied forces, occupied Japan from 1945 to 1952 with the goal of transforming a defeated enemy into a stable democracy. To do this, it devised a host of policies designed to remake Japan in America's image, among the most prominent being the separation of religion—especially Shinto—from the state. Often translated "the way of the gods," Shinto entails the worship of deities called *kami*, many of which inhabit natural features such as trees, waterfalls, and—as in the case of Fuji—mountains. Japan's imperial government (1868–1945) had harnessed this worship by incorporating shrines into a centrally administered hierarchy known as State Shinto. Given its use to justify Japanese militarism, dismantling this bureaucratized religion was one of America's top priorities in early postwar Japan.

Separating religion from the state was, however, a messy affair, forcing Japanese to confront difficult questions for decades to come. One of the thorniest, it turned out, was Who owns Fuji? The imperial government, like the Tokugawa shogunate before it, had recognized Fuji's upper slopes and crater to be the "inner shrine precincts" of Fujisan Hongū Sengen Taisha, a Shinto shrine located at the southwestern base of Fuji that worshiped the mountain as the "god body" (*goshintai*) of its *kami*.[1] It had also registered these precincts, along with those of other Shinto shrines and also many Buddhist temples, as property of the state. Using state land for religious purposes, however, clearly violated both Occupation policy and the country's new constitution. This resulted in a 1947 law that required the Finance Ministry to determine which shrine and temple lands should be deemed religious/private property and which should become secular/public property. The ministry generally affirmed the rights of temples and shrines to the territory they claimed, which in some cases included entire mountains. Fuji, however, was a different story. It was not only the god body of an individual *kami* but also the symbol of an entire nation. To complicate matters, the government operated a meteorological observatory at the highest point on Fuji's summit, Kengamine Peak, which reaches 3,776 meters above sea level (see feature 1 in fig. 46.1).

Shizuoka Prefecture, with jurisdiction over the south slopes of Fuji, backed Sengen Shrine's claim to its inner shrine precincts. Yet neighboring Yamanashi Prefecture, which controlled the north slopes of the mountain, vehemently opposed this. Especially active in the fight against the shrine's claim were the residents of Fujiyoshida, a community located at the volcano's northeastern base that had long challenged Sengen Shrine's authority. They and other opponents, ranging from secularist politicians to real estate developers, coalesced around the argument that Fuji's summit rightly belonged to the Japanese people.

For years Sengen Shrine appealed to the Finance Ministry to recognize its claim, and in November 1951 it submitted the above map, along with an explanatory booklet, to advance its case.[2] This top-down depiction of the summit, with its inward-pointing rock formations (2) and ravines (3), exerts a centripetal force that pulls the viewer's eye toward the shrine's "great inner sanctum"—that is, the crater—and the sacred stones at its heart (4). Located immediately south of the crater is the inner shrine (5). Marked by a building with a pair of crossed finials protruding from the roof—a feature typical of Shinto shrines—it consists of a site for offerings and a site for worship. Conspicuously labeling these sites makes obvious not only what is there but also what is missing: the innermost sanctuary common to many Shinto shrines. By calling attention to this absence, the map emphasizes all the more that the locus of devotion is the crater itself. Also absent from the map is something that *did* overlook the crater—the meteorological observatory—although in a booklet submitted to the Finance Ministry in December, the shrine describes the facility's contribution to the public good as a manifestation of Fuji's "divine virtues."[3]

The crater may have been the focus of Sengen Shrine's claim, but as the line of dashes encircling the map indicates, the inner shrine precincts extended well beyond the crater itself, reaching down to the eighth station (out of ten) for each of Fuji's major climbing routes

FIGURE 46.2 Torii overlooking crater, with meteorological observatory in the distance. Photograph by author.

(6–9). Amid long-standing conflicts over the management of Fuji's summit, the Tokugawa shogunate had ruled that the shrine had jurisdiction at and above the eighth station, and the booklet duly notes this decision in the shrine's favor. The primary goal of the map and its accompanying document, however, is not to make a case based on legal precedent, but to naturalize the shrine's claim by showing that the religious character of the inner shrine precincts derives from the topography itself. The booklet emphasizes at the start that climbing Fuji becomes especially difficult from the eighth station—located between 3,200 and 3,400 meters above sea level, depending on the route—and that the dramatic rock formations, as well as the fields of snow nestled among them, contribute to a feeling of "mystical splendor." The map reinforces the divine status of this terrain by portraying the shrine gates at the eighth station, as well as those that stand either athwart or near the trails that lead to and around the crater. Known as *torii*, these markers of sacred space can be found at shrines large and small throughout Japan (fig. 46.2).

After describing the awe-inspiring nature of the landscape, the booklet proceeds to list the trails that lead to and through it. Although by the 1950s recreation-seeking climbers far outnumbered traditional pilgrims, the heading for this section is *sandō*, meaning a road leading to a shrine, rather than *tozandō*, the common term for a mountain trail. The map, too, uses *sandō* for the major routes leading to the eighth station. For the other trails it uses a word whose characters can be read as *gyōdō*, a term for ascetic discipline and, more specifically, the circumambulation of a sacred object. By using terminology associated with religious practice, the map and the accompanying booklet reinforce the shrine's claim to all the territory above the eighth station. Bolstering this claim is the shrine's careful inventory of the peaks, ravines, rock formations, snowfields (10), and wells (11 and 12) where site-specific forms of worship, purification, and asceticism took place.

The names for a number of sites appearing on the map—the ravine named after Yakushi, the medicine Buddha, for example (3)—are legacies of the Buddhist worship that coexisted with Shinto practice on Fuji until the imperial government took power in 1868. Radical nativists who wanted to purify Shinto of Buddhism's "foreign" influence had powerful allies in the new regime and, in the name of "separating *kami* and buddhas," expunged all signs of Buddhism from shrines around Japan. Most Buddhist images on Fuji's summit were removed or destroyed, while a hall dedicated to Yakushi Buddha was reinvented as an exclusively Shinto shrine (13) and the hall to Dainichi Buddha (the "Great Sun" Buddha) reduced to ruins (14). Yet the continued use of Buddhist names here is evidence that while nineteenth-century nativists may have reshaped their present, they could not obliterate all traces of the past. Moreover, the decision to feature the ruins of the Dainichi Hall rather than ignore them shows that Sengen Shrine was willing to acknowledge that past in order to underscore the historical depth of ritual activity on Fuji.

After years of street protests, Diet hearings, and court battles, the force of precedent eventually won the day. In 1962 the Nagoya District Court ruled that Sengen Shrine should receive title to Fuji's entire summit above the eighth station, with the exception of the meteorological observatory and several other public facilities. The decision was upheld in 1967 by the Nagoya High Court and again in 1974 by Japan's Supreme Court, which cited the constitutional obligation to protect both property rights and religious freedom. By invoking Japan's postwar, American-written constitution to defend a centuries-old claim by a Shinto shrine, the court demonstrated that, like the nativists of imperial Japan, those wanting to build a new order—Japanese as well as Americans—did not transcend the past so much as create new circumstances in which to deal with it.

Notes

1. Today Fujisan Hongū Sengen Taisha uses the title *taisha*, "great shrine," but in the immediate postwar period typically used *jinja*, meaning simply "shrine." The abbreviation Sengen Shrine will be employed for the remainder of this essay. The information presented here comes in large part from Andrew Bernstein, "Whose Fuji? Religion, Region, and State in the Fight for a National Symbol," *Monumenta Nipponica* 63 (1) (Spring 2008): 51–99.

2. Fujisan Hongū Sengen Jinja, *Fujisan okumiya shinkō keitai zu setsumeisho*, November 15, 1951.

3. Fujisan Hongū Sengen Jinja, *Fujisan okumiya keidaichi no shūkyōsei, tsuiho*, December 15, 1951.

Suggested Readings

Bernstein, Andrew. "Whose Fuji? Religion, Region, and State in the Fight for a National Symbol." *Monumenta Nipponica* 63 (1) (Spring 2008): 51–99.

Breen, John, and Mark Teeuwen, eds. *Shinto in History: Ways of the Kami*. Honolulu: University of Hawai'i Press, 2000.

Dower, John. *Embracing Defeat: Japan in the Wake of World War II*. New York: W. W. Norton and Co., 2000.

Earhart, Byron H. *Fuji: Icon of Japan*. Columbia: University of South Carolina Press, 2011.

Hardacre, Helen. *Shintō and the State, 1868–1988*. Princeton: Princeton University Press, 1989.

Ketelaar, James. *Of Heretics and Martyrs in Meiji Japan: Buddhism and Its Persecution*. Princeton: Princeton University Press, 1990.

O'Brien, David M., with Yasuo Ohkoshi. *To Dream of Dreams: Religious Freedom and Constitutional Politics in Postwar Japan*. Honolulu: University of Hawai'i Press, 1996.

Woodard, William P. *The Allied Occupation of Japan, 1945–1952, and Japanese Religions*. Leiden: Brill, 1972.

47 Tange Kenzō's Proposal for Rebuilding Hiroshima

Carola Hein

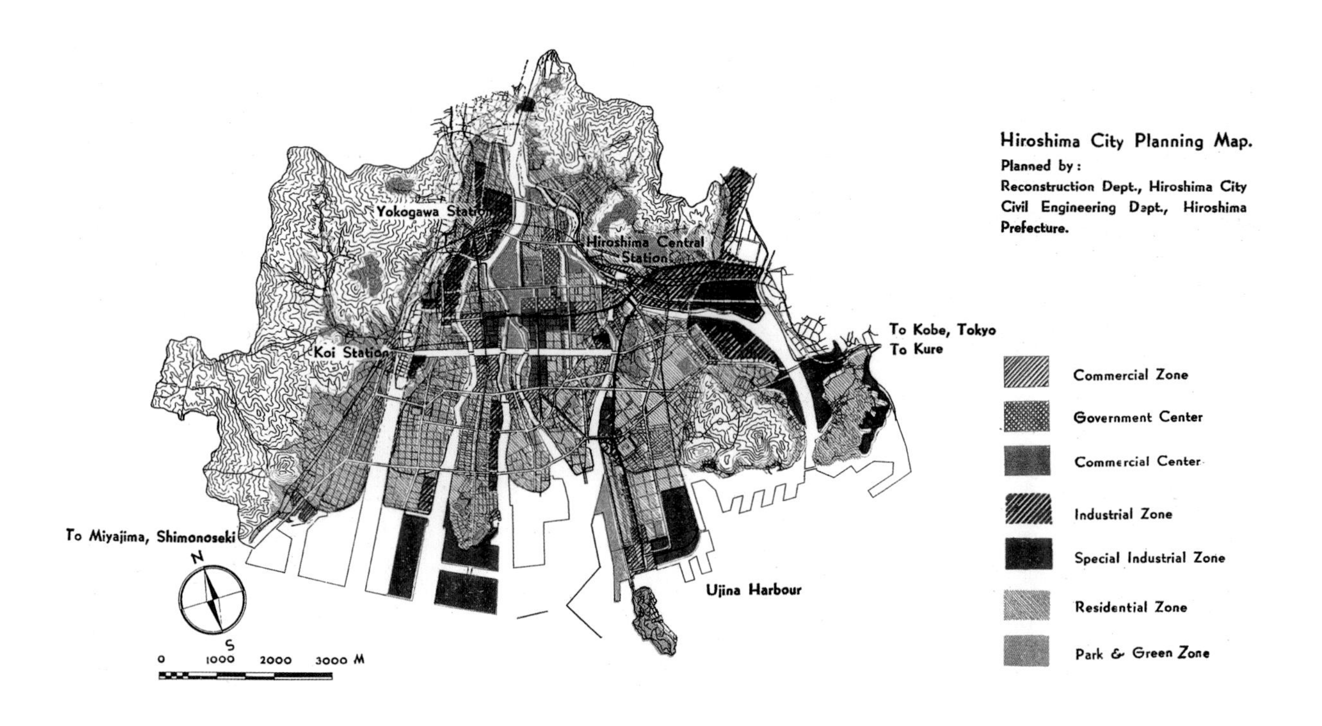

FIGURE 47.1 "Hiroshima City Planning Map," by Tange Kenzō 丹下健三, in *Peace City Hiroshima* (Tokyo: Dai-Nippon Printing), undated (ca. 1948).

The unprecedented nature of the atomic bombing of Hiroshima posed a unique challenge for the city's reconstruction. In the immediate aftermath of the bombing and Japan's surrender, with the city's mayor and half the members of his administration among the dead, Hiroshima initially lacked the manpower needed to tackle the city's rebuilding. In November 1945, however, the city assembly organized a reconstruction committee composed of neighborhood representatives, and following the establishment of the Hiroshima Reconstruction Bureau in 1946, the central government dispatched planning experts and architects to the city.[1] Aware of Hiroshima's status as the first city in the world to have been destroyed by an atomic bomb, in 1949 the city government held a competition to realize a comprehensive project—one that would require both permission of the Allied Occupation authorities and funding from the central government—to rebuild a large urban area in the center of the city for the purpose of memorializing its destruction. Architect Tange Kenzō, who in 1946 had developed a land-use plan for Hiroshima under the auspices of Japan's War Damage Rehabilitation Board, thereafter submitted his design—one of 145 competition entries—for the Peace Memorial Complex to Hiroshima City (fig. 47.1).[2]

According to the competition brief, the project had to be located in the former Nakajima neighborhood, the epicenter of the atomic explosion, which was to become "a symbol of lasting peace and a place suitable for recreation and relaxation for all people."[3] The majority of the entrants limited their proposals to the Nakajima neighborhood, framed by a projected one-hundred-meter-wide Peace Boulevard and two rivers, and featuring the required Peace Memorial Tower, Science Memorial Hall, and cenotaph. Tange, however, was far more ambitious, taking on the larger urban context of Hiroshima as a whole. Additionally, his design exemplifies Tange's particular approach to monumentality and memorializing.[4]

With a Peace Park to run along an axis perpendicular to Peace Boulevard, rather than a monumental Western axis à la Versailles or a Chinese axis, Tange's design appears to have been inspired directly by Shinto religious architecture.[5] In Shinto shrines, visitors enter through a *torii* gate, symbolizing the transition from the physical to the spiritual world, before proceeding to a second gate, which is for prayers. The view beyond this gate is obscured, and the visitor is not allowed to proceed to the inner precinct. Tange uses this same organizational principle in his design for Peace Park. The Peace Memorial Museum serves as the gateway to the inner precinct, while the cenotaph functions as the place for prayers. Beyond, shaded by the trees of Peace Park and separated by the river, is the sacred space, inaccessible to the general public: the A-Bomb Dome. At Ise, bridges—which are among the structures that are ritually rebuilt—lead to the shrine precinct. At Hiroshima, bridges are likewise a central element. Tange, for whom Peace Boulevard signified the "Road to Peace," saw them as symbolizing the link between one culture and another. Perhaps the most surprising part of Tange's Peace Memorial Park design was his proposal for an arch, which appears to be a direct reference to Eero Saarinen's design for the Saint Louis Gateway Arch.[6] For a cenotaph, the centerpiece of his project, Tange eschewed the vertical structure typical of Western monuments, opting instead for a saddle-vault cenotaph inscribed with the names of the dead. While fulfilling the requirement to memorialize the first city destroyed by an atomic bomb, Tange echoed the moniker "workshop for peace" (given to the United Nations Headquarters in New York) in proposing that his design would become a "factory for peace."[7]

Tange's proposal also included facilities spread throughout the city (fig. 47.2). The background of the proposal shows an aerial photograph depicting the larger Hiroshima area with the Ota River's seven river mouths, the six islands that host most of the city, and the surrounding mountains. Green areas indicate proposed locations for new "peace activities."[8] In the center of the city, adjacent to the square of the former castle, Tange proposed an area for international cultural and recreational facilities (including an athletic stadium and a swimming pool, a wrestling arena and theater, a playground, and a children's center). This area is part of a ribbon of riverfront green spaces that starts at the top of the delta and ultimately connects to the Nakajima neighborhood. The triangular-shaped area below the epicenter of the atomic explosion marks the location of Tange's proposed Peace Hall. A one-hundred-meter-wide, tree-lined boulevard running south of the Peace Hall would connect green areas in the mountains (proposed to house international hotels) and, in the east, a building for the Atomic Bomb Casualty Commission. Tange also proposed several leisure amenities for the waterfront, including Ujina Seaside Park, an aquarium, a horse-race track, dormitories for youths visiting from throughout Japan and the world, and a yacht harbor.

Going yet another step, Tange's proposal included infrastructure and institutions to improve the quality of life of the local population: fireproof housing, public health facilities, schools, waterworks, and parks and greenery along the rivers. Tange argued that housing for citizens and the creation of an international memorial complex had to go hand in hand. New buildings, indicated in white with stark black outlines, dot the proposed green spaces. Their function is rendered identifiable through the design: large rectangles surrounded in greenery represent high-rise housing blocks, which reference modernist design proposals like Le Corbusier's 1934 design for Nemours. The more individualistic outlines on the map are proposed cultural and recreational facilities along with hotel buildings. White lines drawn onto the aerial photograph sketch out a comprehensive street network, with several new bridges to facilitate communication across the broader river spaces, tying together the dispersed peace sites and serving as a basis for the postwar redevelopment of the city. The plan further includes, albeit in thinner lines, proposals for urban improvement that are not directly related to

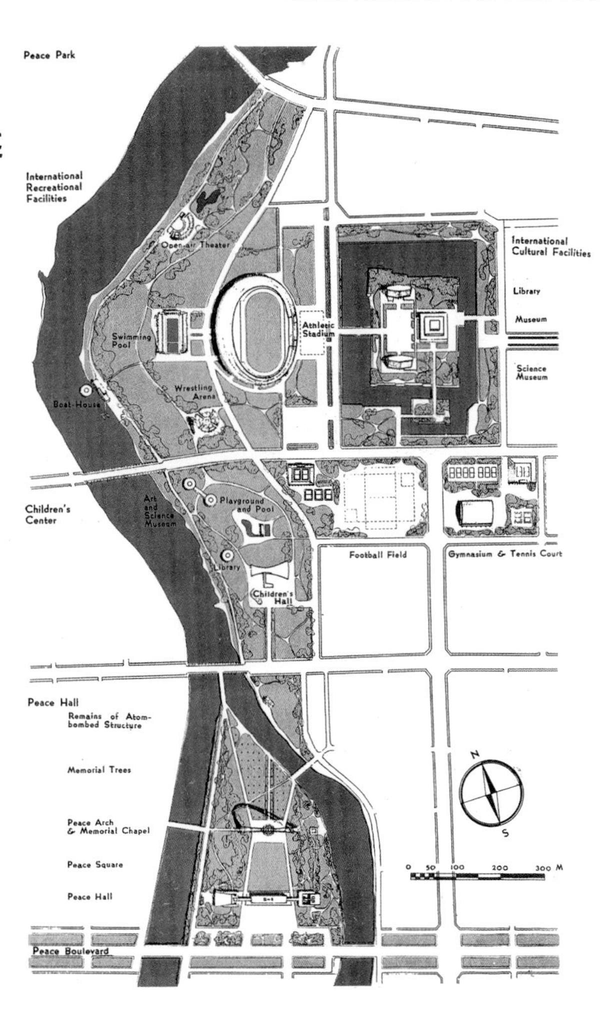

FIGURE 47.2 "Peace Park Project," by Tange Kenzō 丹下健三, in *Peace City Hiroshima*, undated.

the peace function, such as reclamation projects along the waterfront.

According to Tange, his comprehensive approach to rebuilding the city was meant to reflect the officially stated theme of turning Hiroshima into "the symbol of the human ideal for eternal peace as well as a plan for the reconstruction of human Life." The pamphlet accompanying the drawing appropriately argues that Hiroshima now belongs to "the whole human society" and therefore aims to introduce facilities that are of "real service to mankind in its pursuit of peace and happiness."

Tange's proposal for the peace park represents one of the few visionary projects for the postwar reconstruction of Japan and the only one that was realized.[9] In this regard Hiroshima is unique among the many Japanese cities that faced the task of reconstruction. Tange's larger plan for all of Hiroshima, however, was not realized. By 1949, with the city's population having almost doubled, to 270,000, the acute postwar housing shortage made it difficult to clear large urban areas for greenery. Instead, the city government put up public housing on lands west of the castle. In the end, only the heart of Tange's proposal for the Peace Center located in the former Nakajima neighborhood was built. Nonetheless, the realization of only the Peace Memorial Park, completed in 1955, should not be considered a failure. Japan's early postwar economic difficulties meant that almost all visions for transforming Japan's cities would go unrealized. While most Japanese cities were rebuilt without significant memorials or any other obvious traces of their destruction, Hiroshima literally underwent a change of heart, due in large part to the visionary and concrete foundations provided by Tange's proposal.

Notes

1. Hiroshima Peace Culture Foundation, *Hiroshima Peace Reader* (Hiroshima: Hiroshima Peace Culture Foundation, 1994), 19.

2. Tange worked on the project with Asada Takashi, Otani Sachio, and Kimura Tokokuni: "Shinsahyō," *Kenchiku zasshi* 10–11 (Tokyo: Architectural Institute of Japan, 1949), 37–39.

3. Hiroshima Peace Culture Foundation, *Hiroshima Peace Reader,* p. 45.

4. For a further examination of Tange's attitude toward traditional Japanese architecture, see Cherie Wendelken, "Aesthetics and Reconstruction: Japanese Architectural Culture in the 1950s," in *Rebuilding Urban Japan after 1945,* ed. Carola Hein, Jeffrey Diefendorf, and Ishida Yorifusa (New York: Palgrave Macmillan, 2003).

5. Tange et al., "Shinsahyō," 37–39; and Naka Masaki, *Kindaikenchikuka Tange Kenzō ron* (Tokyo: Kindaikenchikusha, 1983), 172.

6. Naka, *Kindaikenchikuka Tange Kenzō ron,* 175.

7. I would like to thank Joan Ockman for drawing this point to my attention. See Naka, *Kindaikenchikuka Tange Kenzō ron,* 167–68.

8. *Peace City Hiroshima* (Tokyo: Dai-Nippon Printing, n.d.).

9. While European planners developed extensive projects for postwar reconstruction, and despite visionary Japanese proposals for colonial endeavors, there were few grand visions for rebuilding in early postwar Japan. Among them are Tange's proposal for Hiroshima, Ishikawa Hideaki's projects for Tokyo, and a competition proposal by Uchida Yoshifumi for Tokyo's Shinjuku district. See Carola Hein, "Visionary Plans and Planners," in *Japanese Capitals in Historical Perspective,* ed. Nicolas Fiévé and Paul Waley (London: RoutledgeCurzon, 2003), 309–46.

Suggested Readings

Hein, Carola. "Hiroshima, the Atomic Bomb and Kenzō Tange's Hiroshima Peace Center." In *Out of Ground Zero: Case Studies in Urban Reinvention,* edited by Joan Ockman and Temple Hoyne Buell Center for the Study of American Architecture, 62–83. Munich: Prestel, 2002.

Kuan, Seng, and Yukio Lippit, eds. *Kenzo Tange: Architecture for the World.* Balden: Lars Müller Publishers, 2012.

Kultermann, Udo, ed. *Kenzo Tange, 1946–1969: Architecture and Urban Design.* New York: Praeger, 1970.

Riani, Paolo. *Kenzo Tange.* London: Hamlyn, 1970.

Yoneyama, Lisa. *Hiroshima Traces: Time, Space, and the Dialectics of Memory.* Berkeley: University of California Press, 1999.

48 Visions of the Good City in the Rapid Growth Period

André Sorensen

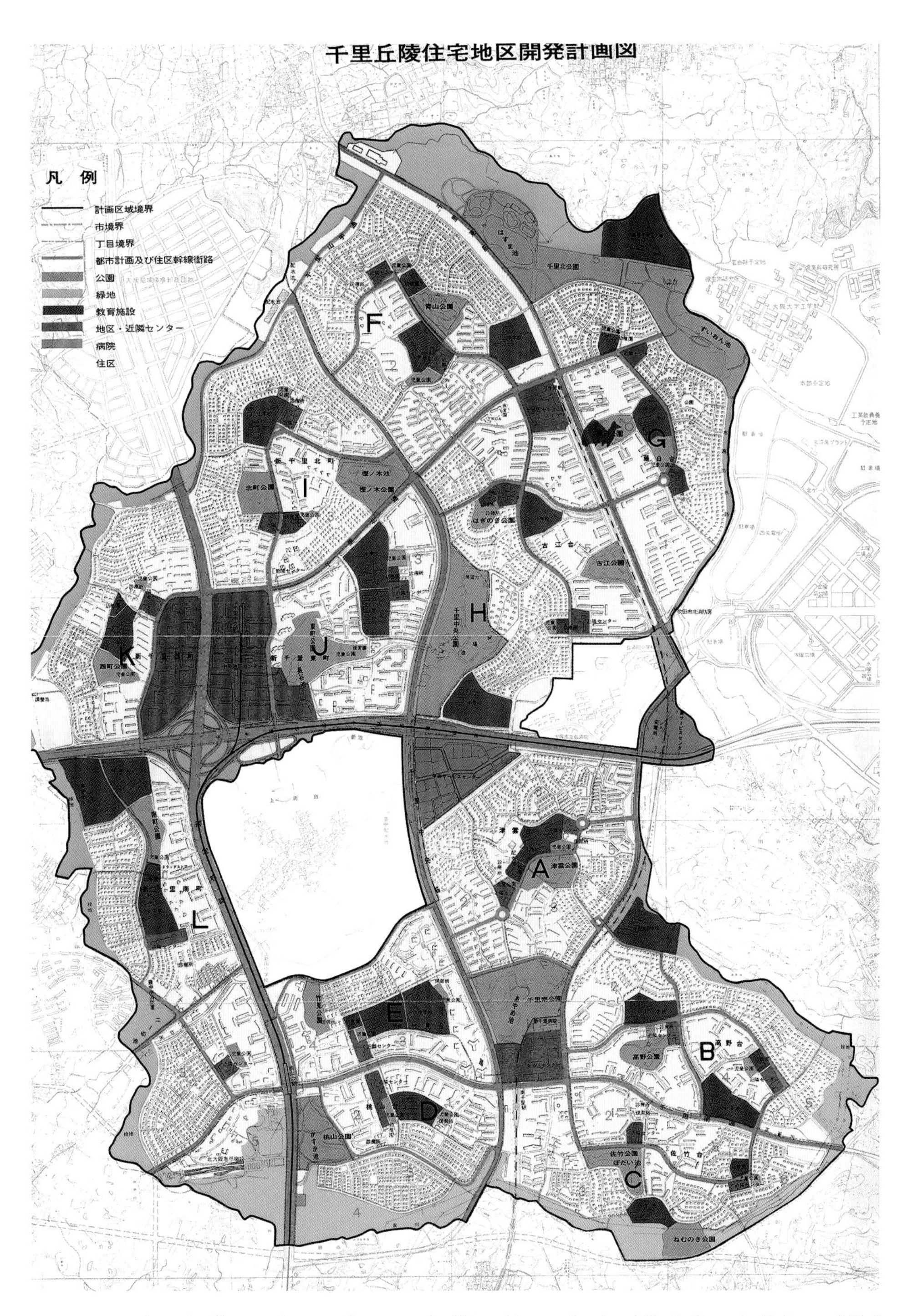

FIGURE 48.1 "Senri Hills Housing Development Plan" [Senri kyūryō jūtaku chiku kaihatsu keikakuzu 千里丘陵住宅地区開発図], from Osaka Prefecture 大阪府, ed., *Building Senri New Town* [Senri Nyū Taun no kensetsu 千里ニュータウンの建設], 1970, 56.

This map comes from a commemorative volume celebrating the completion of Senri New Town, a project of the Japanese Housing Corporation and the Osaka Prefectural Government that was planned in the 1960s and completed the following decade (fig. 48.1).[1] Senri represents a special moment in Japanese postwar housing history: it was built at the height of the economic boom during the decade of rapid growth that followed the frugal period of initial postwar reconstruction, yet before the first land crisis and economic slowdown of the 1970s. Senri is perhaps the best example of the first generation of Japanese new towns that included Tama New Town near Tokyo and Kozoji New Town near Nagoya. Each demonstrates the best practice concepts of "good urban development" as understood by Japanese planners in the early 1960s.

Protests and criticisms leveled at the mass public housing estates of the 1950s constitute a major reason for the design of the new towns. During that period, the government's public housing policy prioritized building the maximum possible number of new units to counter the postwar housing crisis. Large estates were built in areas with cheap land, far from existing built-up areas—and from the schools, shops, and public transit connections found there. This caused serious hardships for the residents of the projects and eventually contributed to a growing opposition from local governments forced to bear the financial burdens of providing schools, water supply, sewers, and other municipal services for thousands of new residents.[2]

The solution was a shift away from the simple focus on large-scale public housing estates to a much more ambitious attempt to build comprehensive new towns. The latter followed British "new town" ideas of the 1950s and the American "neighborhood unit" concept, first proposed by Clarence Perry in 1929 as part of the New York Regional Plan.[3] Japan's New Residential Area Development Act (sometimes referred to as the New Towns Act), passed in 1963, expanded the kinds of development permitted and transformed the Japan Housing Corporation into a "new towns" developer.[4]

In keeping with the ideals of the new-towns movement, Senri is comprehensively designed, with its own town center (shaded in red) providing a range of shopping opportunities and government services, while areas of housing (yellow) include both high-density apartment blocks near railway stations and lower-density single-family detached housing farther from the stations. Residential areas are designed as "neighborhood units" (numbered A to L in the map) centered on elementary, middle, and high schools (shaded in brown), a neighborhood commercial center (red), and a park (green). Two small employment areas are included in the new town (dark blue) near the expressway. The whole town is generously provided with parks and is even girdled by a thin "greenbelt" (light blue) which, although much smaller than those surrounding the English new towns, provides virtually the only green space in the huge metropolitan area filling the alluvial plain north of Osaka.

Unlike the English new towns, however, Senri reflected no attempt to create a self-sufficient, freestanding town with enough jobs for all residents. Instead, it was assumed that most of the population would commute to nearby Osaka or Kobe via three commuter stations on two different railway lines. The town is also bisected by the Chūgoku Expressway, a major limited-access toll road, and by the Shin-Midosuji Road, a northward extension of the famous central-city boulevard through the core of Osaka. Self-containment in terms of jobs would certainly have been an unrealistic goal for Senri, as it is embedded in the middle of one of the major metropolitan regions of Japan. The provision of parks, schools, local shopping, and public transit does, however, answer many of the complaints about the prior model of public housing development.

One distinctive feature of Senri shown clearly on this map is that the main existing settled area, Kamishinden (southeast of the expressway interchange), was excluded from the new-town area, as was a large Osaka municipal hospital. Both appear as white holes in the plan. When the Senri New Town project was designated, the area was sparsely settled, and Kamishinden (which literally translates as "upper new rice paddies") was the only village in the project area. (Most of the incorporated land was hilly and unsuitable for agriculture.) New-town projects were carried out through recourse to a modified land readjustment procedure. Typically, the Japan Housing Corporation would buy a minimum of 40 percent of the project area, then compel all landowners in the project area to participate in the project and contribute a share of their land (usually around 30 percent) toward public facilities such as roads, parks, and land for sale to pay project costs. Not surprisingly, opposition was often encountered from landowners who did not want to participate.[5] The easiest way around such opposition was simply to exclude some areas from the project; as a result, the residents of Kamishinden ended up being surrounded by a new town with which they had few connections. As time went on and opposition to land readjustment grew, new towns became more and more fragmented for this very reason. Both Tsukuba New Town (north of Tokyo) and the Kansai Science City are highly fragmented developments, with many undeveloped and unplanned portions in the middle of town.

Senri represents the most ambitious period of Japanese public housing provision, when the goal was to create well-designed and complete new towns, not just lots of housing units, and when relatively low land prices combined with robust public finances allowed an ambitious approach to the public town-building enterprise. With the land price inflation of the 1980s, the crisis of public finances since the 1990s, and the decline of population since early in the new century, such ambitions now seem ancient history. The Japan Housing Corporation has been rebranded as the Urban Renaissance Agency and, prohibited from building new units, is allowed only to redevelop and renovate its existing stock. The bold plan shown here is emblematic of a bygone era of optimism and confidence in the state as a positive actor in the creation of new urban space.

Notes

1. Osaka Prefecture Kigyō Kyoku, *Senri nyū taun no kensetsu*, 1970.

2. Ishida Yorifusa, *Nihon kindai toshikeikaku no hyakunen* (Tokyo: Jichitai Kenkyūsha, 1987), 296; André Sorensen, *The Making of Urban Japan: Cities and Planning from Edo to the 21st Century* (London: Routledge, 2002).

3. Donald Leslie Johnson, "Origin of the Neighborhood Unit," *Planning Perspectives* 17 (3) (2002): 227–45.

4. Ann Waswo, *Housing in Postwar Japan: A Social History* (London: RoutledgeCurzon, 2002).

5. André Sorensen, "Conflict, Consensus or Consent: Implications of Japanese Land Readjustment Practice for Developing Countries," *Habitat International* 24 (1) (2000): 51–73; André Sorensen, "Consensus, Persuasion, and Opposition: Organizing Land Readjustment in Japan," in *Analyzing Land Readjustment: Economics, Law, and Collective Action*, ed. Yu-Hung Hong and Barrie Needham (Cambridge, MA: Lincoln Institute for Land Policy, 2007), 89–114.

Suggested Readings

Johnson, Donald Leslie. "Origin of the Neighborhood Unit." *Planning Perspectives* 17 (3) (2002): 227–45.

Sorensen, André. "Conflict, Consensus or Consent: Implications of Japanese Land Readjustment Practice for Developing Countries." *Habitat International* 24 (1) (2000): 51–73.

———. *The Making of Urban Japan: Cities and Planning from Edo to the 21st Century*. London: Routledge, 2002.

Waswo, Ann. *Housing in Postwar Japan: A Social History*. London: RoutledgeCurzon, 2002.

49 On the Road in Olympic-Era Tokyo

Bruce Suttmeier

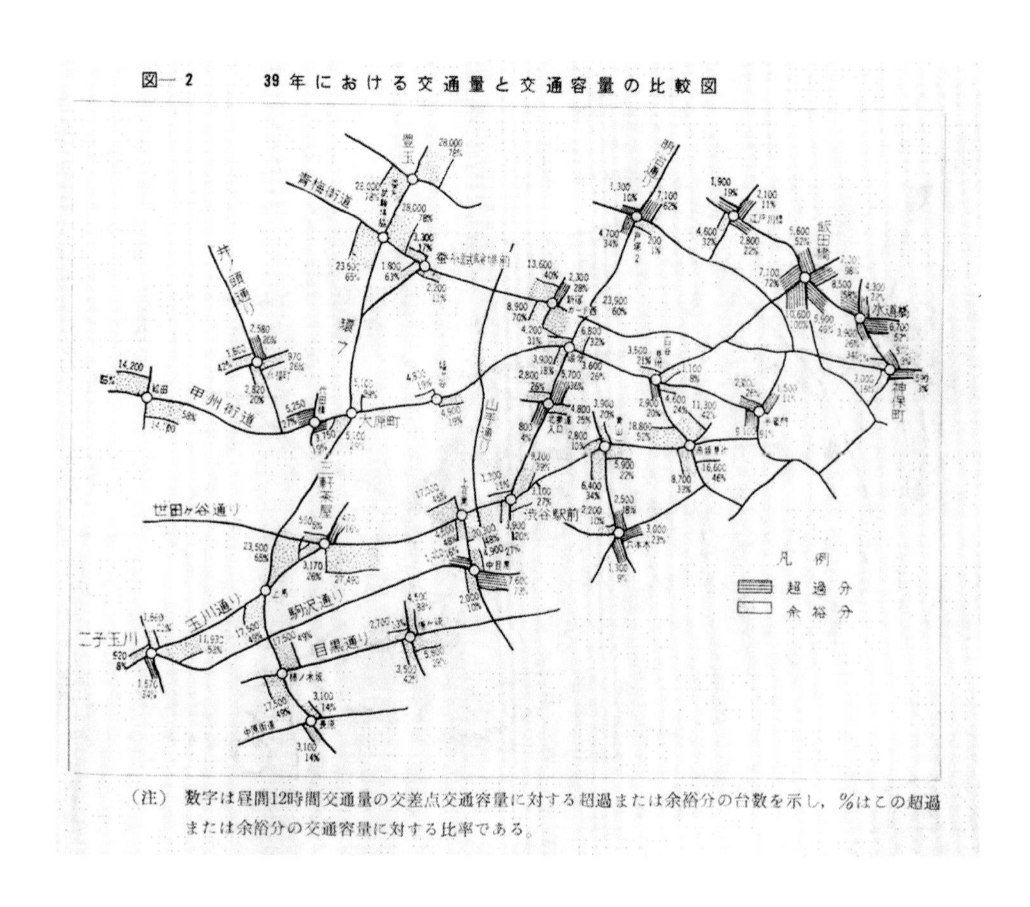

FIGURE 49.1 "A Comparative Map of Traffic Volumes and Traffic Capacities in 1964" [Sanjū kyū nen ni okeru kōtsū to kōtsū yōryō no hikaku zu 三十九年における交通と交通容量の比較図]. Reproduced by permission from *Shintoshi* 新都市18(9), September 1964, 35. City Planning Association of Japan.

In September 1964, with the Tokyo Olympics a mere month away, the streets of Japan's capital were being repaired and refurbished, scrubbed and smoothed, in an effort, as reporters often put it, to prevent the city's already slow-moving streets from suffering total paralysis. The specter of such gridlock spurred massive spending on all kinds of transportation projects, from the Tokyo Monorail to the bullet train to eight new subway routes. Spending on transportation consumed about 75 percent of the city's total one-trillion-yen cost for its five-year Olympic overhaul.[1]

But no project altered Tokyo more than the radically repurposed streets and new highways that emerged on the eve of the Games. Huge elevated expressways, some over forty meters high, began winding their way through the city in the early 1960s, dwarfing nearby buildings, shadowing parks and canals, and towering over large swaths of public spaces. Several modest surface streets nearly doubled in width following the removal of residential buildings and businesses. Such altered streetscapes, while promoted as Olympic projects, were decades in the making. They arose as part of a push by the era's city planners to make Tokyo a more modern (and modernist) metropolis, a place whose slow-moving streets—filled with chatting pedestrians and playing children—were redesigned as "efficient" and "orderly" thoroughfares reserved for the fast-moving motorist. They arose out of the expanding body of research that sought, as Tokyo added over ten thousand new vehicles to its roads each month, to understand and accommodate the extraordinary demands of modern mobility, an imperative essential not only to the city's Olympic success, but also to its long-term viability.

Such research is on stark display in a map published in a special Olympic issue of the journal *Shintoshi* (New city), a publication catering to planning officials and policymakers (fig. 49.1).[2] The map shows traffic volumes around Tokyo in 1964 and catalogs possible trouble spots in the run-up to the games. As readers of this map, what are we to make of this wonky web of lines and circles, crammed with numbers, names, and percentages?

The first thing to notice is its schematic design. Like many topological maps, this one distorts directions, alters scale, and sacrifices strict geographic positioning to illustrate the connective relations between points. Simply put, what matters in this map more than specific city streets is the intersections and their relative position in the network. The map shows a total of twenty-nine intersections, marked with small circles, connected by a network of major boulevards. The shaded sections surrounding each intersection, either light-colored dots or darkened lines, indicate the volume of traffic, from 7 a.m. to 7 p.m., entering the crossing from a particular street. Light dots represent the freer-flowing roads operating under capacity, while dark lines show clogged roads operating above their prescribed thresholds. The figures next to each shaded wedge give the number of cars above (dark lines) or below (light dots) the street's calculated capacity while percentages indicate the proportion this number represents, either over or under the street's recommended traffic volume. Of the ninety-nine shaded wedges, sixty-five display "healthy" movement, and thirty-four show sluggish circulation. That is, about one-third of the streets channeling cars into intersections exceed acceptable limits, averaging 34 percent above capacity. And yet the map, buttressed by the accompanying article (by Hirokawa Yukichi, director of the Metropolitan Police Department's Office of Transportation Planning and Research), stresses the city's readiness, arguing that despite the dire predictions of clogged streets and congested traffic, the city is well prepared to handle the Olympic crowds. How could a map showing so many trouble spots make such a claim?

Answering this question requires knowledge of Olympic geography and midcentury trends in Tokyo development. Looking at the map again, we see a dark-shaded cluster of five intersections in the upper right side of the page, where seventeen of twenty-one streets were overburdened with traffic. These five crossings alone contain half of the map's thirty-four designated trouble spots. What about these sites would make them prone to such congestion? One answer, obscured by the map's topological design, is the cluster's proximity to the center of the city. By the late 1950s, the streets around the Imperial Palace were notoriously congested, containing nine of the ten worst intersections in the city.[3] Yet none of the city's most congested intersections, many of which lay to the east of the palace, even appear on this map. Huge swaths of the so-called low city, the densely populated neighborhoods to the east and north of the palace, are largely ignored by this map, much as they were bypassed in the frenzied building for the games. The map, instead, devotes itself to areas west and south of the old city center. The cluster of trouble spots we see here, sitting north of the palace, are clear outliers, not only because of their geographic position but also because they sit far from any of the important Olympic venues.

These venues were concentrated in the west and southwest portions of the city: the National Gymnasium and athletes' village near Shibuya, the main National Stadium near Shinjuku, and the Olympic Park Stadium Complex out in Komazawa. Shibuya and Shinjuku (near the middle of this map) were becoming the new "centers" of the city, a trend that had started after the 1923 earthquake but vastly accelerated in the 1960s. Most of the streets featured prominently on the map are the so-called "Olympic roads" that, running alongside these venues, were transformed in five short years from modest roads into major thoroughfares. Kannana Avenue (Ring Road 7), running north-south along the entire left side of the page, connected the Komazawa Olympic complex in the far southwest to numerous points farther north. While the full loop circling the city took fifty-eight years to finish, the Olympic portion encompassing the western half of the ring was completed, at enormous expense, in a mere five years. Even more substantial were the changes to Aoyama Avenue, running east-west through the middle of the map. Connecting the city center to Shibuya and Komazawa, Aoyama Avenue was dubbed "Olympic Tokyo's Main Street" for its central role in shuttling visitors to the game venues. Now a center for high-end fashion boutiques and trendy stores, in the late 1950s Aoyama Avenue was a modest street of two-story shop fronts that was expanded to nearly double its original width for the Olympics. Meter for meter, this expansion became the most expensive project in the city's transportation overhaul.[4]

All told, because of the extraordinary expense of compensating landowners, the cost for widening surface streets like Aoyama Avenue ran to about five times that of building the new thirty-two-kilometer elevated highway winding its way up from the airport and out to the Olympic venues. Yet it is impossible to discuss Olympic-era Tokyo, or to fully understand this map, without considering the outsized role of the city's elevated highway system. In the early 1960s, the highways embodied a near-utopian solution to the city's traffic woes. A high-speed road without intersections, it was reasoned, would be free from the gridlock troubling the earth-bound streets below. The most exaggerated expression of such thinking appears in the city plans of the Metabolist school architects, most strikingly in *A Plan for Tokyo 1960*, by Tange Kenzō, the designer of the National Stadium for the 1964 games. Figure 49.2 shows Tange's reimagined city stretching across Tokyo Bay, its design dominated by enormous ten-lane highways structuring

FIGURE 49.2 "Model, Office District in Civic Axis and Residential District on the Sea" [Mokei toshijiku no naka no ofisubiru chiku to kaijō jūkyo chiku 模型都市軸の中のオフィスビル地区と海上住居地区], from Tange Kenzō's *1960 Plan for Tokyo: Toward a Structural Reorganization*. Reproduced by permission from Hirose Mami et al., *Metabolism, the City of the Future: Dreams and Visions of Reconstruction in Postwar and Present-Day Japan* (Tokyo: Mori Art Museum and Shinkenchiku-sha, 2011), 68. Photograph by Kawasumi Akio.

the city through a series of stacked loops. This radical redesign, he predicted, would allow a traffic capacity of two hundred thousand vehicles per hour, or "from ten to thirty times the capacity of high-speed highways now in use." Only by vastly increasing such capacity throughout the city, Tange argued, could Tokyo be saved from its "confused, paralyzed, even moribund state."[5]

It is partly this utopian thinking—the highways' position literally and figuratively "above it all"—that explains their absence from a map devoted to the mundane, exasperating business of congested intersections. The new freeways seemed incapable of such congestion, untroubled as they were by the start-and-stop frustration of surface streets. In the years following the Olympics, amid the excitement of Japan's high-growth era, city planners pushed for an infrastructure commensurate with modern forms of mobility. Indeed, many planners, from avant-garde architects to government officials, stressed that the city's surface streets, with their chaotic mix of pedestrians and vehicles, should function more like the highways by being committed to the high-speed transport of people and goods. For many years, such messianic faith led to the building of ever-larger roadways: an often ill-fated effort to solve gridlock by creating a more motorized Tokyo.

Notes

1. See Igarashi Yoshukuni, *Bodies of Memory: Narratives of War in Postwar Japanese Culture, 1945–1970* (Princeton: Princeton University Press, 2000), 147; and Philip Kaffen, "Ichikawa Kon's 'Tokyo Olympiad: The Olympic Body of Memory,'" in *Olympic Japan: Ideals and Realities of (Inter)Nationalism*, ed. Andreas Niehaus and Max Seinsch (Würzburg: Ergon Verlag, 2007), 50.

2. See Hirokawa Yukichi, "Orinpikku Tōkyō taikai to dōro kōtsū," *Shintoshi* 18 (9) (September 1964): 31–37.

3. See the influential pamphlet *Dōro kōtsū no konran wa sukueru-ka? Kōsoku dōro o chūshin ni Tōkyō wa wakagaeru* (Tokyo: Toshi Keikaku Kyōkai, 1959), 42. Iwaidabashi, listed as the worst intersection in the city, saw 121,911 vehicles a day. The worst intersection shown on the map is Suidobashi, at 35,273. Also see similar figures in Yamada Masao's article "Tōkyō no toshi kōsoku dōro keikaku 1," *Shintoshi* 14 (8) (July 1960): 2–8.

4. The estimated cost of Aoyama Avenue was 21.5 billion yen for 8,199 meters (2.67 million yen per meter), about 90 percent of which was for acquiring land and compensating landowners on either side of the road. The bullet train, by comparison, was on average a "mere" 687,658 yen per meter (380 billion for 552 kilometers). See "Ichi oku en no 'orinpikku seisen,'" *Shūkan Yomiuri* 23 (41) (October 11, 1964): 37.

5. See Kenzo Tange Team, *A Plan for Tokyo 1960: Toward a Structural Reorganization* (Tokyo: Shikenchikusha, 1961), 19, 2.

Suggested Readings

Cybriwsky, Roman. *Tokyo: The Shogun's City at the Twenty-First Century*. New York: John Wiley and Sons, 1998.

Gold, John R., and Margaret M. Gold, eds. *Olympic Cities: City Agendas, Planning and the World's Games, 1896–2016*. 2nd ed. New York: Routledge, 2010.

Lin, Zhongjie. *Kenzo Tange and the Metabolist Movement*. New York: Routledge, 2010.

Seidensticker, Edward. *Tokyo Rising: The City since the Great Earthquake*. New York: Knopf, 1990.

Sorensen, André. *The Making of Urban Japan: Cities and Planning from Edo to the Twenty-First Century*. New York: Routledge, 2002.

50 Traversing Tokyo by Subway

Alisa Freedman

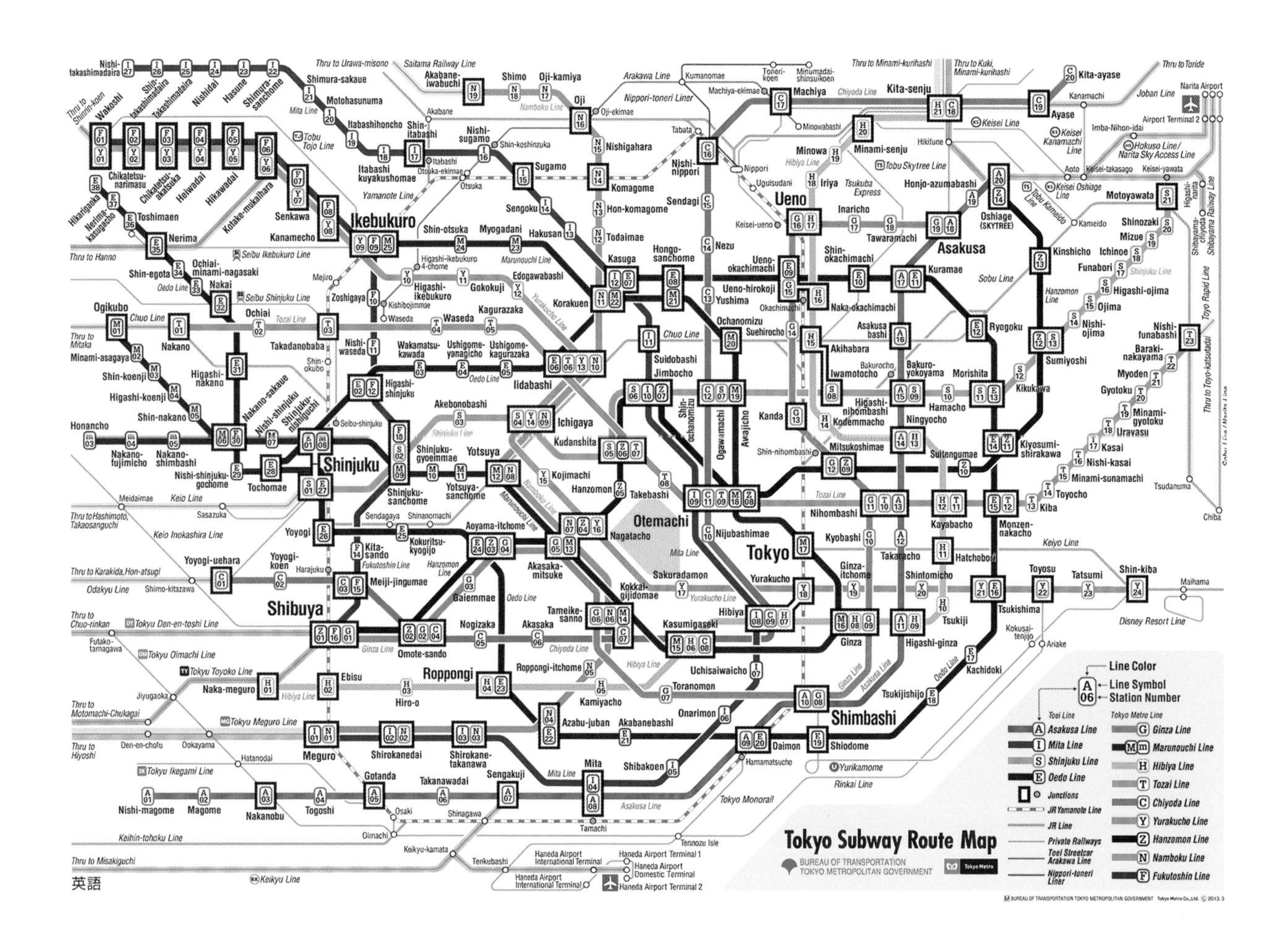

FIGURE 50.1 "Tokyo Subway Route Map," 2013. Bureau of Transportation, Tokyo Metropolitan Government.

The Tokyo Subway Route Map is an indispensable guide to Japan's capital. Invoking the awe and trepidation many people feel in approaching Tokyo itself, its dramatic array of colored lines represent thirteen underground routes joining 285 stations (fig. 50.1). Together with over one hundred train lines, this network constitutes the world's largest and busiest transit system, carrying almost twenty-nine million passengers each day.[1] The system is so extensive that trains and subways are depicted on separate maps. Buses play an important but secondary role, servicing locations hard to reach on rails. More than trains that extend to the suburbs and beyond, subways are arteries connecting nodes of daily life in the heart of Tokyo, and their routes represent flows of twentieth-century urban development.

Subways, like Japan's trains and buses, are owned by government agencies and private companies. Figure 50.1 diagrams the nine routes of the privatized Tokyo Metro (Tōkyō Metoro, 179 stations), transporting roughly six million passengers daily, and seven routes of the Tokyo Metropolitan Board of Transportation (Tōkyō-to Kōtsū), or Toei (106 stations), ridden by nearly two and a half million people each day.[2] In 1970, subway routes were assigned colors, generally based on those of their trains, as well as line numbers. While the designers remain anonymous, the Tokyo subway map might have been based on that for the London Tube, which has used colored route lines since 1908.[3] Lines 1–5 were part of the 1946 plans for Tokyo's high-speed rail network. Toei Mita became line 6 because it connected to Ōtemachi, the subway station closest to Tokyo Station, which opened in 1914 near the Imperial Palace and Marunouchi financial district as the symbolic gateway into the capital.

Downloadable and print versions of the Tokyo subway route map are available in Japanese in three versions: one with station names only, a second showing station names and codes (explained below), and a third depicting rough geometric shapes of stations so that passengers can grasp their layouts. The version with both station names and codes has been translated into seven languages: English, Korean, Chinese, Spanish, German, French, and Russian. Routes are searchable through the Tokyo Metro and Toei websites, along with private transport websites and smartphone apps. Poster-sized maps hung by subway platforms situate routes within the larger Tokyo metropolitan area. Most maps inside passenger cars focus on the route at hand; station stops are usually indicated on small monitors, a practice begun in the Japan Railways (JR) Yamanote Line in 2002. Among the only landmarks announced with station stops are department stores and the Tokyo Subway Museum.

The subway route map requires knowledge to use. To assist passengers who are unable to read Japanese and thus encourage tourism, a new coding system was introduced in 2004. Tokyo Metro replaced the Teito Rapid Transit Authority (Teito Kōsokudo Kōtsu Eidan, Eidan or TRTA for short). Each station was coded with the first letter of its subway line (i.e., G for Ginza) and its sequence en route (i.e., G11 for the eleventh station on the Ginza Line), encircled by the route color (i.e., orange). Stations are generally numbered from west to east, and south to north; those traversed by more than one subway line have multiple codes. For example, in addition to G11, Nihombashi is T10 (Tozai Line, tenth station) and A13 (Asakusa Line, thirteenth station).

Additional features make the subway route map useful for Tokyo residents and tourists alike. Thin lines—gray for JR and light blue for private railways—show how trains connect to subways. Both of Tokyo's airports are prominently noted. The map key and logo marks (blue heart-shaped *M* for Tokyo Metro and green T-shaped gingko leaf of the Tokyo Metropolitan Government for Toei) are found in the lower right corner, which represents Tokyo Bay. The central green expanse designates the Imperial Palace, the symbolic heart of the capital and the sole landmark on the map. The Imperial Palace's closest subway station, Ōtemachi, is serviced by more subway lines than any other and has a direct underground passageway to Tokyo Station. However, the busiest subway station is not Ōtemachi (around 269,800 passengers daily in 2011) but Ikebukuro, through which around 470,300 people pass each day, showing that patterns of work and play deviate from symbolic designations of space.[4] A large typeface is used for names of terminals connecting to the world's busiest train, the JR Yamanote, a loop line constructed around the area considered to be the city center in the years after the 1923 Great Kanto Earthquake.

Tokyo subways developed more than four decades after the Yamanote Line (originally a freight train) and other commuter railways opened in the 1880s. Tokyo's first underground railroad, opened in 1917, carried mail between Tokyo Station and the nearby Tokyo Central Post Office. The Tokyo Underground Railway Company (Tokyo Chika Tetsudō), founded by businessman Hayakawa Noritsugu (1881–1942) in 1920, started construction in 1925 on a 2.2-kilometer route between the Asakusa and Ueno entertainment districts. On opening day, December 30, 1927, over one hundred thousand people waited hours for the five-minute ride on what was celebrated as Asia's first subway.[5] It was a high-fashion and high-tech way to travel; passengers inserted tickets—purchased in vending machines for ten sen, around three sen more than the cost of commuter train tickets—in automatic ticket gates.[6] The Tokyo Rapid Railway (Tōkyō Kōsoku Tetsudō), later part of the Tokyu Railway, started a second route from Shibuya, connecting to the Tokyo Underground Railway at Shimbashi. The companies offered through service from 1939 and in 1941 merged into the Teito Rapid Transit Authority.

The development of routes and distribution of maps show that the first subway was a modern attraction rather than a commuter vehicle. The six-story Asakusa Subway Tower, one of Tokyo's tallest buildings in the 1920s, housed the Subway Store (Chikatetsu Sutoa) and four-story Asakusa Kaminari Gate Cafeteria (Asakusa Kaminari Mon Chokuei Shokudō). Both businesses were open only to ticketed passengers, launching the trend of shopping plazas inside stations. In similar fashion, the Mitsukoshi department store contributed 463,000 yen toward the construction of the opulent 1931 Mitsukoshi-Mae Station (then the only station servicing a company instead of a neighborhood), which led, via one of Japan's first escalators, directly into the first-floor showroom.[7] In the 1930s, subway route maps appeared on commemorative postcards for stations, multi-ride tickets to department stores (*depāto mawari jōshaken*, begun in 1934), and Subway Store matchbook covers. Popular culture captured the subway's modern spirit and critiqued Tokyo's confusing web of transportation. For example, in "Tokyo March" (Tokyo kōshinkyoku)—the theme to a 1929 film directed by Mizoguchi Kenji and the best-selling song of the twentieth century—the last of four verses about love in Tokyo jokes: "Tokyo is big, but when you are secretly in love, it seems small. Let's meet for a tryst in chic Asakusa. You take the subway, and I the bus. But we will never be able to stop at love."[8]

Subways were extended in the 1950s and 1960s to carry the overflow from the JR Yamanote Line and other railways and became integral to the daily commute. In December 1953, Tokyo's first subway was named the Ginza Line in anticipation of the opening of the second subway, the Marunouchi Line, the following month. The most subway expansion to date occurred between 1960 and 1964, concurrent with the construction of Tokyo's highways: the Marunouchi Line reached additional business districts and western suburbs, and the Asakusa, Hibuya, and Tozai Lines opened. At the time, route maps were distributed in various ways. For example, in addition to being available for passengers to pick up inside stations, paper maps were included in the *Metro News* public relations magazine, published from 1960, and electronic route maps were featured on ticket machines during the 1964 Tokyo Olympics.

Subways continue to affect the use of Tokyo urban space. The Ōedo Line, opened in 2000, forms a circle around urban development projects from the beginning of the twenty-first century, most of which were built on reclaimed land south of the old city.[9] Fukutoshin, the newest line (opened in 2008), is so named because it links "secondary city centers," hubs lacking the state symbolism of Tokyo Station. Conspiracy theorists advance the idea that the directions of subway routes have been dictated by military rather than commercial interests and by the need to avoid secret underground shelters for the imperial family.[10]

The route map makes visually apparent several of the ways in which subways have structured Tokyo urban space: their routes trace Tokyo development in the 1920s, 1960s, 1990s, and other key moments of rapid urban growth; placement of their stations both reaffirms the cultural significance of national symbols (i.e., the Imperial Palace) and enables work and play in Japan's capital; and their intersections with railroads are essential to the elaborate design of Tokyo's world-famous mass transit network. Fondness for the subway system—and for the map that functions as its logo—is evidenced by the numerous Tokyo guidebooks based upon it published since the 1930s, and by its appearance over the years on commodities as varied as coffee mugs, handkerchiefs, and smartphone cases.[11] Other Japanese cities—Fukuoka, Kobe, Kyoto, Osaka, Nagoya, Sapporo, Sendai, and Yokohama—have subways, but their maps are simpler and lack the cultural cachet of that of Tokyo.

Notes

1. "Transportation," Tokyo Metropolitan Government, http://www.metro.tokyo.jp/ENGLISH/PROFILE/policy12.htm.

2. "Business Situation," Tokyo Metro, http://www.tokyometro.jp/en/corporate/enterprise/transportation/conditions/index.html.

3. See Claire Dobbin, *London Underground Maps: Art, Design, and Cartography* (London: Lund Humphries, 2012).

4. "Traffic Performance by Station," Tokyo Metro, http://www.tokyometro.jp/en/corporate/enterprise/transportation/ranking/index.html. The second busiest subway station is Kita-Senju (around 281,200 passengers daily).

5. Aoki Eiichi, *Tōkyō no chikatetsu ga wakaru jiten* (Tokyo: Nippon Jitsugyo Shupansha, 2004), 35. Subway service lasted from 6:00 a.m. to midnight, with a train coming every three minutes.

6. See, for example, "Teito kōsokudo kōtsu eidan," *Chikatetsu un'yu gojūnenshi* (Tokyo: Teito Kōsokudo Kōtsu Eidan Eigyōbu Untenbu, 1981), 40–41. Used from 1871 to 1953, the sen equaled one-hundredth of a yen.

7. Hatsuda Tohru, *Modan toshi no kūkan hakubutsugaku—Tōkyō* (Tokyo: Shokokusha, 1995), 152–53 and 225–27.

8. *Shōwa hayari uta shi* (Tokyo: Mainichi Shimbunsha, 1985),

48–49.

9. Kawashima Ryōzo, "Ōedo sen no nori kata daikenkyū," *Tōkyōjin* 161 (January 2001): 60–67.

10. See, for example, Akiba Shun, *Teito Tōkyō chikatetsu no nazo* 86 (Tokyo: Yusensha, 2005); and Ichihara Jun et al., *Chikatetsu no himitsu* 70 (Tokyo: Ikarosu Shuppan, 2009).

11. Examples include journalist Ogawa Takeshi's 1935 *Fashionable Dates: Rendezvous Guide*, a guide to using Tokyo transportation for nights out on the town and secret trysts, and the 2011 *Historical Strolls through Edo and Tokyo by Subway*, by the Tokyo Reikishi Kenkyūkai. Ogawa Takeshi, *Ryūsenkei abekku—Randebū no annai* (Tokyo: Marunouchi Shuppansha, 1935); Tokyo Reikishi Kenkyūkai, *Chikatetsu de iku Edo Tokyo rekishi sanpo* (Tokyo: Kanki Shuppan, 2011).

Suggested Readings

Aoki Eiichi. *Tōkyō no chikatetsu ga wakaru jiten*. Tokyo: Nippon Jitsugyō Shuppansha, 2004.

Freedman, Alisa. *Tokyo in Transit: Japanese Culture on the Rails and Road*. Stanford: Stanford University Press, 2010.

Murakami Haruki. *Underground*. Translated by Jay Rubin. New York: Vintage, 2001.

Toei Transportation Information, available at http://www.kotsu.metro.tokyo.jp/eng/.

Tokyo Metro website, available at http://www.tokyometro.jp/en/index.html.

Tokyo Subway Museum (Chikatetsu Hakubutsukan), available at http://www.chikahaku.jp (in Japanese).

51 The Uses of a Free Paper Map in the Internet Age

Susan Paige TAYLOR

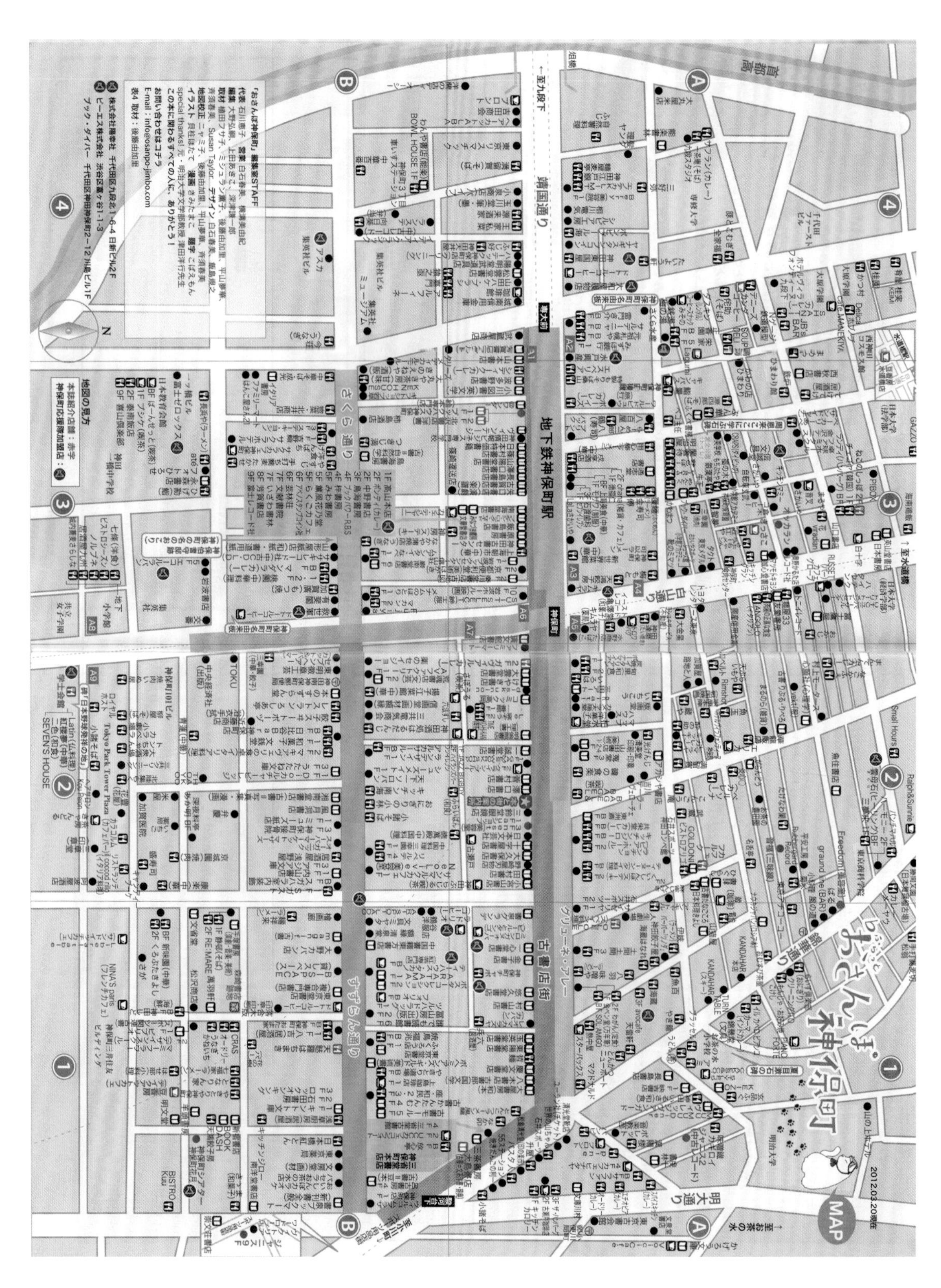

Since the Meiji period, the Kanda/Jinbōchō neighborhood of Tokyo has been central to the Japanese book trade. Many of Japan's oldest universities had their campuses in Kanda, forming an attractive location for bookstores, publishers, and related industries. Now referred to as Jinbōchō after the closest subway station, the area is billed as the world's largest book town, specializing in antiquarian, niche, and used books from Japan, Asia, and the rest of the world. While mapping its numerous bookstores is challenging, several companies and organizations regularly publish maps of the area.[1] As an example of one of Tokyo's "town papers"—free or low-cost magazines that introduce the history, atmosphere, and walking courses of an area—"Walking Guide to Jinbōchō" (fig. 51.1) exhibits a unique strategy for neighborhood-oriented printed ephemera in a digital paradigm: it does not aim to be comprehensive, the points marked on the map change slightly in each issue, and it is not available on the web.[2]

As ephemera, the free paper is intended to be used once and discarded. Some forty thousand copies are printed for each biannual issue and distributed locally through a network of bookstores, cafés, and other shops. The mixture of paid advertisements (which fund this guide), comics, and short articles is designed to create a connection between reader and place. The map occupies the last two pages and is titled in the upper right corner by a cartoonish cat, a book tied to its head with a Japanese headscarf, with the single word "Map" in a speech bubble.[3] Paw prints near the cat suggest the leisurely stroll that this pamphlet is designed to encourage, but few interpretive tools are included. For instance, the largest legend box, instead of clarifying the map, lists the issue's contributors. A smaller legend box labeled "How to read" only explains two symbols. Beyond this, readers are left to make sense of the image on their own.

FIGURE 51.1 "Walking Guide to Jimbōchō" [Osanpo Jinbōchō mappu おさんぽ神保町Map], by Nyamiko ニャミ子, Gotō Yukari 後藤由加里, Hirayama Yumeka 平山夢華, and Saisu Harumi 斉須春美, May 1, 2012. Offset printing, 29.6 × 22.5 cm. Courtesy of Osanpo Jimbōchō, Tokyo, Japan.

The single most striking aspect of the map is its visual density, which actually impedes legibility and results in regular complaints. In two crowded pages, the walking guide identifies 134 bookstores, 165 other shops (ranging from clothing stores to hotels to florists), 270 restaurants, 44 cafés, 15 supporters, 5 local-interest sites, and the purported location of a fictional bookstore that appeared in a popular movie. Famous buildings with multiple bookstores are labeled with orange "speech bubbles," vying for attention against the mass of black symbols for smaller bookstores, cafés, and other shops, all labeled with vertical or horizontal text, depending on space constraints. Commercial considerations, such as the need to include patron shops located in outlying areas while cramming in most of the bookstores and restaurants lining Yasukuni and Suzuran Streets, result in significant distortions of scale. The Jinbōchō intersection and the entrances to the subway station are roughly in the middle of the map, indicating an underlying assumption that people will visit by subway.

The density of points is a reflection of both the abundance of actual shops and complex interests driving the design of the pamphlet. When the map was first created, one contributor insisted on noting identifying landmarks on most corners to aid people who have difficulty finding their way; while not strictly related to the purpose of the map, these wayfinding points have accumulated overtime. Similarly, varying notation styles for businesses occupying the same building have been employed, further complicating the map. Shops that hav purchased an advertisement are shown in red print, creating a color scheme that serves the financial needs of the flyer's production without helping readers navigate its crowded surface. Despite its name, such a "guide" is most useful to those who already have some degree of familiarity with the area. In practice, the map becomes a tool for "locals" to explicate the neighborhood to visitors. Café or shop owners annotate copies for individual customers, directing them to unmarked or hard-to-find locations.[4]

This act of inscription adds a dimension of legibility that the map alone does not offer to the uninitiated.

"Walking Guide to Jinbōchō" is put out by a group of volunteers to coincide with the Suzuran Street Festival. Since 2005, they have relied on local connections to produce it cheaply and regularly. Budgetary concerns influence the legibility of the map, since redrawing it would be costly. In order to showcase Jinbōchō's uniqueness (as well as minimize the workload of volunteers) the group rejects offers from chain stores, such as ubiquitous curry shops and coffee shops, to place advertisements. Because there is a high turnover rate of businesses due to expensive rents, the walking guide's map is updated in each issue. The editors meet new clients interested in purchasing advertisements either through the recommendation of established clients or by walking the area, looking for new shops. They also consult four business owners, each representing roughly one quadrant of the map, regarding changes in the neighborhood. Their method relies on businesspeople who are not organized within a group, unlike publications created by the Used Book Union, which represents its members.

The paper is intended for people who love Jinbōchō and those who will come to love Jinbōchō. For this reason, the makers of the map consider the Internet the wrong medium to introduce it to first-time visitors. If the map is simply uploaded, it may discourage people from visiting in person and instead redirect traffic to the bookstores' online shops. Shopkeepers believe that drawing people into the space will encourage return visits, and that profits generated in-store are an incentive for businesses to stay in a place. If more of their sales move online, the rationale for remaining in a high-rent area diminishes.

The "Walking Guide" website functions as a tool to promote the free paper itself, with prominent text that encourages viewers to "take the walking guide and go out and stroll."[5] The site contains information about the current and upcoming issues, including a table of contents and a list of locations distributing the pamphlet. It aims to motivate visits to the neighborhood rather than encourage content consumption online. Similarly, "Walking Guide to Jinbōchō" utilizes online social media platforms to promote visits to the area.

Chief among the concerns that this flyer addresses is how people circulate through the area. Before the advent of online sales, bookstores lured many pedestrians. However, since students and professors increasingly tend to purchase books online, the bookstores generate less foot traffic, especially on weekdays. Restaurants catering to the lunch crowd have now arguably become the main draw. The location of the subway exits affects the flow of pedestrian traffic, contributing to locations where it is harder to maintain a business. Since Yasukuni Street overtook Suzuran Street as the area's main thoroughfare, pedestrians often bypass the latter altogether and fail to realize that there are more shops on adjacent backstreets. The shops on Suzuran Street, many of which support the walking guide, hope to avoid the fate of nearby Sakura Street, which lost its identity as a commercial venue.

Especially in a digital context, the question of how to encourage pedestrian traffic has become paramount for a place such as Tokyo's Kanda neighborhood. The makers of the walking guide articulate a vision of the relationship between Internet and place by encouraging people to walk and "encounter" the neighborhood rather than searching for isolated shops. The busy surface of the map serves as both PR for the neighborhood businesses and a sign of leisure culture: one that presents a "younger" interpretation of Jinbōchō while still relaying the area's traditional image as Tokyo's "book town." The density of the points also reflects the complexity of the business networks of Jinbōchō, rather than the abstracted points of online map services or customer review websites. We can easily imagine the free map being sandwiched in a bag of purchased books, to be rediscovered later as a memento of one's visit. In that sense, while the flyer is ephemeral, it is meant to facilitate experiences that are more tangible and memorable than those mediated by the ubiquitous Internet.

Notes

1. Maps can be found in the free paper *Hon no machi* (Town of books) and on the Jimbou.info website. The Used Book Union prints a magazine yearly promoting its affiliates, which retails for 1,200 yen.

2. This essay is based in part on interviews conducted by the author with contributors to the free paper. Interviews conducted were as follows: Ōno Hirotsugu, Jinbōchō, Tokyo, May 13, 2012; Ōno Hirotsugu, Shinjuku, Tokyo, July 20, 2013; Saisu, "Harumin," Jinbōchō, Tokyo, June 28, 2013.

3. See S. Mary, P. Benbow, and Bonnie C. Hallman, "Reading the Zoo Map: Cultural Heritage Insights from Popular Cartography." *International Journal of Heritage Studies* 14 (1) (2008): 30–42.

4. This is reminiscent of Roland Barthes's observation that

Japanese clerks often draw maps by hand to help visitors navigate Tokyo's labyrinthine streets. See Roland Barthes, "No Address," in *Empire of Signs*, trans. Richard Howard (New York: Hill and Wang, 1982), 33–37.

5. "Osanpo Jinbōchō WEB—Furiipēpā & Jinbōchō saishin jōhō," http://osanpo-jimbo.com/.

Suggested Readings

Eubanks, Charlotte. "Visual Vernacular: Rebus, Reading, and Urban Culture in Early Modern Japan." *Word and Image* 28 (1) (2012): 57–70.

Kawauchi Yukari and Morimoto Shōichi. "Shōtengai to no renkei ni yoru furī pēpā no sakusei to sono kōka no kenshō." *Senshū daigaku jōhō kagaku kenkyūsho shohō* 80 (June 2013): 25–35.

Tsao, James C., and Stanley D. Sibley. "Readership of Free Community Papers as a Source of Advertising Information: A Uses and Gratifications Perspective." *Journalism and Mass Communication Quarterly* 81 (4) (2004): 766–87.

Yoneno Fumitake. "Machizukuri to mediya no kankeishi." *Koku sōken shiryōshu, machizukuri to mediya kenkyūkai* 218 (http://homepage3.nifty.com/fmeno/lab/PDF/machi_media.pdf).

52 Tsukiji at the End of an Era

Theodore C. Bestor

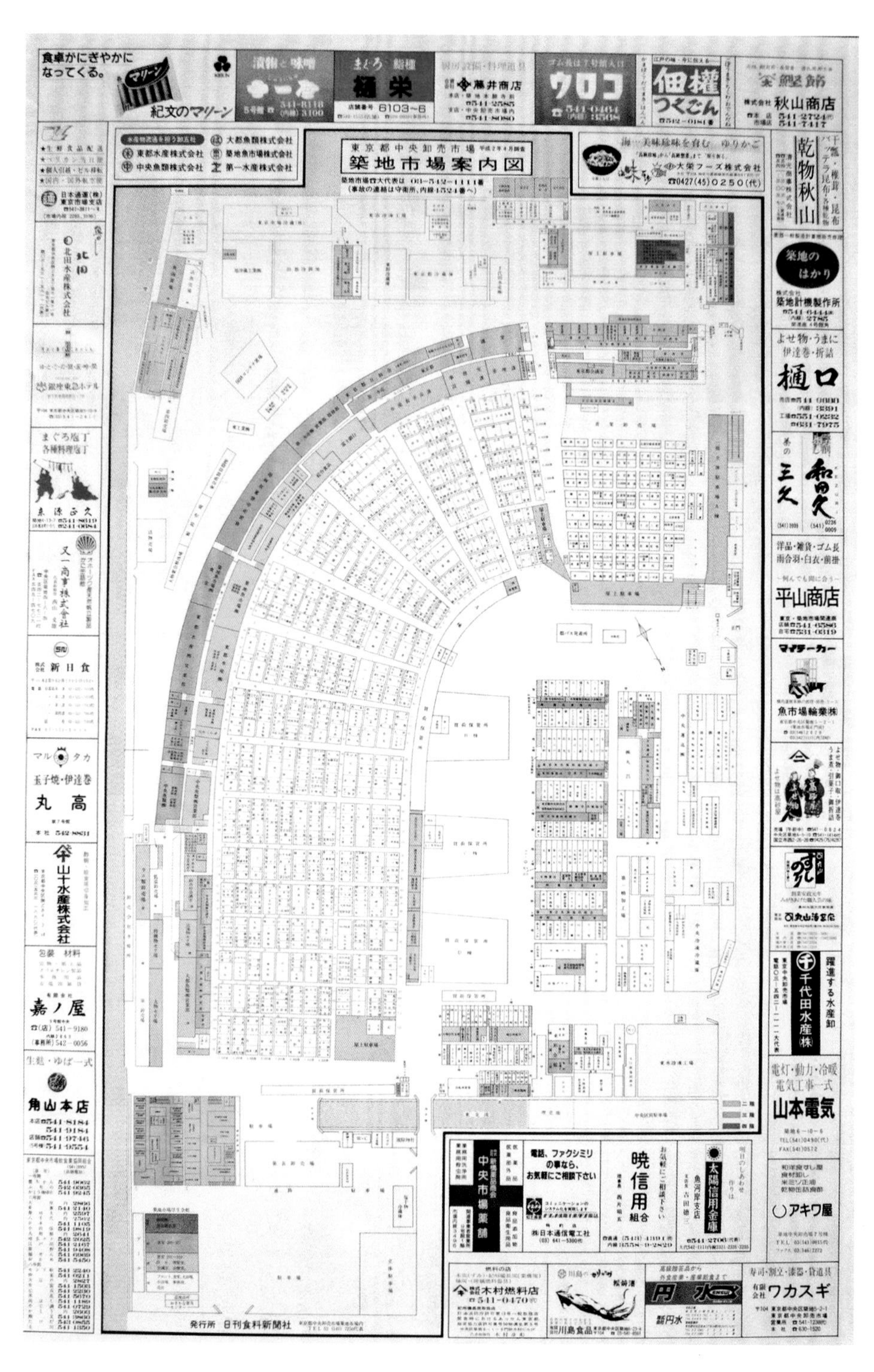

FIGURE 52.1 "Map of the Tsukiji Central Wholesale Market" [Tōkyō-to chūō oroshiuri shijō Tsukiji annaizu 東京都中央卸売市場築地案内図], April 1990. 23.5 × 35.5 cm. Courtesy of Nihon Shokuryō Shinbunsha.

Like all ethnographers, I have come to understand that what I record and report as the "present" becomes historical as soon as I write it. So with this map of the Tsukiji Central Wholesale Market (fig. 52.1). Almost as soon as I began to understand what it could tell me about the present, that present had begun to fade away. The map reveals clues about the origins of the marketplace in the 1930s, and by learning how to read it, I understood aspects of contemporary market organization. Details within the market, though, were constantly changing, causing the map's present-day utility to fade within a couple of years. Beyond the incremental obsolescence of the map, unexpected events in 2013 rendered it a relic of a soon-to-be bygone era.

Near the heart of Tokyo and long the world's largest marketplace for seafood, Tsukiji's inner market contains seven large auction houses and 1,677 stalls operated by about seven hundred firms that collectively cater to thousands of chefs, fishmongers, supermarket buyers, and other trade buyers.[1] The marketplace opened in 1935, after the 1923 Great Kanto Earthquake destroyed the centuries-old fish market in nearby Nihonbashi. The basic physical layout of its marvelous Bauhaus-inspired design endured through the twentieth century with few modifications.[2]

When I began research at Tsukiji, maps helped me to decipher the history and present-day structure of the marketplace. My research drew in part on anthropological perspectives on space and place, focusing on interactions among spatial layout, social institutions, and cultural norms. It also examined the processes through which spaces become culturally meaningful. Having a good map in hand was clearly my starting point.

The most immediate reading of the map shows the marketplace's configuration. The 1,677 stalls are shown in detail in the large curved structure that occupies the top, left, and central parts of this map. In the 1950s each of these was occupied by an independent stall holder. By the 1990s, consolidations had reduced the number of firms to about 1,200. The continuation of this trend left roughly 700 firms in 2013, with some occupying a dozen or more stalls. Reading the map at this level is crucial to comprehending the marketplace's institutional structures.

Less visible are the human networks that help regular clients navigate Tsukiji. Clifford Geertz described the webs of relationships among buyers and sellers in a Middle Eastern suq (market or bazaar) as "grooved channels" in which buyers make their transactional rounds, focusing primarily on familiar trade partners. These kinds of channels operate at Tsukiji as well. Location matters for many reasons, including accessibility for different kinds of clientele, whether established, large-scale buyers (for whom the larger stalls closest to the auction pits are most convenient), or the small-scale walk-in trade (for whom the smaller stalls at the inner edge of the curve are most convenient). Fundamentally, buyers need to know where to find their preferred trading partners.

New maps of Tsukiji have been regularly published over the years, along with directories and guidebooks geared to the chefs, fishmongers, and delivery agents who need to find their regular wholesalers. If "grooved channels" are a feature of Tsukiji, why would new maps be so important? The answer, I discovered, lies in a complicated facet of market governance. In the interests of ensuring long-term equality of access both to auctions and to customers, intermediate wholesalers conduct periodic lotteries among themselves to determine the location of their stalls. Every five to six years, the proprietors of all 1,677 stalls engage in a random lottery that reshuffles their locations. Uncovering this fact, and then asking people at the market to explain how sellers ended up in their current spaces, opened up the complicated structures and norms that govern the daily workings of the marketplace.[3]

At another level, the overall shape of the map reveals the functional logic of an earlier era.[4] When the market opened in 1935, much of the seafood sold at Tsukiji was caught in the waters of Tokyo Bay, and large quantities of fish were landed directly from fishing boats tied up along the Sumida River wharf. The gracefully curving arc of the market sheds, leading west from the wharf, was designed to accommodate rail sidings for special express trains carrying perishable goods that arrived daily from other fishing ports around the country. Neither of these features was in use by the end of the twentieth century. Pollution of the bay's waters had long since brought commercial fishing there to a halt, and express trains were replaced in the 1970s by refrigerated trucks. Still, the curve of the sheds and the little-used wharf remained.

The 1935 layout, based on wharf and rail, created a structure in which seafood originally was unloaded on the eastern edge and moved to the central auction areas (designated for specific kinds of products). Goods purchased at auction could be quickly brought into the stalls and broken down into saleable units for the patrons who entered from central Tokyo to the west. The first iteration of Tsukiji therefore featured a wonderfully logical flow of products from east to west, from bulk commodity to seafood delicacy trimmed to the scale needed by a sushi chef in the Ginza. But the transportation revolution of the 1960s and 1970s turned this on its head. Seafood now arrived by truck from the west, and drivers struggled through traffic to get their cargos to the eastern side of the market. Once through the auction and resale within the market, the products ended up going out the western and northern gates, again by truck!

Since my earliest visits to Tsukiji, the future of the marketplace has been in doubt. In addition to its inefficient design for long-distance refrigerated trucking, the failure to upgrade its electric infrastructure or to adopt digital technologies had isolated Tsukiji from the media that transformed almost all other aspects of Japanese life. Additionally, the economic value of this large plot of land in the center of Tokyo had become astronomically inflated, putting the local government under tremendous pressure to sell the land to developers. Finally, the collapse of Japan's economic bubble in the 1990s fundamentally altered the country's business climate in ways that eroded the viability of Tsukiji. Against relocation stood only neighborhood sentiment, concern for small business, and environmental concerns about the proposed location of the new market. From the 1990s onward, the Tokyo Metropolitan Government, owner and administrator of the marketplace, actively promoted moving it to a new location. Protests by market regulars and environmental activists could not prevent the final decision, made in 2010, to do just that.

Three years later, an even more radical shift rendered this 1990 map a thing of the past. In September 2013, the International Olympic Committee announced its decision to award the 2020 Summer Olympic Games to Tokyo. This announcement sealed Tsukiji's fate. Joining other remnants of Tokyo's *shitamachi* (traditional mercantile neighborhoods) to be obliterated by urban "redevelopment" projects for the Olympics, the current Tsukiji market site was designated to house the international media and broadcast center for the games. The market itself will be moved to a landfill island, Toyosu, on a site previously used by the Tokyo Gas Corporation for processing petrochemicals.

Tsukiji's death knell illustrates how quickly a map can become a historical artifact. The 1923 earthquake cleared the ground for the Tsukiji fish market in the first place by destroying its three-hundred-year-old predecessor in Nihonbashi, which now hosts corporate headquarters, financial institutions, and elite shops. The social earthquake preceding the 2020 Olympiad promises to remake Tokyo's economic and cultural geography once more, again relocating the fish market. Like the last move, so the next one will create new spatial environments for the seafood trade, generating new patterns of institutional governance, human interactions, and cultural understandings of place. Grooved channels will take new shapes, and new maps will be made.

Acknowledgments

The author thanks Victoria Lyon Bestor and Yukari Swanson for their generous assistance in the preparation of this article.

Notes

1. In 2011, Tsukiji handled almost 550,000 tons (497 million kilograms) of raw, frozen, and processed seafood, with a total wholesale value of 423.8 billion yen (US$5.4 billion). Tōkyō-to Chūō Oroshiuri Shijō Tsukiji Shijō, *Tsukiji shijō gaiyō*, 2012. The dollar value is calculated at a rough average of the 2011 exchange rate: one US dollar equaling 79 yen.

2. Tsukiji's history, social structure, impact on food culture, the architectural and transactional layout of the marketplace, and efforts (up through 2002) to relocate Tsukiji are detailed in Theodore C. Bestor, *Tsukiji: The Fish Market at the Center of the World* (Berkeley: University of California Press, 2003).

3. See Bestor, *Tsukiji*, especially chap. 7.

4. The map is roughly oriented with north at the bottom. The Sumida River (on the left edge of the map) is to the east of the marketplace and flows south (toward the top of the map). The shopping and entertainment districts are beyond the map's scope to the right-hand side (toward the west).

Suggested Readings

Bestor, Theodore C. *Neighborhood Tokyo*. Stanford: Stanford University Press, 1989.

———. *Tsukiji: The Fish Market at the Center of the World*. Berkeley: University of California Press, 2003. (Japanese translation, *Tsukiji*. Tokyo: Kirakusha, 2007.)

Geertz, Clifford. "The Bazaar Economy: Information and Search in Peasant Marketing." *American Economic Review* 68 (2) (1978): 28–32.

Low, Setha M., and Denise Lawrence-Zuniga, eds. *The Anthropology of Space and Place: Locating Culture*. Oxford: Blackwell, 2003.

53 Probabilistic Earthquake Hazard Maps

Gregory Smits

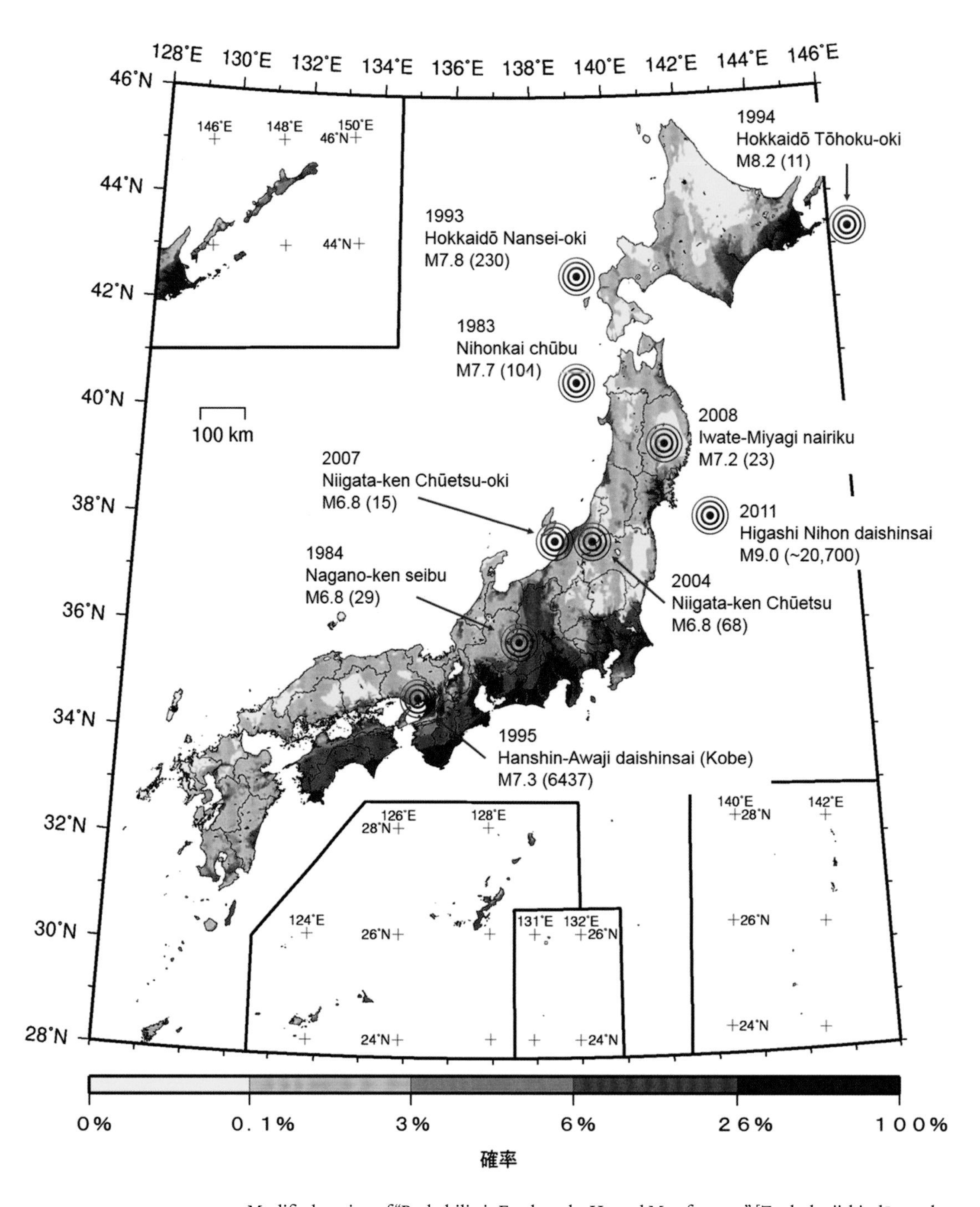

FIGURE 53.1 Modified version of "Probabilistic Earthquake Hazard Map for 2010" [Zenkoku jishindō yosoku chizu 全国地震動予測地図2010年版], published online by the Jishin chōsa iinkai of the Japanese Ministry of Education, Culture, Sports, Science, and Technology, accessible at http://www.jishin.go.jp/main/chousa/10_yosokuchizu/. Darker areas correspond to higher probabilities of a major quake. Epicenters of earthquakes causing ten or more fatalities since 1978 have been superimposed on the map by the author, along with magnitude and fatality data. Personal collection of the author.

In 2010 the Japanese government published a probabilistic earthquake hazard map. That map is reproduced here with an important modification: the epicenters of actual earthquakes that killed ten or more people between 1978 and 2011 have been superimposed onto it, with fatalities given in parentheses (fig. 53.1).[1] The color scale used in the original indicates the calculated probability of a major earthquake ("weak 6 or stronger" on the Seismic Intensity Scale of the Japan Meteorological Agency [JMA]) occurring within thirty years.[2] The lighter-colored areas of the map are presumably safer owing to a lower probability of strong ground motion. The map purports to say something useful about earthquake forecasting and appears to embody the authoritative, objective expertise of the scientific community. It may be, however, that this map is a manifestation less of established scientific principles than of social and historical forces.

The current focus on earthquake forecasting in Japan is a direct result of its postwar history of earthquake prediction, which in turn is the result of a complex interplay among politics, public expectations, mass media, and actual seismicity. A formal earthquake prediction program in postwar Japan has its roots in a 1962 document, *Prediction of Earthquakes: Progress to Date and Plans for Further Development,* subsequently known as "the Blueprint." It proposed that the state provide funds for accumulating data on possible earthquake precursors for ten years. At the end of that period, enough data would exist to determine whether earthquake prediction was possible. It was a modest proposal and might not have gone anywhere, but the earth intervened.

The 7.5 magnitude Niigata earthquake struck on June 16, 1964. Although it only caused twenty-six fatalities, damage to buildings and infrastructure was extensive. Moreover, Niigata was Japan's first televised earthquake. One result was money in the 1965 budget for earthquake prediction research. These funds arrived just in time for the Matsushiro swarm, consisting of thousands of small earthquakes between 1965 and 1970. In an interview, the mayor of Matsushiro famously pleaded for "more science." Seismologists set up instruments throughout the region, and local residents nervously anticipated a big shock that never happened. In 1967, the earthquake prediction budget doubled.

Two broader forces in society had also become significant by the end of the 1960s. One was the acceptance (despite some holdouts) of the theory of plate tectonics. This breakthrough in knowledge raised hopes in the scientific community that earthquake prediction might soon be possible. At the same time, there was a rise in unrealistic public expectations that in an age when travel to the moon was possible, surely experts could predict earthquakes. In this context, Transportation Minister Nakasone Yasuhiro consulted with members of the seismological community and explained that their agreeing to drop the world "research" from publicly funded earthquake prediction projects and initiatives would result in a roughly tenfold increase in funding. In other words, earthquake prediction funding would vastly increase if seismologists agreed to wording that implied earthquake prediction was possible. They took the money.[3] One possible rationalization was that with such an increase in funding, earthquake prediction would indeed soon become reality. Seismologist Susan Hough describes a similar push for government funding in the United States relying on promises of earthquake prediction as having been "somewhere between optimistic and totally irresponsible."[4]

Meanwhile, Japan was about to encounter a major earthquake panic. In 1977, seismologist Ishibashi Katsuhiko warned of an imminent Tōkai earthquake, and the mass media amplified his message. It was a doomsday scenario whereby a powerful megathrust earthquake originating in the Nankai Trough would shake Osaka and the surrounding area violently and generate a massive tsunami that would wash away vast areas of the densely populated coast, damage nuclear reactors, and cause vast loss of life. Ishibashi urged the immediate start of "concentrated observation." The state took him seriously, and the earth intervened. The 7.0 magnitude Izu-Ōshima earthquake of January 14, 1978, occurred just outside the eastern edge of the Tōkai earthquake zone. One effect of this event was to exacerbate public and government fears of the dreaded Tōkai earthquake. Moving at record speed, the Diet passed major earthquake legislation later that same year.

The Special Measures Law for Large-Scale Earthquakes (abbreviated "Daishinhō" in Japanese) is mainly concerned with the implementation of martial law in the likely area of Tōkai earthquake devastation. The law also stipulates that the JMA will predict the next Tōkai earthquake. The makers of the law did not seriously consider *whether* JMA personnel could indeed predict a Tōkai earthquake. Instead, the Special Measures Law took such prognosticating powers as a given. The law also opened the doors to massive influx of government money, not only for the concentrated observation that Ishibashi advocated, but also for the building of seawalls and other infrastructure. Owing to the Special Measures Law, Japan is the only country in which earthquake prediction, a feat that most seismologists regard as impossible, is legally mandated.

When we zoom out to survey the larger picture of Japanese seismicity, the 7.1 magnitude Fukui earthquake of June 28, 1948, killed nearly four thousand, a grim capstone to wartime devastation and several deadly earthquakes earlier in that decade. For nearly fifty years afterward, Japan experienced very few earthquake-related fatalities. The relative calm ended suddenly and without warning on January 17, 1995, when the Rokkō-Awaji fault zone—unknown at the time—produced an earthquake that killed roughly sixty-five hundred and did extensive damage to Kobe and nearby areas.[5]

This event dramatically revealed the inability of seismologists to predict destructive earthquakes. It also revealed a lack of basic knowledge about the locations of the many faults under the Japanese islands. The event resulted in changes in Japan's earthquake bureaucracy. First, many agencies quickly dropped the term "prediction" from their names. Second, a systematic effort to discover, map, and assess the characteristics of active faults began. In connection with these two changes, the seismological community in Japan shifted focus from short-term "prediction" (*yochi*) to longer-term "forecasting" (*yosoku*). Probabilistic earthquake hazard maps are the main visual expression of this changed focus.

Even the most optimistic proponents of earthquake forecasting do not claim that precision similar to that of weather forecasting is possible. Indeed, there are no fixed standards for assessing earthquake forecasting, and current practices are the result of arbitrary choices. The preferred interval is thirty years, and the typical predictive statement gives a range of probability of an earthquake of a certain magnitude or intensity within a region specified by a color on the map. Retroactively calculating such a statement for Kobe in 1995 results in a thirty-year probability of 0.4 to 8.0 percent, because the relevant fault appears to rupture once every 1,800–3,000 years.

Stated probability ranges are often large, such as 0.9 to 9.0 percent. Even these wide ranges are the result of arbitrary starting assumptions and spotty data about past earthquakes in a region. Changing the assumptions or the historical data can result in greatly altered probabilities.[6] Even if probabilistic forecasting of earthquakes were somewhat accurate and based on firm geophysical principles, it is difficult to imagine how such information would be broadly useful. It might help inform building codes, but a prudent builder anywhere in Japan should assume that a destructive earthquake is possible. Moreover, there is no way to know in advance when such an event will occur. The dreaded Tōkai earthquake from the late 1970s, for example, is still pending.

Notice that the dark regions of the map do not correspond well to actual deadly earthquakes, most of which have occurred in apparently safe areas. In creating a hazard map, an investigator first employs an overall judgment for each region to ascertain earthquake scale and frequency, often using historical data. There is no standard or common methodology for arriving at this impressionistic assessment. These estimates of recurrence intervals for regions translate into probabilities of the future occurrence of strong ground motion.[7] Recently, the international seismological community has debated the usefulness of probabilistic earthquake hazard maps, seeking ways to make such maps and the science behind them more rigorous and realistic.[8]

Notes

1. This modification of the map follows that of seismologist Robert Geller. See Robert Geller (Robaato Geraa), *Nihonjin wa shiranai "Jishin yochi" no shōtai* (Tokyo: Futabasha, 2011), 163. The map was issued by the Jishin Chōsa Kenkyū Suishin Honbu (Headquarters for Earthquake Research Promotion), an agency within the Ministry of Education, Culture, Sports, Science, and Technology.

2. The current iteration of the scale, dating from 1995, classifies all earthquakes into seven categories by intensity, with levels 5 and 6 subdivided into "weak" and "strong."

3. For details see Geller, *Jishin yochi*, 90; and Shimamura Hideki, *Jishin yochi wa uso darake* (Tokyo: Kōdansha, 2008), 103–4.

4. Susan Hough, *Predicting the Unpredictable: The Tumultuous Science of Earthquake Prediction* (Princeton: Princeton University Press), 69.

5. The official name for this event is the Hyōgo-ken nanbu earthquake, and it is better known outside Japan as the Kobe earthquake.

6. Shimamura, *Jishin yochi*, 164–68.

7. Geller, *Jishin yochi*, 156–69.

8. See, for example, Seth Stein, Robert J. Geller, and Mian Liu, "Why Earthquake Hazard Maps Often Fail and What to Do about It," *Tectonophysics* 562–63 (2012): 1–25.

Suggested Readings

Clancey, Gregory. *Earthquake Nation: The Cultural Politics of Japanese Seismicity, 1868–1930*. Berkeley: University of California Press, 2006.

Smits, Gregory. *Seismic Japan: The Long History and Continuing Legacy of the Ansei Edo Earthquake*. Honolulu: University of Hawai'i Press, 2013.

———. *When the Earth Roars: Lessons from the History of Earthquakes in Japan*. Lanham: Rowman and Littlefield, 2014.

54 Citizens' Radiation Mapping after the Tsunami

Jilly Traganou

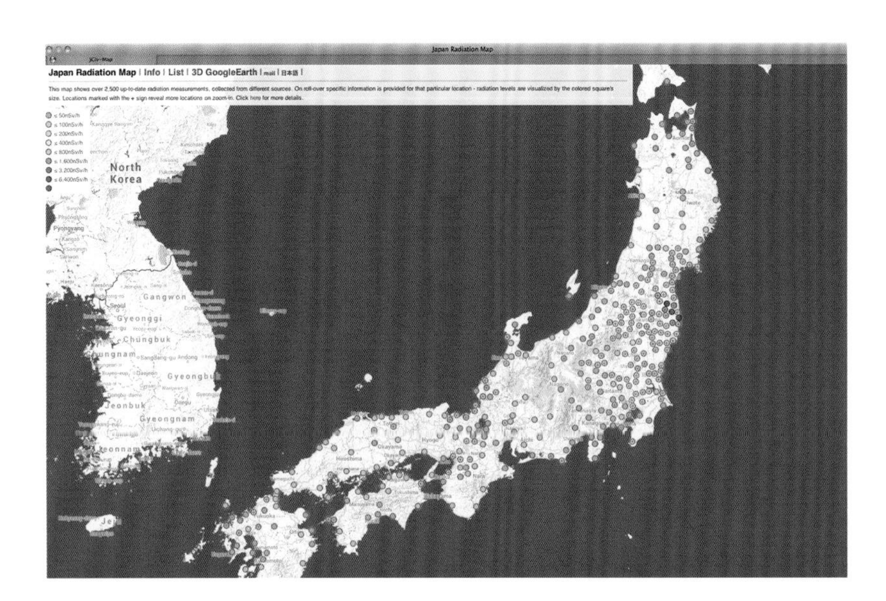

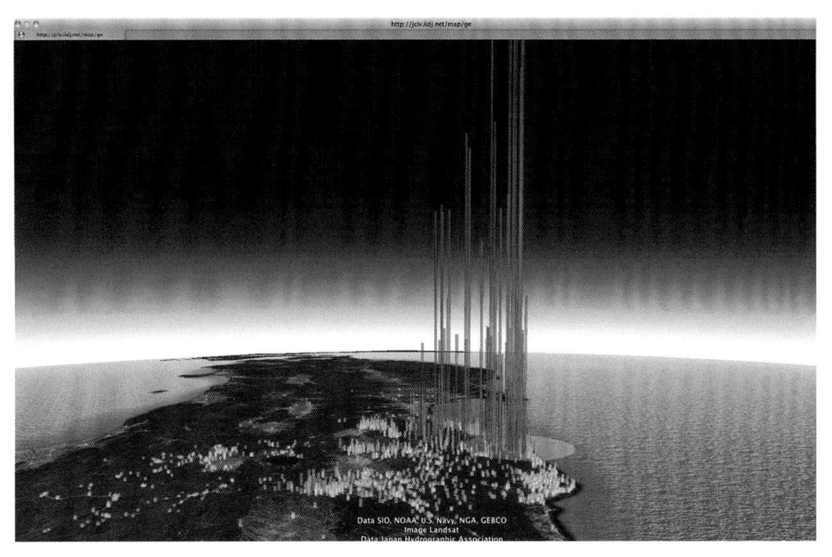

FIGURE 54.1 "Japan Radiation Map," digital map designed by the Institute of Information Design Japan (IIDj), 2011. This map is continually updated on the web at http://jciv.iidj.net/map/. Courtesy of IIDj.

FIGURE 54.2 "Japan Radiation Map," three-dimensional view. On the web at http://jciv.iidj.net/map/. Courtesy of IIDj.

Radiation maps produced by volunteers proliferated after the triple disaster—earthquake, tsunami, and nuclear meltdowns—that hit Japan's Tohoku region beginning on March 11, 2011. This cartographic phenomenon is both an outcome of a distrust of political institutions that may follow natural disasters and an indication of the appearance of a "citizen science" that attempts to bridge expert knowledge with social infrastructure. Immediately after the disaster, and amid contradictory government reports on radiation measurements, many

Japanese citizens became acutely aware of their inability to access reliable information. Moreover, most nonexperts were unable to comprehend the provided data and use them in ways that would help them make informed decisions. With public safety potentially threatened across the country, citizens throughout much of Japan felt insecure about the radiation levels in their environment and the degree of danger that the accident posed to their health.

Citizen mapping projects stepped into the breach. Open-source technologies and Internet-based communication have made do-it-yourself cartography possible in unprecedented ways. Constituting the interface for community learning, the Internet provides the possibility for constant updates of content that is user generated or derived from official sources. The posting of information on resources and the capacity for direct users' exchange allowed Japanese citizens to undertake their own radiation measurements and to improve their skills in the process. These efforts are a reflection of the belief in the value of social capital in recovery after disaster. As Daniel Aldrich has observed, "High levels of social capital—more than such commonly referenced factors as socioeconomic conditions, population density, amount of damage or aid—serve as the core engine of recovery."[1] While most online radiation maps are aimed at empowering the public, however, they also differ from each other. Some citizen mappers in the wake of the Fukushima nuclear accident were motivated primarily by an interest in rendering official information legible. Others were more ambitious, aiming at building social infrastructure: a capacity of civil society that did not need to be mediated by governmental authorities or the market. "Japan Radiation Map," by the Institute of Information Design Japan (IIDj) and the Safecast global sensor network, both emerging a very short time after the disaster, exemplify some of these differences.

The maps created by IIDj (figs. 54.1 and 54.2) bridge professional expertise with citizen initiative. While not experts in radiation, designers at the IIDj felt compelled to produce radiation maps that would convert available data into legible online information for the general citizenry. An interest in having comprehensive and accessible information that would cover the wider geographic territory of the country (and indeed of the whole world, for comparative purposes) was another premise of the map, which was published just twelve days after the disaster.

A small team of designers, editors, and programmers of IIDj collected material from scattered prefectural sources. Having no common format, the data had to be normalized and localized to their specific geocoordinates. The team also identified and translated location names into English, a clear distinction from the official sites which provided only Japanese versions. Combining two- and three-dimensional visualizations, IIDj attempted to provide accurate, legible information that would make viewers aware of the radiation magnitude in a comparative perspective. One important question involved how to visually code different radiation levels. While IIDj wanted to avoid explicit data interpretation, according to Andreas Schneider—a founding partner of IIDj who developed the "Japan Radiation Map"—the group "needed to provide people with visual cues for capturing the significance of the mapped data points on a global as well as local level."[2] In addition to using color coding established by the International Atomic Energy Agency (IAEA) to show increases in measured radioactivity, the group added location markers indicating the direction and speed of wind, which were particularly important in the first phase after the accident.

When looking at the IIDj's planar map (fig. 54.1), viewers can immediately understand radiation levels in locations across the country as well as detailed measurements, updated hourly, in over twenty-five hundred locations. This differs from the official "Environmental Radioactivity and Radiation Information Map"—produced by the Nuclear Safety Technology Center of Japan's Ministry of Education, Culture, Sports, Science, and Technology—which only lists maximum values of each prefecture and omits measurements of the accident site itself. Converting the radiation data to vertical cylinders (fig. 54.2) provides an immersive view that shows the tremendous difference between the nuclear accident's location in Fukushima and other areas of Japan. Comparisons with other countries are necessary, according to IIDj, in order to give an understanding of what might be an average daily value of radioactivity, which is determined by various natural and human factors. The institute noted that radiation levels in Tokyo before the accident were lower than the radiation levels in some parts of Germany, for example, where due to the geological structure the natural background radiation is consistently higher than the Japanese average.[3]

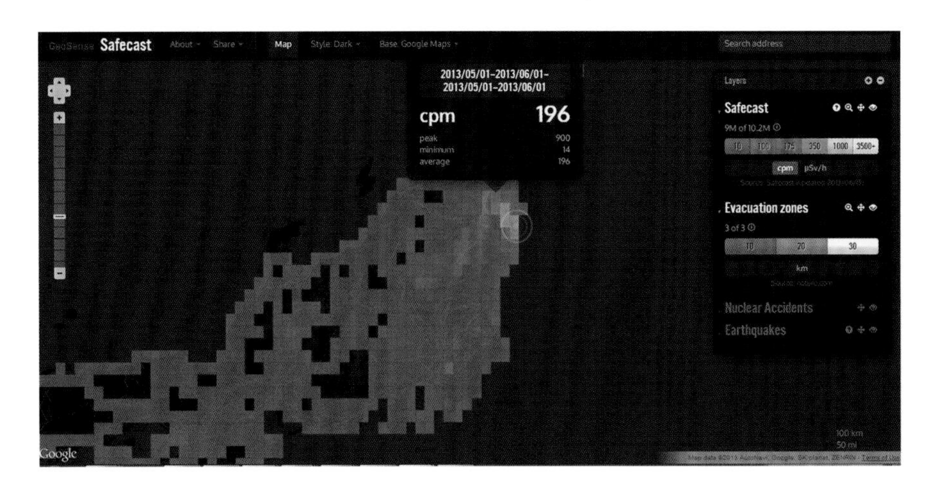

FIGURE 54.3 "Safecast Map," depicting over 4 million radiation data points collected by the Safecast team. Captured online on October 20, 2013 (http://map.safecast.org/map/140.1726041766169,36.94429246653189,7).

IIDj's approach to collecting information from selected official sources was different from crowdsourcing, which aggregates input from a large number of users. Characteristic of the latter approach are maps by the global sensor network Safecast, a volunteer nonprofit organization dedicated to collecting and sharing radiation measurements (fig. 54.3). Safecast's efforts covered a wide range of activities that included the construction of new Geiger counters, training of field members, and design and online posting of constantly updated, interactive maps. Their data aggregated a variety of sources, deriving from measurements of their own field members in addition to those from governmental and nongovernmental agencies. Going beyond data presentation alone, Safecast's website is an interface for communication and dialogue on issues that vary from monitoring methods to data interpretation.

The major goal of Safecast's effort is to amplify social capacity that already exists and expand it to empower wider populations. According to Joi Ito, Safecast founding member and director of the MIT Media Lab, the group's aim is to create "rapid, agile and resilient systems,"[4] which are necessary for overcoming situations of crisis. Immediately after the accident, Safecast started a Kickstarter campaign and turned to the collective Tokyo Hacker Space to tap civilians' knowledge of radiation. Giving Geiger counters to citizens and showing them how to construct their own was a priority for this project, especially in areas where coverage was sparse. By April 2011, Safecast had created a network of interdisciplinary experts and volunteers who both improved technical equipment (online maps and devices) and took measurements in the field.

Safecast realized that it was important to take as much information as possible (at multiple points, rather than at designated sensors), because, with the possibility of data substantially differing even between points that were in proximity to each other, regional averages were not necessarily useful. Frequency of measurement was also important due to the possibility of environmental factors such as weather conditions affecting the results. Their monitoring effort is thus both massive and ongoing. Mobile devices, which are mounted outside cars, perform readings every five seconds, while field members report readings to the website eight to ten times per day. The compact, GPS-equipped Geiger "bento" device designed by Safecast members is an important part of the project (fig. 54.4). Geiger devices sold out internationally just a few days after Japan's triple disaster, leaving citizens with no way to acquire their own. Being able to construct one's own equipment, rather than depend on the market, is seen as particularly critical.

During the first fifteen months, Safecast collected more radiation data—3.5 million data points—than all prior projects in history.[5] This gives Safecast maps a granularity that most others lack, allowing them to be searched by inputting an address in Japan or elsewhere. They can be particularly useful after a nuclear accident, when they can indicate that radius-based evacuations that are usually mandated by governments might be flawed, since radiation moves in nonconcentric patterns.[6] This is crucial for making informed decisions on population evacuation and destination after a nuclear accident.

Crowd-sourced maps, however, are not flawless. Andreas Schneider, in a review of online radiation cartog-

FIGURE 54.4 Homemade "bento" Geiger device created by Safecast. Photograph courtesy of Safecast.

raphy, questioned the validity of some user-generated data in the Safecast maps. Although the crowd-sourcing of data has received great publicity, "the non-qualified, non-expert nature of the many contributions," he pointed out, "may only add noise to the overall mass of information."[7] Additionally, studies show that some of the unmoderated networking technologies that appeared after the Fukushima accident (such as the "Person Finder") were vulnerable to malicious hoaxes and spam.[8] For Schneider, what is needed is "publicly available observation data provided by neutral and certified organizations, [using] internationally agreed conventions or measurement units, thresholds and classifications of criticality."[9] On the other hand, for Safecast, what is valued the most is a bottom-up approach, through which scientific infrastructure is built in civil society in an independent way. Despite their differences, however, both of these citizen-based initiatives agree on the importance of making data open to the public. Schneider and his team offer their professional expertise pro bono to their fellow citizens, ameliorating serious deficits in governmental and media communication practices. Safecast aggregates already existing expertise and invests in its expansion into a self-reliant social infrastructure. Both suggest that citizen cartography—a dispersed, nonauthoritarian practice—is a powerful tool in today's civil society.

Notes

1. Daniel Aldrich, *Building Resilience: Social Capital in Post-disaster Recovery* (Chicago: University of Chicago Press, 2012), 15.
2. Institute for Information Design Japan, "Japan Radiation Map," http://jciv.IIDj.net/map/. Aldrich, *Building Reslilience*, 4.
3. http://jciv.IIDj.net/map/germany/.
4. Safecast trailer, http://vimeo.com/69002438.
5. *Safecast Blog*, http://blog.safecast.org/about/.
6. Sean Bonner, "Safecasting inside the Evacuation Zone," *Safecast Blog*, September 20, 2011, http://blog.safecast.org/2011/09/safecasting-inside-the-evacuation-zone/.
7. For Schneider, the Safecast map fully visualizes the great quantity of available measurements. But he also has design-related objections with the legibility of the map, due to its color scheme, which makes distinctions difficult to draw. Andreas Schneider, "Reviewing Online Maps about Radioactivity in Japan," *Information Design Journal* 192 (2011): 195.
8. Aldrich, *Building Resilience*, 162.
9. Schneider, "Reviewing Online Maps," 197.

Suggested Readings

Aldrich, Daniel. *Building Resilience: Social Capital in Post-disaster Recovery*. Chicago: University of Chicago Press, 2012.

Davies, Gemma, Rebecca Ellis, Amy Fowler, and Duncan J. Whyatt. "How Reliable Are Citizen-Derived Scientific Data? Assessing the Quality of Contrail Observations Made by the General Public." *Transactions in GIS* 17 (2013): 488–506.

Havens, Kayri, and Sandra Henderson. "Citizen Science Takes Root." *American Scientist* 101 (2013): 378–85.

Institute for Information Design Japan. "Japan Radiation Map." http://jciv.IIDj.net/map/.

Schneider, Andreas. "Online Maps about Radioactivity in Japan." *Information Design Journal* 19 (2011): 188–97.

55 Run and Escape!

SATOH Ken'ichi 佐藤賢一

FIGURE 55.1 A settlement site that was washed away by the tsunami on March 11, 2011, with only a guide map of the town remaining. Numerous houses used to stand in rows behind this guide map just before the tsunami hit the town. The photograph was taken by the author in April, 2012, in Yamamoto Town, Miyagi Prefecture.

The Japanese archipelago sees frequent volcanic activity and earthquakes, being located at the junction of the Eurasian Plate, the North American Plate, and the Philippine Plate. Weather events can be catastrophic as well. Typhoons from the South Pacific hit Japan almost every summer, frequently causing strong winds and torrential rain as well as flood damage, while northern Japan suffers from heavy snow in the winter. All in all, Japan has a long history of quite harsh natural disasters. Tsunami events in particular can devastate coastal areas following earthquakes and have often taken a serious toll on human life.

The Tohoku coast of Honshu has been hit by major tsunami four times since the dawn of the modern period: in 1896, 1933, 1960, and 2011. The greatest damage of all came in the wake of the recent Great East Japan Earthquake in March 2011. This unprecedented tsunami triggered a catastrophic disaster, causing some twenty thousand casualties. Although various issues with disaster prevention and the evacuation guidance system were pointed out in 2013, the affected area still struggles to envision a way forward as reconstruction continues.

As previous essays have shown, Japan's natural disasters and major fires have repeatedly destroyed built enironments, topography, and landscape in the blink of an eye. Each time, humans have recorded the disaster, leaving records for people in later ages to use. Those records naturally include maps. Today, however, maps have come to play a new role in natural disasters, connected not with recording the past but with forecasting the future. The process of planning for future catastrophes now includes creating hazard maps that allow people to visualize and avoid potential danger zones. In other words, rather than merely documenting damage that has already happened, hazard maps are now being used to plan for the next event, a completely new function.

As discussed in chapter 53, government agencies at the local and national level in the 1990s began to make official predictive hazard maps. Many different types arose, with special maps being created for fire, flood, landslide, earthquake, tsunami, and other dangers. Then in the wake of the Great East Japan Earthquake of 2011, communities across the country began to step up their own disaster preparations. Along with the creation of new hazard forecast maps, this prompted a reexamination of historical hazard maps as well as the compilation of documents taking their lessons into account. The bulk of this work was undertaken by groups charged with preparing for disaster, aided by engineers specializing in damage-minimizing construction techniques, with citizens generally playing only the passive role of receiving the experts' recommendations.

Unfortunately, a number of difficulties emerged in trying to use the resulting maps. Because they were ultimately designed to forecast the scope of future disasters, they were inevitably presented in such a way as to exaggerate the likelihood that disasters would occur. Ordinary people found it difficult to understand this, tending to take their information at face value. Moreover, the public did not have a high degree of familiarity with probabilistic hazard maps as a genre, and doubts were raised in disaster-prone areas about the challenge of making them accessible to local residents.

It was in this context that an initiative to create a different sort of disaster-related map emerged after the Great East Japan Earthquake of 2011. The Run-and-Escape Map Project proposed a new model: one where local residents would evaluate their own residential areas and plot evacuation routes from a potential disaster.

It was the volunteer club at the architectural firm Nikken Sekkei that suggested the idea. The project was born as a result of volunteer activities launched after the earthquake by employees who were normally engaged in designing high-rise buildings. A desire to verify the best evacuation route from a tsunami—which can be expected again in the future—surfaced during a dialogue

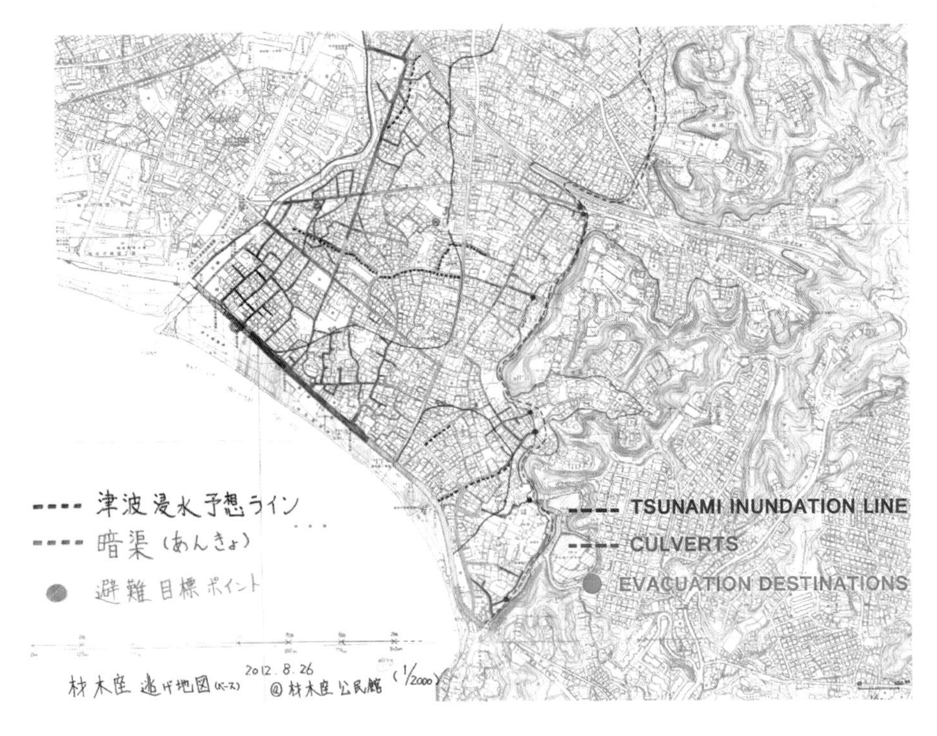

FIGURE 55.2 "Run-and-Escape Map, Zaimokuza District, Kamakura City, Kanagawa Ken" [Kanagawa ken Kamakura shi Zaimokuza chiku no nigechizu 神奈川県鎌倉市材木座地区の逃げ地図], with added English labels. Map made in August 2012 by the People/Place Kamakura Network [Hito machi Kamakura nettowāku ひと・まち・鎌倉ネットワーク]. The colors on the map and in the scale gradient refer to the minutes required to walk to an evacuation point: green, 0–3; yellow green, 3–6; yellow, 6–9; orange, 9–12; red, 12–15; purple, 15–18; black, 18 and over. Accessed online at http://www.nigechizuproject.com/?p=167. Courtesy of Nikken Activity Design Lab.

with the residents in the tsunami-affected area. That desire was materialized through the "run-and-escape" map.

The steps in creating a run-and-escape map are as follows:

1. Obtain a map of the community in question and identify the areas that sustained damage from tsunami in the past.
2. Designate evacuation sites that lie outside those areas.
3. Identify routes through which people can reach each evacuation site on foot and color-code them based on the time it takes.
4. Once all escape routes within the area are identified, clarify the shortest routes for residents of a given neighborhood, as well as hazardous spots to avoid at the time of evacuation.

Project organizers say that the idea is based on designing safe evacuation paths from large buildings, a requirement for megastructures throughout Japan. The run-and-escape map essentially scales up this procedure, substituting the setting of the disaster-stricken region for the inside of a megastructure. As a novel attempt to actively incorporate disaster preparedness measures into mapping, it constitutes a noteworthy landmark in the history of Japanese cartography.

The Run-and-Escape Map Project has characteristics and values that are not found in the usual hazard map projects described above. The first difference is the fact that the map is not created in a top-down manner, with disaster prevention experts preparing the map and residents being merely on the receiving end. Instead, local residents participate in the process, working on tasks that range from selecting evacuation sites to searching for the shortest routes by applying their own experience and knowledge of the area. They of course seek advice from disaster prevention experts when necessary. This process produces a map that can be used as safety education material while increasing awareness of disaster prevention across a given locality. In practice, residents of every age including children and the elderly have participated in the production of these maps, deepening disaster awareness dramatically.

In addition, run-and-escape map projects have another benefit: once the maps are made, local residents can readily see where the nearest safety zones are located, and how to reach those zones most effectively. Once this kind of information is compiled cartographically, it is a simple matter to mark escape routes on the landscape with street signs.

By preparing a run-and-escape map, local residents can gain new information about where safe places are within their community. Additionally, since it is safer and cheaper to designate everyday roads as evacuation routes than to build special bypass roads for evacuation, it now becomes feasible at the design stage of a new urban plan to anticipate disasters. In fact, such instances have been reported as a useful by-product of the Run-and-Escape Map Project. While the process of compiling such maps is very simple, the fact that the information is presented from the point of view of someone trying to survive a catastrophic event makes them easy for ordinary people to read and use compared to their conventional counterparts. Not only for tsunami but for other types of hazards as well, the future application of the run-and-escape process is promising, showing a value and utility distinct from those of traditional disaster cartography.

Humans must continue to live in the face of natural disasters. These events unfold at geographic scales that an individual human being simply cannot comprehend through direct experience; if we want to grasp them, at least at this stage in history, we have no method other than painstakingly gathering up widely scattered data and pulling them together in cartographic form. The daily weather forecast map may be a useful reference point. Although now so commonplace that their origins are all but forgotten, weather maps too were created by aggregating masses of local data to make sense of something as variable over space and time as the weather. Now, cartography has taken its latest turn, giving rise to a novel mapping practice with practical value for clarifying the best escape routes in the event of a natural disaster. Having demonstrated a whole new utility for maps, run-and-escape maps are sure to become an intimate part of our everyday cartographic culture in the future.

(Translated by Cactus Communications)

56 Postmortem Cartography: "Stillbirths" and the Meiji State

Fabian Drixler

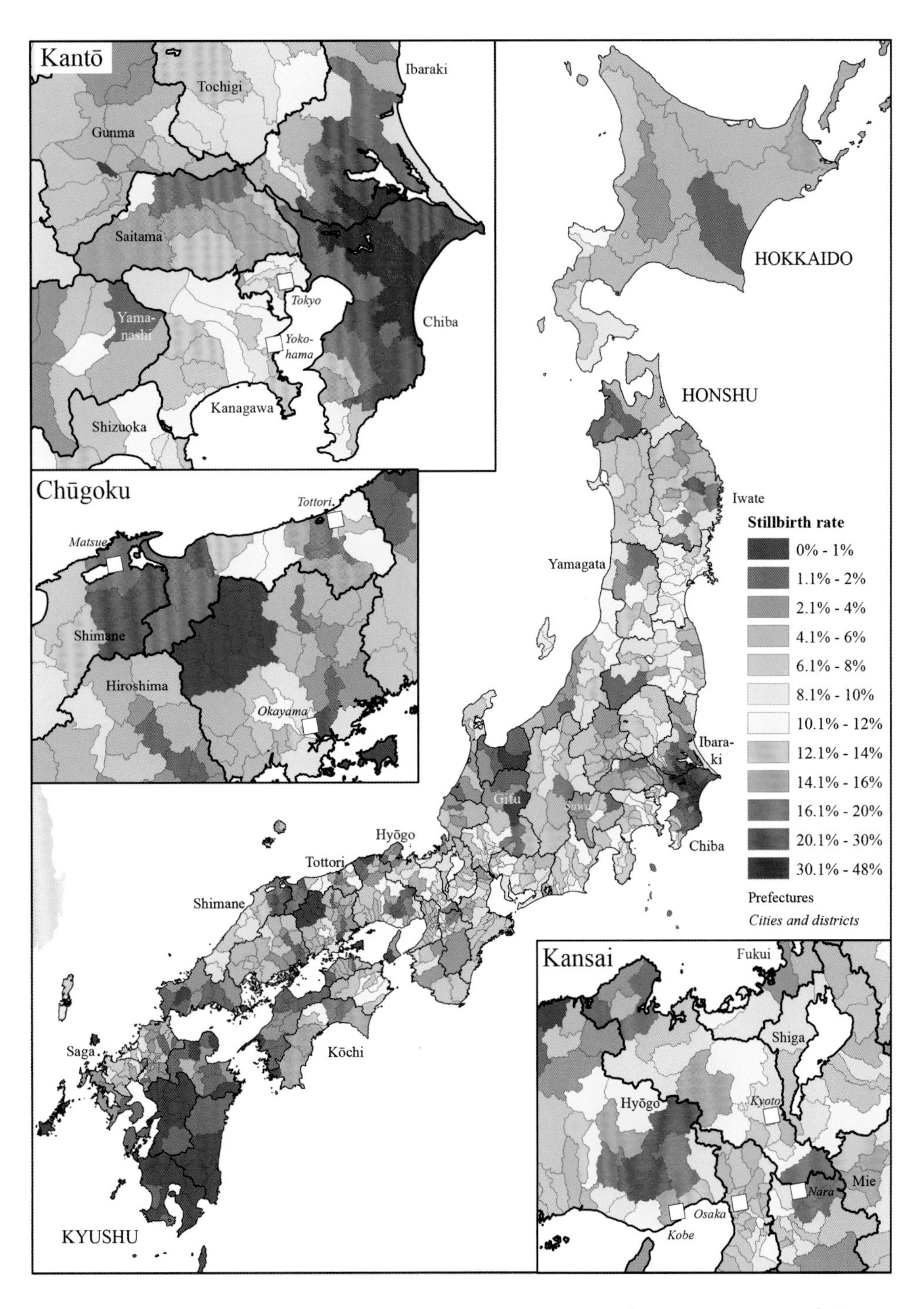

FIGURE 56.1 "Reported Stillbirths in the Districts of Japan as a Percentage of All Births, 1890–1894." © 2013 Fabian Drixler.

A map can be a powerful investigative tool, uncovering phenomena that lie hidden in innocuous national averages. Consider stillbirths. Near the beginning of the twentieth century, Japan's national rate stood above 8 percent. This was much the world's highest, but as several other countries reported about 5 percent of their children stillborn, this Japanese superlative was not necessarily suspicious. After all, diets in Japan were often meager, pregnant women often performed hard physical labor, and syphilis—a peril to unborn children—was a common affliction.

Enter computer-assisted cartography. Mapping vital statistics of the early 1890s shows that, in fact, Japan was a quilt of widely divergent stillbirth rates (fig. 56.1).[1] In regions such as southern Kyushu, anomalously few fetal deaths were reported (*dark green*). In others (*mid-green*), their incidence was similar to that in Europe at the time. But in a belt that stretched with some interruptions from northern Kyushu to southern Hokkaido, stillbirth rates were twice or thrice as high as elsewhere in the world (*light green and yellow*). Finally, in eight or nine hotspots (*red*), the statistics claimed that more than a fifth, and in one case fully 48 percent, of children came into this world without pulse or breath.

No biological explanation for these hotspots is persuasive. They were not vales of starvation, nor were they exceptional for the drudgery of their pregnant women, and their incidence of syphilis did not come close to accounting for such stillbirth rates. Instead, the hotspots document that under a modern statistical system, an old reproductive culture endured in which "thinning out" offspring was common practice. After an abortion or an infanticide, a small corpse and the end of a pregnancy needed to be explained. A stillbirth report accomplished both ends. With individual reports, it was difficult to tell truth from fiction, especially once the body of the child could no longer be examined. Taken together, the reports add up to such stupendous rates and such skewed sex ratios as to document the persistence of routine infanticide into Japan's modern age.

The map that suggests this conclusion is not a physical object curated in a museum or library. Alongside the great archive of old maps that has inspired most of the essays in this book, Japan's past has also left us a hoard of spatially specific data; with the help of geographic information systems (GIS), these figures can unfurl into a second cartographic archive of Japanese history. Numbers project apparent factual authority, but like any map, a GIS visualization of historical statistics is not an imprint of objective reality. It instead reflects the concerns of the statisticians who gathered the information and defined the categories, as well as the uses to which the governed put those categories—such as Japanese parents who found it convenient to file a stillbirth report for a child that was alive when it emerged into the world of light and air.

In the mid-Meiji period, Japan's prefectures published prodigious quantities of spatially specific statistics. Kären Wigen has called the resulting yearbooks "arguably the most important medium of geographical description" for these local governments.[2] Stillbirths first appeared in most of these yearbooks from 1886, usually tabulated by sex, legitimacy, and district. This platoon of numbers, marching quietly amid an entire field army of other figures on acres sown and inkstones mined, children schooled and silkworms bred, was not designed to attract attention. To the extent that we can take the paucity of evidence as evidence of paucity, the stillbirth statistics inspired surprisingly little comment and concern at the time. When their geographic patterns are revealed in a map, however, their implication is much harder to ignore.

Districts with frequent stillbirths were not sprinkled randomly across the Japanese landscape but formed distinct and stable clusters. The location of those clusters belies the idea that infanticide was the practice of impoverished, backward peripheries. The largest contiguous area with very high stillbirth rates lay on the very outskirts of Tokyo, an area that was certainly not noted for its poverty. The hinterlands of Kobe and Nagoya were also home to stillbirth clusters, as were the environs of Nara and the plains around Matsue on the Japan Sea coast.

Within a cluster, the stillbirth rates typically faded

as one moved away from its geographic center. In the case of the hotspot just east and north of Tokyo, a core of eight districts with stillbirth rates above 30 percent was surrounded on all sides by districts with rates above 15 percent. This geography is suggestive of the processes by which a tolerance of infanticide was perpetuated. If a couple's sense of which circumstances could justify an infanticide was shaped by the opinions of their relatives and neighbors, we would expect infanticides to cluster in space, just as those salient relationships did. A woman who lived at the edges of a stillbirth cluster would be more likely to know people who disapproved of infanticide than somebody living at its center, and she might therefore feel more constrained in her choices when giv-

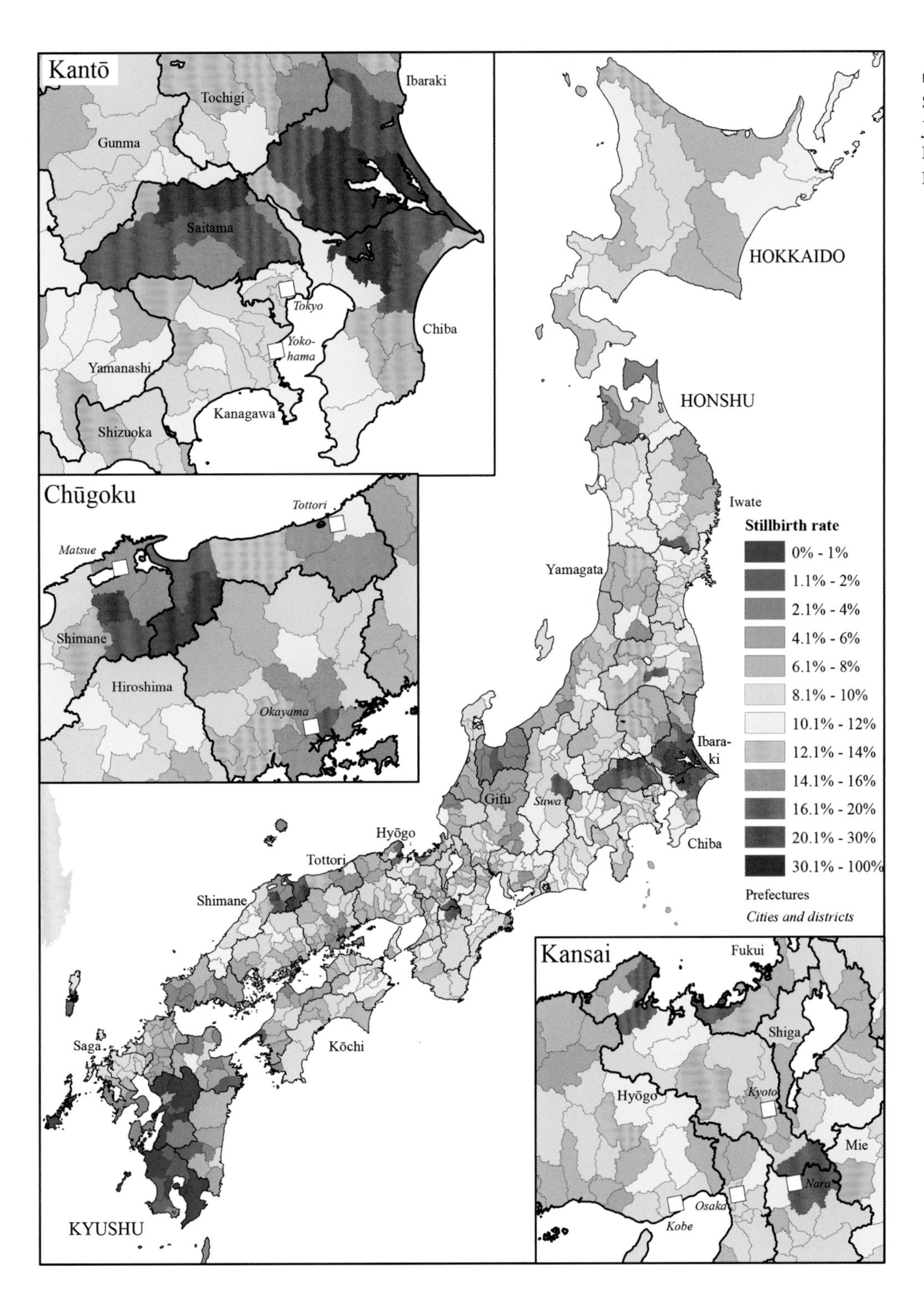

FIGURE 56.2 "Reported Stillbirths in the Districts of Japan as a Percentage of All Births, 1900–1904." © 2013 Fabian Drixler.

ing birth to another baby. The resulting geography of reported stillbirths suggests that understandings of life, death, and responsibility remained highly localized at a time when Japan was supposedly in the advanced stages of its nation-building process.

Suwa, a district in Nagano where stillbirths were between two and five times as frequent as in the neighboring districts, is a clear exception to the cluster pattern, but its peculiar geography lends support to the network theory of infanticide. Suwa was a mountain basin, and for all that a booming silk industry connected it to national networks of commerce and industry, the terrain may have isolated Suwa socially from its immediate neighbors in ways that district boundaries elsewhere often did not. For the same reason, most clusters spilled across prefectural boundaries, unless these coincided with physical features that separated communities. The cluster around Matsue, for example, faded out very gradually as one traveled east or west along the Japan Sea coast, taking no account of the boundary of Tottori and Shimane, but ended abruptly on its mountainous interior flank, where these two prefectures met Okayama and Hiroshima. With some notable changes, these patterns persisted into the twentieth century (fig. 56.2).

When the stillbirth reports that give color to our two maps were filed, the law treated abortion as a felony and infanticide as a form of homicide. That at the height of its international prestige—a time when it won hard-fought victories over Chinese and Russian armies—the Meiji state allowed tens of thousands of infanticides to occur every year puts into perspective the extent to which the modern state penetrated the private lives of individual subjects. Although the central government may well have been unaware of just how many infanticides its subjects committed, local officials often knew that newborns were killed in large numbers. It seems that those who tried to intervene remained exceptional. In the mountainous hinterland of Nara, one such man, Kuwabara Asajirō, so impressed the locals with his fervor that they, perhaps half in jest, called children that they let live under his influence "Kuwabara kids."[3] By implication, Kuwabara was an exception. As late as the 1920s, when stillbirths were reported more rarely but still in suspicious patterns, one observer described the situation in a village in Yamanashi as follows: It had no police officer, so a forestry official based in the village was in charge of law enforcement. If such an officer, whose term in the village was typically short, aspired to popularity, he would "pretend not to hear anything about the villagers' infanticides. . . . Naturally, babies were few in such years. In years when an unpopular officer resided in the village, babies would conversely be numerous." Apparently, employees of the municipality were aware of this dynamic, which they observed with "wry smiles."[4]

The Tokugawa order has been described as a "flamboyant state," uncompromising in its pronouncements but flexible in their implementation, and a regime in which the "performance" (rather than reality) of obedience maintained the Great Peace.[5] Two million feigned stillbirths between 1886 and 1940 and the maps that make them visible raise the possibility that at a local level at least, this habit of making a show of accepting the political order while quietly ignoring some of its laws was very much alive under the modern state.

Notes

1. The two maps in this chapter are based on 311 prefectural statistical yearbooks *(tōkeisho)*, published on microfilm as *Meiji nenkan fuken tōkeisho shūsei* (Tokyo: Yūshōdō Firumu Shuppan, 1963–79), with missing years supplemented with yearbooks made available on prefectural websites, as well as with *tōkeisho* published by some districts. For harvesting much of this information from hundreds of reels of microfilm, I am indebted to the cheerful meticulousness of Zhao Zixiang. Since figures do not survive for every year in every district, the maps are based on the following imputation technique: If a year was missing in a district, it was projected forward from the last year available by multiplying the district stillbirth rate in year t with the prefectural rate in year $t+1$ divided by the prefectural rate in year t. It was then projected backward from the next available year by the same logic. The two projection results were then averaged, weighted in proportion to the proximity of the year of departure to the year being estimated. District boundaries follow Mark Henderson et al., *Japan Meiji Gun (Beta)* (Cambridge, MA: G. W. Skinner Archive, 2010), http://skinner.hmdc.harvard.edu/?q=content/japan-meiji-gun-beta, corrected and modified for subsequent boundary changes. For grayscale and colorblind-proof versions of the two maps, see fabiandrixler.net.

2. Kären Wigen, *A Malleable Map: Geographies of Restoration in Central Japan, 1600–1912* (Berkeley: University of California Press, 2010), 139.

3. Onshi Zaidan Boshi Aiikukai, eds., *Nihon san'iku shūzoku shiryō shūsei* (Daiichi Hōki Shuppan, 1975), 166.

4. Ono Takeo, *Zōtei nōson shakaishi ronkō* (Ganshōdō Shoten, 1932), 101–3.

5. Philip C. Brown, *Central Authority and Local Autonomy in the Formation of Early Modern Japan* (Stanford: Stanford University Press, 1993), 25–27, 229–33; and Luke Roberts, *Performing the Great Peace: Political Space and Open Secrets in Tokugawa Japan* (Honolulu: University of Hawai'i Press, 2012), esp. pp. 3–5, 18.

Suggested Readings

Burns, Susan. "When Abortion Became a Crime." In *History and Folklore Studies in Japan*, ed. David Howell and James Baxter. Kyoto: International Research Center for Japanese Studies, 2006.

Drixler, Fabian. *Mabiki: Infanticide and Population Growth in Eastern Japan, 1660–1950*. Berkeley: University of California Press, 2013.

———. "Hidden in Plain Sight: False Stillbirths and Family Planning in Imperial Japan." Article manuscript in preparation.

Hillier, Amy, and Anne Kelly Knowles, eds. *Placing History: How Maps, Spatial Data, and GIS Are Changing Historical Scholarship*. Redlands, CA: ESRI Press, 2008.

Roberts, Luke. *Performing the Great Peace: Political Space and Open Secrets in Tokugawa Japan*. Honolulu: University of Hawai'i Press, 2012.

Taeuber, Irene. *The Population of Japan*. Princeton: Princeton University Press, 1958.

Wigen, Kären. *A Malleable Map: Geographies of Restoration in Central Japan, 1600–1912*. Berkeley: University of California Press, 2010.

57 Reconstructing Provincial Maps

Nakamura Yūsuke 中村雄祐

FIGURE 57.1 Scene of collective research on the "Genroku Provincial Map of Higo Province" [Genroku kuniezu, Higokokuzu 元禄国絵図 肥後国図]. Photograph courtesy of the Historiographical Institute, the University of Tokyo.

In 2006, an interdisciplinary research project entitled "Building Geographical Historiography" was initiated, with a focus on analyzing provincial maps drawn under the direction of the Tokugawa shogunate.[1] Over the ensuing six years, a total of thirty-three provincial maps, together with many related maps and documents, were investigated by scholars of Japanese cartography, analytical scientists specializing in cultural properties, and artists familiar with traditional techniques of Japanese painting. The members jointly examined these artifacts from multiple perspectives, looking at their bibliographic and archival characteristics, their physical properties, and the tools and techniques employed to create them. At the culmination of the project, a provincial map was reconstructed by a member of the team in order to examine experimentally the process of creating such large maps.

As explained in chapter 10 above, the provincial maps of the Edo era are exceedingly rich cultural resources. On the one hand, they are famous for their overwhelming size (over three meters on each side) and their gorgeous construction and finish. Such lavishly painted and carefully preserved artifacts effectively serve as a comprehensive catalog of techniques and materials used in Japanese painting at particular moments in the past. On the other hand, provincial maps were not just decorative artifacts but administrative documents, prepared under the strict direction of the shogunate. Careful research has revealed that complicated political negotiations between the shogun and feudal lords, as well as among neighboring lords, lay behind the colorfully drawn images of villages, mountains, and rivers.

Until recently, the unwieldy size of these provincial maps has made systematic comparative research a formidable challenge. Digital information technology has brought about a breakthrough with respect to this problem, and digital photos are now used extensively to examine these maps in detail. Scholars are rightly wary, however, of losing the most important aspects of more traditional approaches in the midst of enthusiasm for new technologies. The search for a creative combination of digital and traditional approaches in research on cultural artifacts takes time, calling as it does for close collaboration and serious discussion among experts in various fields.

With these considerations in mind, a diverse research team was formed for this recent project. Historians specializing in Japanese cartography conducted basic bibliographic and archival research on the maps under examination and scrutinized them using traditional methods including direct, minute observation with magnifying glasses. Meanwhile, art conservation experts who specialize in scientific measurement techniques (such as X-ray fluorescence spectrometry and visible reflection spectrometry) analyzed the color materials (pigments and dyes) used in the maps. Artists who work in the style of traditional Japanese painting joined the team to examine the materials and techniques used to make the provincial maps. In the final phase of the project, a team of artists reconstructed a Genroku-era provincial map of Bizen Province with a view to experimentally re-creating the actual map production process (fig. 57.1).

Whereas each member had already worked in various research settings, the project provided the first opportunity for everyone to work together at the same site. Tackling large provincial maps provided us with a unique experience of collaborative thinking. Once a map was spread out on the floor, the researchers surrounded it, and each began to carry out his or her research agenda, save for the art conservation experts, who needed separate spaces because of the technical conditions required for measurements. About five to ten researchers would cluster around each map, eagerly straining to see the points of interest while ensuring that they did not accidentally touch it. As a natural consequence of this approach, members saw what others were doing, exchanged comments and ideas, and gave assistance when needed. Conservation experts joined the discussion during breaks. The various topics discussed included the characteristics of the areas depicted, the drawing techniques and tools, the use of pinholes and traces of spatulas to apply paint in particular positions, and the methods for joining pieces of paper into irregular shapes.[2]

One of the key interests shared among the researchers was the association among the colors, materials, and symbols used in the maps. Through detailed observation of provincial and other Edo-era maps, historians have demonstrated that different colors were used systematically to express specific landscape features. Red lines, for example, were commonly used for routes, blue lines for waterways, and white squares for castles. Artists and art historians, meanwhile, knew that a variety of natural materials were used to express subtle color variations during the period. This is a significant difference from the modern age, when synthetic paints and digital pixels are believed to be able to express infinite color variations.

Accordingly, an interdisciplinary network of research activities evolved following the trajectory of the name-color-material association:

1. *Selection of points for measurement.* Historians and artists chose specific spots on the map (five to ten millimeters in diameter) for the testing of color materials based on their knowledge of the Japanese cartographic tradition. The points were chosen on the areas painted thick enough and as close as possible to the edges for safe measurement. The labels of the points were chosen following the naming traditions in Japanese art or, when unavailable, by giving modern names (fig. 57.2)
2. *Examination and estimation of color materials.* The selected points were examined by multiple methods including direct observation aided by magnet scopes, comparison with color guide slips published by a Japanese printing company, infrared photography, X-ray fluorescence spectrom-

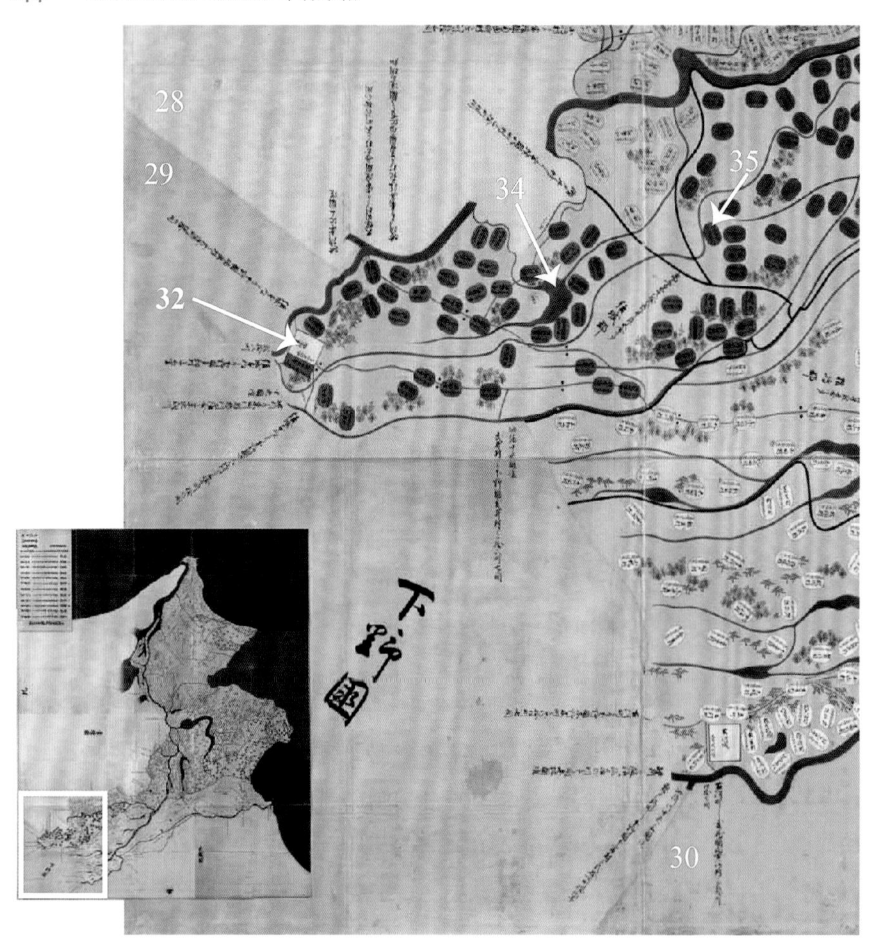

FIGURE 57.2 Examples of selected points for measurement, "Tenpō Provincial Map of Shimofusa Province" [Tenpō kuniezu, Shimofusa-koku-zu 天保国絵図 下総国図]. Courtesy of the National Archives of Japan, Digital Archive. Image reproduced by permission from *Research Report for the University of Tokyo's Historiographical Institute, 2011–2: New Developments in "Building Geographical Historiography,"* pt. 2: *Report on Original Research*, 38, 40.

etry, visible reflection spectrometry, and digital cameras. The results of the above examination were compared with each other to estimate the color materials used. While scientific analysis can detect chemical characteristics of pigments and dyes, it alone is not enough to determine the color materials, particularly when different materials are mixed. In those cases, direct observation with magnet scopes by experts in Japanese painting proved helpful.

3. *Historical interpretation of color use for map depiction.* Once the estimates of color materials were obtained, we could ask how and why they were employed on the maps. At this stage the expertise of different team members was tapped, and the discussion shed light on the identification of color materials.

Here we can see that collective thinking emerged from close examination of the provincial maps from multiple perspectives and methods. Some of the results have already been published in coauthored articles and discussed in symposia. This kind of intellectual interplay would have been impossible from simply analyzing digital images on computer displays.

As the final stage of the project, team member Arai Kei decided to re-create a provincial map to examine the actual process of map creation (see chap. 58). This could be called "experimental history," modeled after experimental archaeology. The experience was highly stimulating. As Sugimoto Fumiko, the principal investigator, memorably noted, "One of the most exciting moments in this project was when I could sit on a brand-new reconstructed provincial map." Indeed, the tactile experience of working with these large documents opened the way to new discoveries for many members of the team. In particular, the trials and errors during our attempt at reconstruction stimulated new research questions about how the original large-size provincial maps were actually produced, including the division of labor, the configuration of the workspace, and cost management. Furthermore, the team's struggles with such huge maps had the effect of putting our purportedly "traditional" techniques in sharper historical perspective. What, we asked, have we in fact kept intact from the past? What have we improvised on the basis of our ancestral tradition? What have we lost forever? Finally, we began to debate whether our methods could have broader application in the future. Might other organizations, such as local governments or schools, reconstruct provincial maps of their own areas? If this were to prove difficult, what kinds of support might be useful? Such questions open the door to further interdisciplinary collaboration with experts from an expanding range of fields.

In retrospect, the research project, which ended in 2012, unfolded in accordance with the material characteristics of the provincial maps. To some extent this was a logical consequence of the project design. But the actual development of the research process was influenced by the physical state of these unwieldy maps in a more profound sense than we had anticipated. The systematic, collective attention to the material aspects of historical maps has expanded the horizons of the traditional approach significantly. Through the scientific measurement and re-creation of a provincial map, researchers were able to reexperience the way material conditions and human skills conditioned visual expressions in the Edo period. Sugimoto states that these findings offer invaluable insights for future research into historical maps. Our experience encourages us to reconsider the importance of materiality in an age when "big data" and computational analysis often seem to trump human beings' analytical performance in information space.

However, this attention to materiality should not be seen as negating the importance of digital technologies. We have been using them extensively with this new approach. In fact the project has left us with the formidable challenge of systematically archiving a great body of data most of which are digitally formatted. We also need to remember the cautionary experience of how easily the "cutting-edge" technology of one era rapidly becomes obsolete, leaving a legacy of unusable data. Therefore, the appropriate application of digital technologies is truly within the scope of our collective endeavor. We warmly welcome the participation of history-minded digital experts to our research network.

Notes

1. The project was sponsored by a Japan Society for Promotion of Science grant-in-aid for scientific research and led by Dr. Sugimoto Fumiko as principal investigator.

2. Experienced researchers, by meticulously examining the physical condition of old maps, are sometimes able to recognize traces of past actions such as making holes and pressing spatulas onto the paper before painting and folding.

58 The Art of Making Oversize Graphic Maps

Arai Kei 荒井経

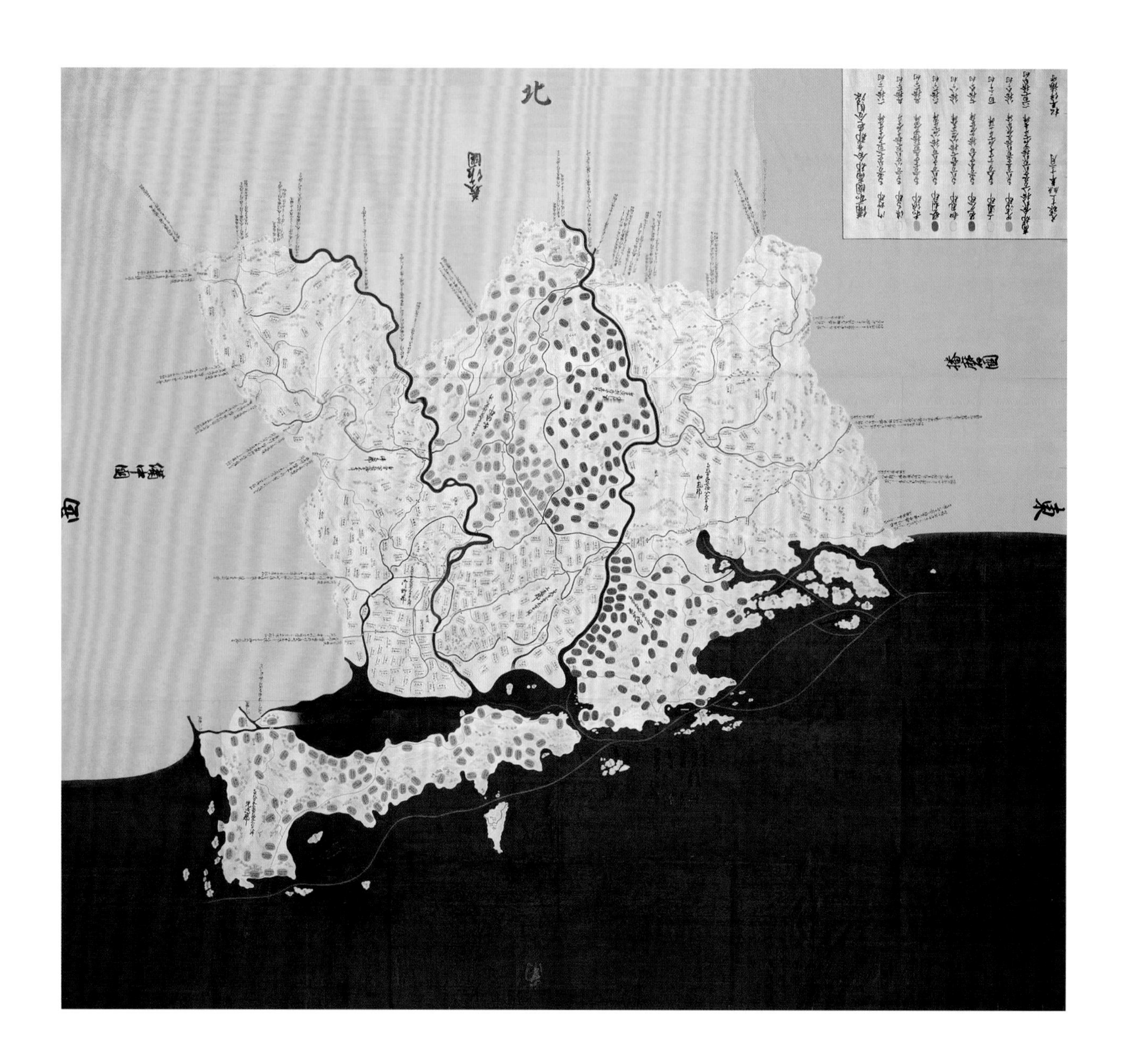

FIGURE 58.1 "Reconstructed Genroku Bizen Province Map" [Fukugenzu Genroku Bizen kuniezu 復元図・元禄備前国絵図], completed 2010. Manuscript, 316 × 348 cm. Based on "Genroku Bizen Provincial Map" [Genroku Bizen kuniezu 元禄備前国絵図], ca. 1700s. Courtesy of Okayama University Library.

In September of 2010 I completed work on an oversized map of Bizen Province (present-day Okayama Prefecture) with a group of graduate students in the Graduate School of Fine Arts at Tokyo University of the Arts (Tokyo Geijutsu Daigaku).[1] Measuring over 10× 11 feet (316 ×348 cm), our model was one of many such giant maps requisitioned by the Tokugawa shogunate at the end of the Genroku period (1688–1704) in the first few years of the eighteenth century.[2] Over three hundred years later, our team took up the challenge of reconstructing the map at its

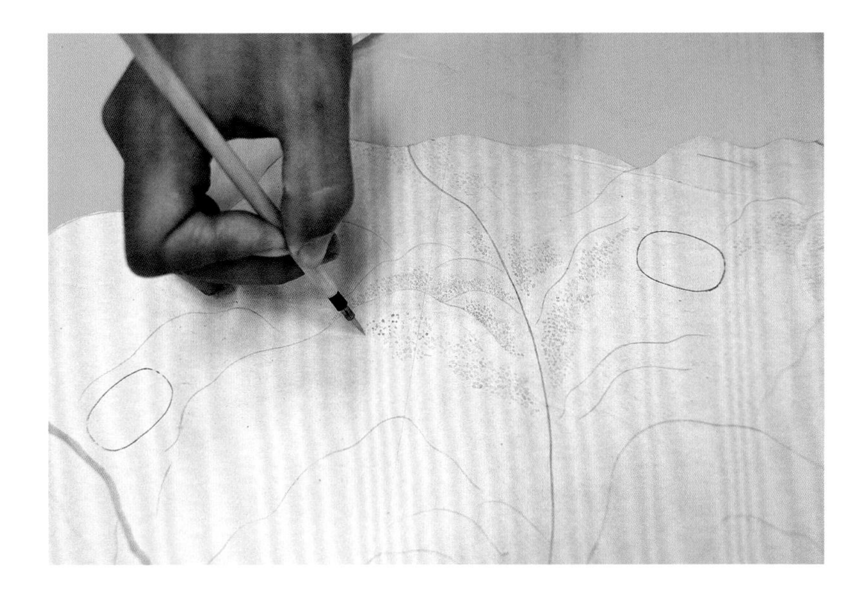

FIGURE 58.2 Detail showing dotting technique for "Reconstructed Genroku Bizen Province Map." Photograph by the author.

original size, using the same techniques and materials originally used (fig. 58.1). Like myself, the graduate students who carried out the day-to-day work on the project specialize in the practice of *nihonga*, or Japanese-style painting. This was crucial to the success of the project, since *nihonga* techniques were originally used to make this kind of map.

The political transformation brought about by the Meiji Restoration in 1868 ushered in major changes in all areas of everyday life and learning in Japan due to the introduction of modern ideas and practices from the West. At the same time, however, Japan's traditional culture underwent its own modernization. A double culture was the result, with Western and Japanese clothing or food, for example, coexisting but remaining distinct from each other. So it was in the case of painting too, with *nihonga* developing in parallel with *yōga*, or Western-style painting. Furthermore, after 1868, a new Western-inspired system of exhibitions and institutional education incorporated the practice of *nihonga*, and under those circumstances it has endured as a distinct form of modern Japanese painting to the present day. In Japan today, students who aspire to become painters must choose between *nihonga* and *yōga* upon taking entrance exams for art schools. Most students who opt for *nihonga* will ultimately show their work in exhibitions dedicated to *nihonga* and sell their work in a market dedicated to this art form.

Nihonga artists continue to employ the same materials and related techniques utilized to make eighteenth-century province maps, such as Chinese ink (*sumi*), natural mineral pigments, soft brushes, handmade paper, animal glue, and wheat starch paste.[3] Be that as it may, our reconstruction project for the Bizen map did not merely entail making a facsimile replica using period techniques. It was also necessary for us, as contemporary practitioners of *nihonga*, to account for a rupture brought about in Japanese history by modernization. This exercise in "experimental history," whereby we were to re-create from scratch a historically accurate artifact, required us to address the cultural gaps in our own understanding of maps and mapmaking.

Working primarily with a team of four doctoral students, we finished the project in forty days, working seven hours a day.[4] The most difficult problem we encountered was the time-consuming dotting technique necessary for painting the landscape of the region depicted on the map (fig. 58.2). Using continuous solid colors to represent the space of Bizen Province proved simple enough. The challenge arose with the depiction of mountains and trees in a homogeneous landscape without recourse to perspectival techniques. Each mountain and tree needed to be rendered minutely at a diameter of one to two millimeters. To accomplish this required 250 hours of work in a nine-day period. While the painstaking process of using tiny marks to represent thousands of mountains and trees was joyless work for the graduate students, we learned an important lesson by persevering with this historically accurate technique.

The particular Japanese word that comes to my mind

when thinking about this technique is labor (*tema*), which is composed of the characters for "hand" and "time." Work involving the hands and requiring copious amounts of time is typically associated with industrial arts such as welding or crafts like woodworking. In order to make our map, a similar kind of labor was required. We learned that the most refined techniques of *nihonga* were not needed to depict landscape elements in a historically accurate way, since an actual view of topography was not evidenced in the original map. The main requirement instead was simply the execution of thousands of tiny brush marks on a huge surface. The doctoral students grew weary of having to expend so much seemingly wasted labor on a landscape without engaging in any craftsmanship. I tried telling them that they should not feel upset, since if nothing else they were going to be paid the hourly wage we agreed on no matter how long it took to mark all the dots. There was no need for them to rush. But this did not raise their spirits, so I decided to hire a group of master's-degree students to make the remaining marks for the landscape portions of the map.

Ever since completing our map I have been thinking about the role of labor in painting. If modern maps had been our standard, I could have reduced the total cost of the project by 20 percent without compromising the map's cartographic functionality simply by eliminating the dotting technique. From this same modern point of view, the luxuriant paper and rich color pigments we used seemed to have no bearing on the map's functionality either, so in theory they also seemed expendable. The use of the simplest of *nihonga* techniques and such valuable materials amounted to little more than excessive ornamentation. The map seemed to have been made to achieve a certain appearance. Together with whatever objectives people at the time may have had for making the map, it looked to us as though the map had value insofar as it was an expression of expended labor.

Labor and craft are not necessarily equivalent. Craft does not always require labor, and labor is a wasteful extravagance if unaccompanied by craft, as my four doctoral students experienced firsthand. And yet since the cost of labor is human effort, the value of labor (and the objects that result from it) can be measured quantitatively, nowadays usually in the form of money. Perhaps, then, the dotting technique used in the Bizen map had value after all, since it represented, in abstract terms, the labor required to produce the object. So many tiny brush strokes allowed one to gauge the human cost of making the map.

Painting in the modern world is a private matter that does not take place in the context of duty to royalty or religion, as it did in the past. Rather, it is premised on amateurism. Somerset Maugham's novel *The Moon and Sixpence*, whose main character is modeled on Paul Gauguin (1848–1903), presents the image of the artist holding fast to his autonomy while living a life of poverty. This reminded me of my doctoral students. In reconstructing the map, these students of *nihonga* disdained the dotting technique regardless of any monetary value associated with it. They are indeed the children of modern art.

Before the advent of the modern Japanese state in 1868, painters in Japan worked in a vocation that guaranteed them a fixed income from those they served, whether this was the central government in Edo or daimyo in domainal governments. Such painters made provincial maps like the Bizen map. They did not construct it as a representation of their own individuality as autonomous painters, but rather as an offering of the work required of them to paint it. In other words, they engaged in making presentations, not simply of painted objects, but of the painstaking effort it took to execute the order to paint the map. The countless dots in the enormous Bizen map are nothing more or less than the traces of labor expended by painters whose anonymous service means that their names were not preserved for posterity.

(Translated by Robert Goree)

Notes

1. The reconstruction of the map was linked to the following project, which was carried out in 2010: "New Developments in Building Geographical Historiography" (JSPSK KAKENHI grant no. 21242018, principal investigator Sugimoto Fumiko, the Historiographical Institute, University of Tokyo).

2. Okayama University Library's Ikeda Collection (T1-20-1).

3. See Sugimoto Fumiko's introduction to part I of the present volume for details about the production and distribution of pigments and paper during the Edo period.

4. Most of the work was carried out by four doctoral students at Tokyo University of the Arts, who did the tracing, coloring, and shading (Koga Umito, Takeda Hiroko, Peng Weixin, and Miyako Emi); six master's students, who did the dotting (Igarashi Yuki, Ueda Mari, Suzuki Hiroo, Han Myongook, Hirao Rina, and Yasuhara Shigemi); and three students majoring in calligraphy at Tsukuba

University, who copied the calligraphic text (Nakamura Yumiko, Anjo Narumi, and Baba Ayaka). For more details, see our blog (in Japanese): http://maruta.be/kuniezu.

Suggested Reading

Tokyo University of the Arts. *An Illustrated Dictionary of Japanese-Style Painting Terminology*. Tokyo: Tokyo Bijutsu, 2010.

Epilogue

SUGIMOTO Fumiko 杉本史子

What was the reaction of people in the past who first came into contact with maps that showed extensive lands beyond the sphere of daily life? Maybe they were surprised and bewildered by the representation of distances they didn't experience in everyday life and the new perspective this gave them as they looked down on the places where they were actually standing.

COLLEGE FRESHMAN, literature major

My sense is that it was indispensable to record on paper an objective construction for conveying the verifiable existence of things people couldn't actually see.

COLLEGE FRESHMAN, pharmacology pre-med major

By making maps according to consistent surveying methods and uniform scale, it became possible for people inside as well as outside Japan to view maps with the same gaze. Scientific progress is significant not only for making life more convenient, but also for introducing shared ways of seeing.

COLLEGE FRESHMAN, pharmacology major

I went to England for a home stay several weeks after 3/11 [the Great East Japan Earthquake]. While there I saw a map on the TV news that showed the magnitude of the earthquake and the spread of radiation, which made it easy for those around me in England to imagine the disastrous circumstances into which Japan had fallen. Based on such accurate information, the high school I attended raised funds to send to Japan. I realized that maps can make accurate and easy-to-understand information available to people in other places because of their simple shapes (i.e., by conveying forms and circumstances with points, lines, and planes), not as objects to be used only in daily life but also in facing global crises. I think that when it comes to depicting space, the two dimensions of maps are much easier to understand than words.

COLLEGE FRESHMAN, literature major

(Excerpts from essays written by Keiō University students in spring 2013)

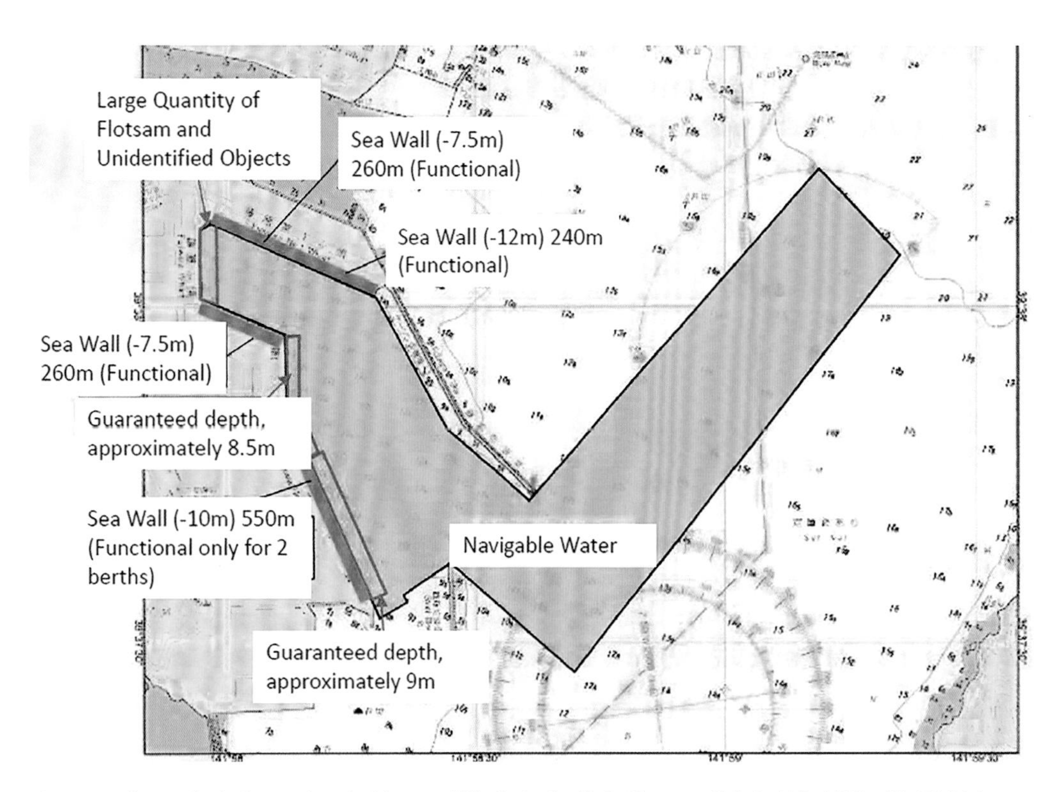

Figure 59.1. "Map of Miyako Bay (Navigable Water)" [Miyako-kō (kōkai kanō ryōiki) 宮古港（航海可能領域）], by Kokudo Kōtsūshō Kōwankyoku (Port and Harbor Bureau, Ministry of Land, Infrastructure, Transport and Tourism) and Kaijō Hoanchō (Japan Coast Guard), posted to the web on March 17, 2011. Accessed online at http://www.kaiho.mlit.go.jp/info/kouhou/h23/k20110317/k110317-2.pdf. English labels added by author.

Some four centuries after maps began to take off in early modern Japan, an earthquake of unprecedented size struck northern Honshu at 2:46 p.m. on March 11, 2011, precipitating a tsunami that would destroy coastal towns in the Tohoku region as well as the nuclear power plant at Fukushima. As we have seen, maps played a pivotal role in documenting the damage, as well as in relief and rescue efforts. Prominent among the thousands of cartographic documents to emerge from this disaster was an emergency survey along the damaged Pacific coastline, conducted from March 15 to March 26 by the Japan Coast Guard, designed to quickly assess the degree to which protective sea walls remained functional.

We conclude this volume with a sea map of Iwate Prefecture's Miyako Bay, based in part on the results of that survey (fig. 59.1).[1] Using a base map made before the disaster, the image published on the Japan Coast Guard website on March 17, 2011, now showed viable navigation routes in real time. The emergency surveys supplying the new information dealt only with sea routes and seawalls and did not account for the circumstances of harbor facilities or the coastline. Just as in the predisaster version, yellow indicated land while blue indicated water depths of 5 meters, where it was not safe for large vessels to pass. The updated image also revealed where flotsam and unidentified objects were accumulating in the bay. One section of the seawall was now rated to handle surges of

8.5–9 meters, not 10 meters, as it had before the tsunami struck. The indications on the map told a story of grave damage.

But the map also attempted to indicate with pale orange the areas that had been determined by emergency surveys to be safe for ship navigation. And it indicated in dark orange those undamaged sections of the sea wall that had been confirmed. Using this new information, the enormous oil tanker and dredger *Hakusan* was able to navigate the bay and deliver relief supplies. At the height of danger, the map's simple shapes—points, lines, and planes—revealed the valuable functionality of maps, conveying a sense of powerful reassurance by aggregating information in real time to the extent possible for the sake of rescuing people and helping them survive.

For cartographic historians whose relationship to maps is often remote, this recent naval survey serves as a jarring reminder that our very lives have come to depend on maps. Cartography has never been an abstract or disinterested enterprise; what is at stake in the case of disaster maps like this one may be nothing less than life itself. If this seems a fitting place to end our journey through Japanese cartographic history, it is because of the compelling way it testifies to the continuing centrality of maps in an uncertain world. We may not be able to predict precisely what form Japanese maps will take in the future, but we can be confident that they are not going away any time soon.

Note

1. This map was brought to my attention by Imai Kenzō of the Ippan Zaidanhōjin Nihon Suiro Kyōkai.

Acknowledgments

The idea for this project was sparked by the 2011 publication of *Mapping Latin America: A Reader* (University of Chicago Press). The editors gratefully thank Jordana Dym, coeditor of that volume, for her encouragement and support. In addition, we are glad to have a chance to publicly thank Nancy Hamilton for her cheerful and sharp-eyed editorial help; David Rumsey for his generous financial support; three anonymous reviewers for their astute suggestions; the many public institutions and private collectors who graciously shared their maps with us; and the outstanding staff at the University of Chicago Press, especially Mary Laur, Logan Smith, and Erik Carlson, for transforming the manuscript into the book you hold. Finally, the editors express our tremendous gratitude to Sayoko Sakakibara. Her organizational skills and commitment to the project, all while carrying and then caring for her newborn daughter Maia-Paulina, helped bring this book into being.

About the Authors

ARAI Kei is associate professor at Tokyo University of the Arts, specializing in conservation techniques in Japanese painting. He is simultaneously an active artist in the traditional *nihonga* style and a scholar of its materials and techniques, having studied contemporary Japanese painting at Tsukuba University before undertaking formal graduate training in *nihonga* techniques and conservation at Tokyo University of the Arts. His dissertation, "From the Techniques of the Kano School to the Techniques of Modern Japanese Paintings," explores the influence of nineteenth-century European synthetic pigments on modern Japanese painting. Arai has since published numerous articles on Japanese painting materials and techniques and was a contributor to the English-language *Illustrated Dictionary of Japanese-Style Painting Terminology* (Tokyo Bijutsu, 2010).

Andrew BERNSTEIN is associate professor of history at Lewis and Clark College, specializing in the religious and environmental history of Japan. His first book, *Modern Passings: Death Rites, Politics, and Social Change in Imperial Japan* (University of Hawai'i Press, 2006), was awarded the Oregon Book Award for General Nonfiction in 2006. He is currently writing a comprehensive "biography" of Mount Fuji that treats the celebrity volcano as an actor in, and product of, both the physical world and the human imagination.

Mary Elizabeth BERRY, Class of 1944 Professor of History at the University of California, Berkeley, is the author of *Hideyoshi* (Harvard University Press, 1982), *The Culture of Civil War in Kyoto* (University of California Press, 1994), and *Japan in Print: Information and Nation in the Early Modern Period* (University of California Press, 2006). She is now working on a book that tries to answer the question: How did Japan make the revolutionary leap from a manorial economy to a market economy in the seventeenth century? She has served as president of the Association for Asian Studies and is a member of the American Academy of Arts and Sciences.

Theodore C. Bestor is Reischauer Institute Professor of Social Anthropology at Harvard University. His book *Tsukiji: The Fish Market at the Center of the World* (University of California Press, 2004; published in Japanese in 2007) is based on extensive fieldwork at Tokyo's vast Tsukiji wholesale market, the world's largest marketplace for seafood and the center of Japan's sushi trade. As Tsukiji prepares to move to a new site, he is continuing research on the market's end and the rebuilding of the area in preparation for the 2020 Tokyo Olympics. Currently, he is also examining the global popularity of Japanese food, and the recent UNESCO designation of *washoku* (traditional Japanese cuisine) as a Global Intangible Cultural Heritage item. Bestor served as president of the Association for Asian Studies (AAS) during 2012–13.

Daniel Botsman is professor of history and East Asian Studies at Yale University. A graduate of the Australian National University, he received his PhD from Princeton in 1999. His publications include a translation of the autobiography of the postwar Japanese foreign minister, Okita Saburō, and *Punishment and Power in the Making of Modern Japan* (Princeton University Press, 2005). His current book project considers the impact of nineteenth-century liberalism on Japanese society with particular reference to the experiences of outcaste communities, prostitutes, and other marginal social groups.

Philip C. Brown is professor of history at the Ohio State University, Columbus, Ohio. A graduate of the University of Pennsylvania, he is a specialist in early modern and modern Japanese history and focuses on developments affecting rural Japan. He is author of *Central Authority and Local Autonomy in the Formation of Early Modern Japan: The Case of Kaga Domain* (Stanford University Press, 1993) and *Cultivating Commons: Joint Ownership of Arable Land in Early Modern Japan* (University of Hawai'i Press, 2011). His current research examines Japan's changing response to flood and landslide risk in the nineteenth and twentieth centuries. Combining study of both technological and environmental history, his research for this project extends to a case study of Japanese civil engineering in Taiwan.

Fabian Drixler is associate professor of history at Yale University. In *Mabiki: Infanticide and Population Growth in Eastern Japan, 1660–1950* (University of California Press, 2013), he argues that changing images, metaphors, and worldviews brought about a reverse fertility transition in lands where people once raised only a few children. He is currently trying to understand how communities responded to climatic shocks during the eighteenth and nineteenth centuries, and why starvation was such a prominent feature of the Tokugawa order. He is also gathering materials to reconstruct Japan's seventeenth-century demography to find out if and how the Tokugawa realm transitioned from growth to stability.

David Fedman is assistant professor of East Asian history at the University of California, Irvine, specializing in the environmental history and historical geography of Japanese imperialism. His publications on the history of cartography include "Triangulating Chōsen: Maps, Mapmaking, and the Land Survey in Colonial Korea" (*Cross-Currents: East Asian History and Culture Review*, 2012); "A Cartographic Fade to Black: Mapping the Destruction of Urban Japan During WWII" (*Journal of Historical Geography*, 2012); and "Mapping Armageddon: The Cartography of Ruin in Occupied Japan" (*The Portolan* 92, Spring 2015). His dissertation concerns the history of forestry and natural resource management in colonial Korea. Fedman has also studied the history of Japanese alpinism and environmental ethics as a Fulbright Fellow in Hokkaido.

Nicolas Fiévé, professor at the École Pratique des Hautes Études in Paris, teaches the history of premodern Japanese urbanism, architecture, and gardens. He is presently director of the East Asian Civilizations Research Center in Paris (http://www.crcao.fr). Major publications include *L'architecture et la ville du Japon ancien* (Maisonneuve et Larose, 1996); *Historical Atlas of Kyōto: Spatial Analysis of the Memory Systems of a City* (UNESCO/L'Amateur, 2008); with Paul Waley, *Japanese Capitals in Historical Perspective: Power, Memory and Place in Kyōto, Edo and Tokyo* (Routledge-Curzon, 2003); and with B. Jacquet, *Vers une modernité architecturale et paysagère: Modèles et savoirs partagés entre le Japon et le monde occidental* (Collège de France, 2013).

Alisa FREEDMAN is associate professor of Japanese literature and film at the University of Oregon. Her major publications include *Tokyo in Transit: Japanese Culture on the Rails and Road* (Stanford University Press, 2011), an annotated translation of Nobel Prize winner Kawabata Yasunari's *The Scarlet Gang of Asakusa* (University of California Press, 2005), and the coedited volume *Modern Girls on the Go: Gender, Mobility, and Labor in Japan* (Stanford University Press, 2013). She has authored articles and edited collections on Japanese versions of international children's culture, modernism, urban studies, youth fashions, media discourses about gender norms, television history, humor as social critique, and intersections of literature and digital media. She has also translated Japanese novels and short stories.

Robert GOREE is assistant professor of Japanese at Wellesley College, whose teaching and research interests fan out from prose and poetry to visual art, print culture, cultural history, geography, film, and translation. His forthcoming book, *Illustrated Gazetteers and the Mapping of Culture in Early Modern Japan*, explores the history and enduring cultural significance of illustrated guidebooks published in Japan during the eighteenth and nineteenth centuries. Goree has taught at Harvard, Columbia, and Boston University, and holds a PhD in East Asian languages and literatures from Yale University.

Carola HEIN is professor of architectural history and urban planning at the Delft University of Technology. She has published widely on topics in contemporary and historical architectural and urban planning in Europe and Japan. A Guggenheim Fellowship supported her research on the global architecture of oil, while an Alexander von Humboldt fellowship allowed her to investigate large-scale urban transformation in Hamburg between 1842 and 2008. Her current research interests include the transmission of architectural and urban ideas as part of international networks, focusing specifically on as part of port cities and landscape of oil. She serves as IHPS editor for the Americas for the journal *Planning Perspectives* and as book review editor for Asia for the *Journal of Urban History*. Her books include *Port Cities: Dynamic Landscapes and Global Networks* (Routledge, 2011), *Rebuilding Urban Japan after 1945* (Palgrave Macmillan, 2003), and *Cities, Autonomy and Decentralisation in Japan* (Taylor and Francis, 2006).

Todd A. HENRY is associate professor of modern Korean and East Asian history at the University of California at San Diego, where he also serves as an affiliate faculty member of the Program in Critical Gender Studies. In addition to articles published in the *Journal of Asian Studies, Positions: Asia Critique, Asea yŏn'gu (Journal of Asiatic Studies),* and *Sitings: Critical Approaches to Korean Geography*, he is the author of *Assimilating Seoul: Japanese Rule and the Politics of Public Space in Colonial Korea, 1910–1945* (University of California Press, 2014). He is currently working on a comparative and transnational study of contemporary queer Korea with a focus on sexualized labor, colonial/military occupation, and the entertainment industry. His research has been supported by grants from the Fulbright Foundation and the Korea Foundation.

Cary KARACAS, associate professor in the Department of Political Science and Global Affairs at the College of Staten Island, City University of New York, is a cultural geographer who specializes in urban Japan. His research focuses on the civilian experience of aerial bombing during wartime, and how memories of catastrophic loss are inscribed upon the urban landscape. He is co-creator of JapanAirRaids.org, an online digital archive dedicated to the dissemination of primary documents, analysis, and remembrances related to the firebombing of urban Japan. His current book projects include *Voices from the Ashes: The Firebombing of Tokyo in Memory and History*, which will feature translated accounts of civilians who experienced the major air raids on Japan's capital in 1945, and *Tokyo at War: The Unmaking of a Metropolis*, which examines the profound changes that the city underwent between 1936 and 1945.

KOMEIE Taisaku is a historical geographer working as associate professor in the Department of Geography, Graduate School of Letters, Kyoto University. His research topics includes two major concerns: (i) environment-human relationship in Japanese mountain areas from medieval to modern times, especially focusing on forestry and swidden agriculture; and (ii) Japanese representation and experience of geographies in the modern colonies of the Japanese empire in the early twentieth century, especially focusing on forestry, heritage, and tourism in Korea and northeastern China. Early articles on the first topic were collected in *Chū-kinsei sanson no keikan to kōzō* (Structure and landscape of medieval and early modern mountain settlements; Azekura Shobō, 2002), based on his PhD dissertation at Kyoto University. His research experience in the historical geography and historical resources of Japan are also represented in his contributions to *Seto-shi shi* (A history of Seto City; 2003–10) and *A Landscape History of Japan* (Kyoto University Press, 2010).

Joseph Loh is director of public programs and engagement at Bard Graduate Center in New York. He holds a PhD in art history and archaeology from Columbia University. His 2012 dissertation explored folding screens that feature painted images of Western European maps of the world, produced during the 1540s–1640s, when Japan encountered the West for the first time. He has had fellowships at the University of Tokyo and the British Library and has held positions at Columbia University, the University of British Columbia, the University of Oregon, and the Metropolitan Museum of Art, New York.

Matsui Yoko is professor in the Section for Overseas Materials of the Historiographical Institute, University of Tokyo, specializing in the relationships between Japan and foreign countries in the early modern period. Engaged in compiling and translating the diaries kept by the heads of the Dutch factory in Japan, she has been studying Japan's relationship with the Dutch. Her recent research focuses on the people who had direct contact with the foreigners in the port city of Nagasaki, such as Dutch residents in Deshima, interpreters, suppliers, prostitutes, and other women. She is also interested in the descriptions of early modern Japan by such European observers as Kaempfer, Titsingh, and Siebold. She is the author of *Kenperu to Siboruto: "Sakoku" Nihon o katatta ikokujin tachi* (Kaempfer and Siebold: Foreigners who narrated Japan under "Sakoku"; Yamakawa Shuppansha, 2010).

Miyazaki Fumiko is professor emerita at Keisen University. She is interested in analyzing religious and cultural developments in premodern Japan in their social, political, and economic context. She has published articles on various aspects of the cult of Mount Fuji, including the aspiration for world renewal developed by the believers, their criticism of the predominance of men over women in society, and their attempts to abolish the exclusion of women from holy places based on traditional taboos. She is also interested in other aspects of the history of religion in Tokugawa Japan, including developments in popular pilgrimage sites and governmental control over unauthorized religious associations and movements.

D. Max Moerman is professor of Asian and Middle Eastern cultures at Barnard College, Columbia University. His research interests lie in the visual and material culture of Japanese Buddhism. He is the author of *Localizing Paradise: Kumano Pilgrimage and the Religious Landscape of Premodern Japan* (Harvard University Asia Center, 2004), "Demonology and Eroticism: Islands of Women in the Japanese Buddhist Imagination" (*Japanese Journal of Religious Studies*, 2009), and the forthcoming *Japanese Buddhism and the World Map: Cartography, Cosmology, and the Epistemology of Vision*.

Tessa Morris-Suzuki is Distinguished Professor of Japanese History and holds an Australian Research Council Laureate Fellowship at the Australian National University. Her research focuses on aspects of modern Japanese and East Asian regional history, particularly cross-border movement between Japan and its Asian neighbors; issues of history, memory, and reconciliation in Northeast Asia; grassroots social movements in Japan; and the modern history of ethnic minorities and frontier communities in Japan. Her most recent works include *Exodus to North Korea: Shadows from Japan's Cold War* (Rowman and Littlefield, 2007); *To the Diamond Mountains: A Hundred Year Journey through China and Korea* (Rowman and Littlefield, 2010); *Borderline Japan: Foreigners and Frontier Controls in the Postwar Era* (Cambridge University Press, 2010); and the coedited *East Asia beyond the History Wars: Confronting the Ghosts of War* (Routledge, 2013). In 2013 she was awarded the Fukuoka Prize for contributions to the study of East Asia.

Nakamura Yūsuke is professor in the Department of Cultural Resources Studies at the University of Tokyo, specializing in the study of literacy and cognitive artifacts. He earned his PhD from the Graduate School of Arts and Sciences of the University of Tokyo in 1995 with a dissertation on the transformation of oral history in West Sudan (the present-day Republic of Mali). Since then he has participated in various development projects in Latin America (Bolivia, Guatemala, and elsewhere) and has conducted field research on the use of documents and other cognitive artifacts in those projects. He has also joined interdisciplinary academic research projects on historical documents such as cadastral maps in Latin America and provincial maps in early modern Japan. Recently, he has been interested in the impact of information and communication technology in both development projects and humanities research, in collaboration with computer scientists.

Oka Mihoko is assistant professor in the Historiographical Institute at the University of Tokyo. Her specialty is the socioeconomic and maritime history of Japan during the period of exchange with Portugal and Spain during the sixteenth and seventeenth centuries. Recently she has published on Japanese Christians and foreign missionaries, focusing on the diplomacy of Nobunaga, Hideyoshi, and early Tokugawa shoguns. Her PhD dissertation (Kyoto University, 2006) was published as *Shōnin to senkyōshi: Nanban bōeki no sekai* (Nanban trade: The merchants and the missionaries; University of Tokyo Press, 2010). Her current project is on the mercantile society of Nagasaki, looking at both the domestic and the international commercial networks of merchant families.

Catherine L. Phipps is associate professor of East Asian history at the University of Memphis. She received her PhD from Duke University in 2006 and specializes in the history of modern Japan, maritime empires, and historical geography. Her first book, *Empires on the Waterfront: Japan's Ports and Power, 1858–1899* (Harvard University Asia Center, 2015), analyzes Japan's foreign trade ports and their networks to reveal how local citizens and the government alike navigated the limits, dangers, and opportunities of informal empire in East Asia. Her current research projects examine Japan's coastal security, policing, and smuggling in the nineteenth century, and the physical, material, and cultural impact of Japan's empire on the metropolis of Osaka.

Sayoko Sakakibara is lecturer in the Department of History at Stanford University, specializing in the religio-political history of premodern Japan. After earning her first PhD from Osaka University of Foreign Studies and working as a postdoctoral fellow at the Historiographical Institute, the University of Tokyo, she acquired a second PhD at Stanford. Her 2012 dissertation, "Domesticating Prince Shōtoku: Tokugawa Sacred Geography and the Construction of a National Landscape," redefines the concept of the state in the Japanese historical context by focusing on the development of the national cult of Prince Shōtoku in the late medieval and early modern periods. Her current project involves interrogating religious texts from the Tokugawa period to understand the geography of religious networks within the early modern national landscape. Her recent publications include contributions to *Taishi shinkō to Tenjin shinkō: Shinkō to hyōgen no isō* (Shōtoku cult and Tenjin cult: The phases of worships and narratives; Shibunkaku Shuppan, 2010) and *Kōsaku suru chi: Ishō, shinkō, josei* (Checkered intelligence: Attire, cults, and gender; Shibunkaku Shuppan, 2014).

Satoh Ken'ichi is associate professor at the University of Electro-Communications in Tokyo, specializing in the history of early modern sociology, science, and technology. He has investigated the state of archival preservation for materials related to Edo-era mathematics in various regions of the country and has analyzed treatises by the Japanese mathematician Seki Takakazu. Since the devastating Great East Japan Earthquake and tsunami of 2011, he has been compiling historical sources related to natural hazards in Japan. Satoh is chief editor of *Kagakushi kenkyū* (The journal of the History of Science Society of Japan). His publications include "Kinsei Nihon sūgakushi" (The history of premodern Japanese mathematics; *Kagakushi kenkyū*, 2012), among others.

J. Charles Schencking is professor of history at the University of Hong Kong. He has published widely on modern Japanese history, the history of natural disasters, and war, state, and society in Japan. His most recent book, *The Great Kantō Earthquake and the Chimera of National Reconstruction in Japan* (Columbia University Press, 2013), explores how people experienced, interpreted, and attempted to use the Great Kantō Earthquake of 1923 not only to rebuild Tokyo so as to reflect a new urban modernity, but also to reconstruct society. His current research project—"America's Tsunami of Aid: Charity, Opportunism, and Delusion following Japan's Great Kantō Earthquake"—investigates the political, social, and cultural factors that shaped one of America's largest humanitarian aid projects in the pre-1945 world.

Peter D. Shapinsky is associate professor of history at the University of Illinois, Springfield. He is the author of *Lords of the Sea: Pirates, Violence, and Exchange in Late Medieval Japan* (Center for Japanese Studies, University of Michigan, 2014). He is currently working on a project on cross-cultural cartography in fifteenth- and sixteenth-century East Asia.

Henry D. Smith II, professor emeritus of Japanese history at Columbia University, has a special interest in the cultural history of the city of Edo-Tokyo, particularly in the many ways in which it was represented in pictures and maps. His books include *Hiroshige: One Hundred Famous Views of Edo* (George Braziller, 1986); *Hokusai, One Hundred Views of Mt. Fuji* (George Braziller, 1988); *Kiyochika: Artist of Meiji Japan* (Santa Barbara Museum of Art, 1988); and *Ukiyo-e ni miru Edo meisho* (The famous places of Edo in ukiyo-e; Iwanami Shoten, 1993). He has also written widely on the story of the Forty-Seven Samurai and is currently at work on a project to explore the print technology, colorants, and meanings of color in nineteenth-century Japanese woodblock prints. A list of publications is available at http://www.columbia.edu/~hds2/CV.htm.

Gregory Smits is associate professor of history and Asian studies at Pennsylvania State University. His work focuses on intellectual history, including the history of science. His recent research examines aspects of the history of earthquakes in early modern and modern Japan. With Bettina Gramlich-Oka he is coeditor of *Economic Thought in Early Modern Japan* (Brill, 2010). His earthquake-related publications include "Conduits of Power: What the Origins of Japan's Earthquake Catfish Reveal about Religious Geography" (*Japan Review*, 2012), *Seismic Japan: The Long History and Continuing Legacy of the Ansei Edo Earthquake* (University of Hawai'i Press, 2013), and *When the Earth Roars: Lessons from the History of Earthquakes in Japan* (Rowman and Littlefield, 2014).

André Sorensen is professor of human geography at the University of Toronto Scarborough. He has published extensively on Japanese cities, urban planning, and planning history. His monograph *The Making of Urban Japan: Cities and Planning from Edo to the 21st Century* (Routledge, 2002) won the book prize of the International Planning History Association. In 2007 he was elected a fellow of the University of Tokyo School of Engineering in recognition of his research on Japanese urbanism and urban planning. His current research examines institutions, urban space, and temporal processes in urbanization and urban governance from an institutionalist perspective, with a focus on property development, infrastructure management, and the creation of increasingly differentiated property rights in urban settings.

Sugimoto Fumiko is professor in the Section for Early Modern History at the Historiographical Institute, University of Tokyo. She specializes in legal, political, and cultural history from the seventeenth to the first half of the nineteenth century, using not only textual documents but also maps, bird's-eye views, and other spatial representations. In 2011, she published *Ezugaku nyūmon* (Ezu, society, and the world: An introduction to early modern Japanese maps; University of Tokyo Press), serving as first editor. This is the first book to provide a comprehensive overview of interdisciplinary research on early modern Japanese maps, encompassing perspectives from history, the history of science, geography, architecture, and art history, among other fields. She has also written *Ryōiki shihai no tenkai to kinsei* (Capturing spaces in early modern Japan; Yamakawa Shuppansha, 1999) and coedited two other volumes: *Rekishi o yomu* (Reading history; University of Tokyo Press, 2004) and *Ezu to chizu no seijibunka-shi* (Maps in political culture; University of Tokyo Press, 2001).

Bruce Suttmeier is associate professor of Japanese at Lewis and Clark College. He has published on several postwar writers, including Kaikō Takeshi and ōe Kenzaburō, as well as on travel writing in the 1960s and on war memory in the postwar period. His recent work includes "Speculations of Murder: Ghostly Dreams, Poisonous Frogs and the return of Yokoi Shōichi," in *Perversion and Modern Japan: Experiments in Psychoanalysis*, edited by Keith Vincent and Nina Cornyetz (Routledge, 2011). His current project involves public space and the rise of the highway in pre-Olympic-era Tokyo.

Suzuki Junko is a long-time curator of cartographic materials at Japan's National Diet Library. Her research concerns the history of cartography and the evolution of map archives in Japan. On the latter subject, she compiled a guide to the cartographic materials held by the National Diet Library (*Chizu shiryō gaisetsu*; Kenkyū Shirōzu, 1996). Her research in cartographic history concerns the modernization of Japanese maps, with an emphasis on the pioneering surveys of Inō Tadataka as well as the innovations in maritime charts during the late Tokugawa and early Meiji decades. She is currently conducting research on the Home Ministry's efforts to compile maps and other geographic materials from the former daimyo during the early Meiji years, with an eye to understanding the goals, methods, and principles of ministry officials. Publications include "Bakufu kaigun kara kaigun suirobu e" (From the Tokugawa shogunate navy to the Hydrographic Office of the Admiralty; *Tokyo Daigaku Shiryō Hensan-jō kenkyū kiyō*, 2013) in addition to numerous coedited and cowritten works on the surveys of Inō Tadataka.

Tamai Tetsuo holds the position of professor emeritus at both the National Museum of Japanese History and the Chiba University Department of Engineering. His original specialty is Japanese architectural history, with an interest in clarifying the way architecture and land-use plans have shaped urban space. While his research has ranged from the ancient to the contemporary periods, he has focused on the turning point between the late medieval and the early modern era, a period for which both cartographic documents and artifacts are particularly rich. This era spanning the sixteenth and seventeenth centuries is when the foundations of modern Japanese cities were largely put in place. While Edo is his chief focus, he has also explored such castle towns as Osaka and Sendai, along with the famous ports of Sakai and Hakata. Recently his research interests have taken him to other areas in Asia, including Viet Nam, Thailand, and India. Publications include *Edo—Ushinawareta toshi kūkan o yumu* (Edo: Reading the lost urban space; Heibonsha, 1986) and the coedited *Ajia kara miru Nihon toshi shi* (Japan's urban history as seen from Asia; Yamakawa Shuppansha, 2013).

Susan Paige Taylor earned her BA from Georgetown University and her MA from the Graduate School of Interdisciplinary Information Studies at the University of Tokyo. She is currently pursuing a PhD in anthropology at Harvard University. Her research focuses on urban used book markets in Japan and South Korea, specifically regarding the creation of and perpetuation of bibliographic knowledge through booksellers' networks. She has done significant fieldwork in Japan and South Korea on the international antiquarian used book markets there. Her research interests include visual media, cartography, cognitive mapping, urban space, and used markets. She is interested in photography, oral history, and collaborative history writing as methodological approaches in ethnographic research. Her publications include "Bookstore Owners' Cognitive Maps in Networked Booktown Jimbocho," in *Japan Studies: The Frontier* (International Christian University, 2014).

Ronald P. Toby is professor emeritus of history and East Asian studies at the University of Illinois. He received his PhD in Japanese history from Columbia University in 1977. Toby publishes in both English and Japanese on premodern Japanese international history and cultural, social, cartographic, and visual history. His books include *State and Diplomacy in Early Modern Japan* (Princeton, 1984), *The Emergence of Economic Society in Japan, 1600–1859* (coedited, Oxford, 2004), *Japan and Its Worlds: Marius B. Jansen and the Internationalization of Japanese History* (coedited, I-House Press, 2007), *"Sakoku" to iu gaikō* (The politics of "seclusion"; Shōgakukan, 2008), and *Ezugaku nyūmon* (Ezu, society, and the world: An introduction to early modern Japanese maps; University of Tokyo Press, 2011). In 2012 he was named the first recipient of the National Institutes for the Humanities Prize in Japanese Studies.

Jilly Traganou is an architect by training and associate professor in spatial design studies in the School of Art and Design History and Theory at Parsons the New School for Design. She is the author of *The Tōkaidō Road: Traveling and Representation in Edo And Meiji Japan* (RoutledgeCurzon, 2004), and coeditor with Miodrag Mitrašinović of *Travel, Space, Architecture* (Ashgate, 2009). Traganou is reviews editor of the *Journal of Design History*. She is currently working on a book provisionally titled *Designing the Olympics* (contracted by Routledge). Her broader research interests include the study of spatial controversy, design as critical pedagogy, and the role of design in social movements.

Uesugi Kazuhiro is associate professor of history at Kyoto Prefectural University. His specialty is the historical geography of information in early modern Japan. After earning his PhD from Kyoto University, he was appointed to the Kyoto University Museum, where he managed and conducted research on old maps in the museum's collection. His book *Edo chishiki-jin to chizu* (Maps and intellectuals in 18th-century Japan; Kyoto University Press, 2010) discussed the role of maps in the networks among literati in the eighteenth century. He also explored the history of maps from early modern through modern times in his coedited book *Nihon chizu shi* (A history of cartography in Japan: Yoshikawa Kōbunkan, 2012). An outline history of maps in Japan was also published in English in his paper (with Akihiro Kinda) "Landscape and Maps," in *A Landscape History of Japan* (Kyoto University Press, 2010).

Constantine N. Vaporis teaches Japanese and East Asian history at the University of Maryland Baltimore County. He has received numerous fellowships for his research, including a Fulbright Scholar's Award and an NEH Fellowship for College Teachers. He is the author of *Breaking Barriers: Travel and the State in Early Modern Japan* (Harvard University Council on East Asian Studies, 1995); *Tour of Duty: Samurai, Military Service in Edo and the Culture of Early Modern Japan* (University of Hawai'i Press, 2008); *Nihonjin to sankin kōtai* (The Japanese and alternate attendance; Kashiwa Shobō, 2010); and *Voices of Early Modern Japan: Contemporary Accounts of Daily Life during the Age of the Shoguns* (ABC-CLIO/Greenwood Press, 2012; paperback edition, Westview Press, 2013). Vaporis is the founding director of the Asian Studies Program at UMBC and was recently appointed the 2013–16 UMBC Presidential Research Professor.

Paul Waley is senior lecturer in human geography at the University of Leeds. He has written widely on the historical geography of Tokyo, as well as on contemporary issues of concern in the city. One strand of recent work has involved a study of the role of memory and commemoration in the urban landscape, developed in a chapter entitled "Who Cares about the Past in Today's Tokyo?" published in *Urban Spaces in Japan: Social and Cultural Perspectives*, edited by Christoph Brumann and Evelyn Schulz (Routledge, 2012).

Brett L. Walker is Regents Professor and Michael P. Malone Professor of History at Montana State University, Bozeman. His research and teaching interests include environmental history, the history of science and medicine, Japanese history, and comparative world history. He is the author of *The Conquest of Ainu Lands: Ecology and Culture in Japanese Expansion, 1590–1800* (University of California Press, 2001), *The Lost Wolves of Japan* (University of Washington Press, 2005), *Toxic Archipelago: A*

History of Industrial Disease in Japan (University of Washington Press, 2010), and *A Concise History of Japan* (Cambridge University Press, 2015). He also coedited two volumes: *JAPANimals: History and Culture in Japan's Animal Life* (Center for Japanese Studies, University of Michigan, 2005), with Gregory Pflugfelder, and *Japan at Nature's Edge: The Environmental Context of a Global Power* (University of Hawai'i Press, 2013), with Julia Adeney Thomas and Ian Jared Miller.

Anne Walthall, professor emerita of history at the University of California, Irvine, has written several books on Japanese and East Asian history. Recent publications include *Japan: A Cultural, Social, and Political History* (Cengage, 2006), *The Weak Body of a Useless Woman: Matsuo Taseko and the Meiji Restoration* (University of Chicago Press, 1998), and the edited volume *Servants of the Dynasty: Palace Women in World History* (University of California Press, 2008). With Patricia Ebrey she coauthored *East Asia: A Cultural, Social, and Political History* (3rd ed., Cengage, 2013). With Sabine Frühstück, she coedited *Recreating Japanese Men* (University of California Press, 2011). Her latest article is "Shipwreck! Akita's Local Initiative, Japan's National Debt, 1869–1872" (*Journal of Japanese Studies*, 2013). Walthall is currently working on a manuscript tentatively titled "Practicing Faith: The Hirata Family, 1800–1945," drawing on the Hirata Atsutane archives.

Watanabe Miki is associate professor in the Department of Interdisciplinary Cultural Studies at the University of Tokyo. She received her PhD from the University of Tokyo and has published *Kinsei Ryūkyū to Chū-Nichi kankei* (Early modern Ryukyu and China-Japan relations; Yoshikawa Kōbunkan, 2012), based on her dissertation. The book examines how Ryukyu established and maintained its position between Ming-Qing China and Tokugawa Japan, and how Ryukyuans assimilated and internalized relations with both China and Japan and linked them with their local identity.

Kären Wigen teaches Japanese history and the history of cartography at Stanford University. A geographer by training, she earned her doctorate at the University of California at Berkeley in 1990. Her first book, *The Making of a Japanese Periphery, 1750–1920* (University of California Press, 1995), mapped the economic transformation of southern Shinano during the rise of the silk industry. *A Malleable Map: Geographies of Restoration in Central Japan, 1600–1912* (University of California Press, 2010) returned to the ground of that study, exploring the roles of cartography, chorography, and regionalism in the making of modern Shinano. Wigen's interest in world history has taken the form of collaborative projects, including *The Myth of Continents* (University of California Press, 1997), coauthored with Martin Lewis, and the coedited volume *Seascapes* (University of Hawai'i Press, 2007).

Steven Wills is assistant professor of history at Nebraska Wesleyan University. He specializes in the social and cultural history of Edo-Tokyo, with an emphasis on the reciprocal relationship between urban development and conflagration in the eighteenth and nineteenth centuries. He coauthored a chapter in *Flammable Cities: Urban Conflagration and the Making of the Modern World* (University of Wisconsin, 2012), and he is currently investigating the history of fire and firefighting in early modern Osaka and Kyoto.

Roderick Wilson earned a PhD in East Asian history from Stanford University in 2011 and is an assistant professor with a joint appointment in the Department of History and the Department of East Asian Languages and Cultures at the University of Illinois at Urbana-Champaign. In 2013–14, he was awarded a Faculty Fellowship by the Illinois Program for Research in the Humanities. His research focuses on people's interactions with their local environments in early modern and modern Japan with a particular focus on rivers, cities, and science and technology. He has recently authored articles about the Edomae fishery and Tokyo's waterways and bay. Currently, he is working on a book manuscript entitled *Turbulent Stream: Reengineering Environmental Relations along the Rivers of Japan, 1750–2000* as well as a second book-length project about the urban and environmental history of Tokyo.

Marcia Yonemoto is associate professor of history at the University of Colorado, Boulder. Her research concerns the cultural history of Japan's early modern period (ca. 1590–1868). She is the author of *Mapping Early Modern Japan: Space, Place, and Culture in the Tokugawa Period* (University of California Press, 2003). Her scholarly articles have appeared in the *Journal of Asian Studies*, *Japan Forum*, the *Geographical Review*, *East Asian History*, the *U.S.-Japan Women's Journal*, and other venues. Yonemoto is currently completing a book on the history of women and family in the late seventeenth through early nineteenth centuries, entitled *The Problem of Women in Early Modern Japan.*

Index